LINEAR ALGEBRA

LINEAR ALGEBRA

WALTER NEF

Professor of Mathematics
University of Berne, Switzerland

Translated from the German by

J. C. Ault

DOVER PUBLICATIONS, INC.
NEW YORK

Published in Canada by General Publishing Company, Ltd., 30 Lesmill Road, Don Mills, Toronto, Ontario.
Published in the United Kingdom by Constable and Company, Ltd., 10 Orange Street, London WC2H 7EG.

This Dover edition, first published in 1988, is an unabridged, corrected republication of the English translation (by J. C. Ault), first published by McGraw-Hill Publishing Company Limited, Maidenhead, Berkshire, England, and the McGraw-Hill Book Company, New York, 1967, of *Lehrbuch der Linearen Algebra*, first published by Birkhäuser Verlag AG, Basel, Switzerland, 1966. The translation was originally included by McGraw-Hill in the "European Mathematics Series" and the "International Series in Pure and Applied Mathematics."

Reprinted by special arrangement with Birkhäuser Verlag AG, Ringstrasse 39, CH-4106 Therwil, Basel, Switzerland.

Manufactured in the United States of America
Dover Publications, Inc., 31 East 2nd Street, Mineola, N.Y. 11501

Library of Congress Cataloging-in-Publication Data

Nef, Walter, 1919–
    [Lehrbuch der linearen Algebra. English]
    Linear algebra / by Walter Nef ; translated from the German by J. C. Ault.
        p.   cm.
    Reprint. Originally published: New York : McGraw-Hill, 1967.
    (International series in pure and applied mathematics)
    Translation of: Lehrbuch der linearen Algebra.
    Bibliography: p.
    Includes index.
    ISBN 0-486-65772-8
    1. Algebras, Linear. I. Title.
QA251.N4413   1988
512′.5—dc19   88-11847
                               CIP

CONTENTS

PREFACE

This book is based on an introductory course of lectures on linear algebra which I have often given in the University of Berne. This course is intended for students in their second year and is for both specialist and supplementary mathematicians. These latter usually include actuaries, astronomers, physicists, chemists and also, sometimes, mathematical economists. Because of the wide diversity of the audience, it was necessary to develop the discussion from as few basic assumptions as possible and this is a feature of the book. It is assumed that the reader has had a grammar school education in mathematics and that he has the ability to think abstractly. It is also helpful if he has some knowledge of vector geometry because this will provide him with visual illustrations of the general theory.

It was also necessary to restrict the course to the most important topics so that in the book, with the exception of the last chapter, only real and complex vector spaces will be considered.

In addition, it was necessary to consider the applications of linear algebra in accordance with the needs of the majority of the audience. Apart from details concerning the choice of material, the main effect of this requirement has been the inclusion of simple techniques for the solution of the most important types of numerical problem. Naturally, it has not been possible to give a complete discussion of these.

It is not usual practice to deal with linear programming, Tchebychev approximations, and game theory in a textbook of linear algebra but these topics are becoming increasingly important, so I have presented them in the form of introductory chapters.

The publication of this book would hardly have been possible without the very considerable help of several co-workers. Especially, I would like to thank Miss Dora Hänni for typing the manuscript and Messrs. H. P. Bieri, H. P. Blau, D. Fischer, and N. Ragaz for their help with the proof-reading and also H. P. Blau for drawing the diagrams.

Lastly, I thank the Birkhäuser Verlag and the McGraw-Hill Publishing Company, Limited for their always friendly co-operation and their careful preparation of the book.

W. NEF

Notes for the Reader

1. The book is divided into fourteen chapters and the chapters into sections. The numbering of definitions, theorems, examples and individual formulae starts afresh in each section. For example a reference to Theorem 2.4;9 would mean the Theorem 9 in section 2.4. Similarly, 3.3;(11) would mean the formula (11) in section 3.3. The number of the section is omitted whenever the reference is made within the same section.

2. In contrast with the usual notation, the matrix which has just one column consisting of the components ξ_k of a vector x will be denoted by ξ (instead of x). The transpose of ξ is denoted by ξ'. This avoids the problem of having different things denoted by the same symbol. (See 5.2.2.)

3. At the end of some of the sections, there are 'Exercises' and 'Problems'. The exercises are numerical examples and the problems are theoretical examples which are often extensions of the material in the text.

4. Readers who are not interested in linear programming, Tchebychev approximations, and game theory may omit the following sections and chapters: 2.6, 3.4, 6.3, 7.3, 7.4;8, 9, 10.

CHAPTER 1

SETS AND MAPPINGS

In this chapter we state and discuss the properties of sets and mappings which will be needed in the subsequent chapters. We will do this at an intuitive level without attempting to carry out a strictly axiomatic approach.

1.1 Sets

A *set* is a collection of objects which is thought of as an entity in itself. The objects x in a set M are usually referred to as the *elements* of the set and this relationship is written formally as

$$x \in M.$$

This is usually read as 'x is an element of M' or 'x belongs to M' or simply as 'x is in M'. A set M is said to 'contain' its elements. If the object y is not an element of M, we write $y \notin M$.

A set is completely determined by its elements. Consequently two sets M and N are considered to be equal if and only if they contain the same elements, i.e., $M = N$ if and only if

$$x \in M \quad \text{implies} \quad x \in N$$

and $\qquad\qquad x \in N \quad \text{implies} \quad x \in M.$

A set is said to be *finite* if it contains only a finite number of elements.

If every element of the set A is also an element of the set B, we say that A is a *subset* of B and express this in symbols by

$$A \subseteq B \quad \text{or} \quad B \supseteq A$$

These are usually read as 'A is a subset of B' or 'A is contained in B' or 'B contains A'.

In order to prove that the set A is a subset of the set B it is therefore necessary to show that

$$x \in A \quad \text{implies} \quad x \in B.$$

1

Clearly $A = B$ if and only if $A \subseteq B$ and $B \subseteq A$. Every set A is a subset of itself. All other subsets M of A are referred to as *proper subsets* of A and we express this in symbols by $M \subset A$.

In many cases it is convenient to be able to consider the set which contains no elements at all. We call this the *empty set* and denote it by the symbol $\varnothing$. The empty set is a subset of every other set.

In order to characterize a set, it is necessary to specify its elements. For a finite set A, this can be done by writing out a list of its elements in the form $A = \{x_1, \ldots, x_n\}$. Thus, for example, $\{1, 3, 5\}$ is the set which contains the numbers 1, 3 and 5 as its elements. If A is countably infinite, we will extend this notation to write $A = \{x_1, x_2, x_3, \ldots\}$.

More generally, we will use the notation

$$A = \{x; \phi(x)\} \tag{1}$$

where $\phi(x)$ denotes a statement by which the elements of A are characterized. In words (1) reads, 'A is the set of all those elements x for which the statement $\phi(x)$ is true'. For example, in 3-dimensional space, if x is an arbitrary vector and y is a fixed vector, then

$$A = \{x; x \text{ is orthogonal to } y\}$$

is the set of all vectors x which are orthogonal to the vector y. If ξ is a real number and N is the set of all natural numbers (positive non-zero integers), then

$$B = \{\xi; \xi^2 \in N\}$$

is the set of square roots of the natural numbers.

1.2 Families of Sets

Many of the sets studied in mathematics have sets as their elements, that is to say, we will need to deal with 'sets of sets'. If we wish to emphasize that the elements of a set $\mathscr{S}$ are again sets, we will call $\mathscr{S}$ a *family of sets* and use capital script letters.

Example 1. For each positive non-zero real number ρ, let A_ρ be the set $A_\rho = \{\xi; -\rho \leqslant \xi \leqslant \rho\}$. Then $\mathscr{S} = \{A; \text{ there is a } \rho > 0 \text{ such that } A = A_\rho\}$ is the family consisting of all the sets A_ρ.

We define the *intersection* of a family of sets $\mathscr{S}$ by

$$\bigcap(\mathscr{S}) = \{x; x \in A \text{ for all } A \in \mathscr{S}\}$$

and the *union* of $\mathscr{S}$ by

$$\bigcup(\mathscr{S}) = \{x; \text{ there is an } A \in \mathscr{S} \text{ such that } x \in A\}$$

i.e., the intersection $\bigcap(\mathscr{S})$ is the set of all elements x which belong to all the sets $A \in \mathscr{S}$, and the union $\bigcup(\mathscr{S})$ is the set of all elements x which are in at least one of the sets $A \in \mathscr{S}$.

In Example 1, $\bigcap(\mathscr{S}) = \{0\}$—only the number 0 lies in all the sets A_ρ. $\bigcup(\mathscr{S})$ = the set of all real numbers.

In the following special cases, the intersection and union will be denoted as shown. When $\mathscr{S}$ consists of just two sets A, B,

$$\bigcap(\mathscr{S}) = A \cap B \quad \text{and} \quad \bigcup(\mathscr{S}) = A \cup B.$$

When $\mathscr{S}$ consists of n sets $A_1, \ldots, A_n$

$$\bigcap(\mathscr{S}) = A_1 \cap A_2 \cap \ldots \cap A_n = \bigcap_{k=1}^{n} A_k$$

and

$$\bigcup(\mathscr{S}) = A_1 \cup A_2 \cup \ldots \cup A_n = \bigcup_{k=1}^{n} A_k.$$

It will sometimes also be convenient to use a generalization of this notation to write $\bigcap_{A \in \mathscr{S}} A$ and $\bigcup_{A \in \mathscr{S}} A$ instead of $\bigcap(\mathscr{S})$ and $\bigcup(\mathscr{S})$.

If $A \cap B = \varnothing$, we say that A and B are *disjoint*.

Example 2. Let A, B, C be the sets of points of a plane represented by the following diagram. Then $A \cap B \cap C$ and $A \cup B \cup C$ are the shaded areas.

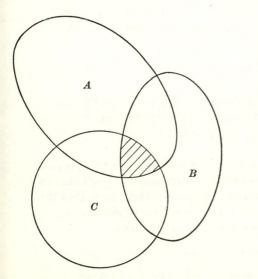

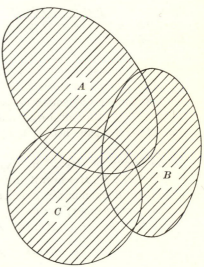

$$A \cap B \cap C \qquad\qquad A \cup B \cup C$$

Fig. 1.

The first theorem follows directly from the definitions.

Theorem 1. If $\mathscr{S}$ is a family of sets and $A \in \mathscr{S}$, then $\bigcap(\mathscr{S}) \subseteq A \subseteq \bigcup(\mathscr{S})$.

Theorem 2. If $\mathscr{S}_1 \subseteq \mathscr{S}_2$, then $\bigcap(\mathscr{S}_1) \supseteq \bigcap(\mathscr{S}_2)$ and $\bigcup(\mathscr{S}_1) \subseteq \bigcup(\mathscr{S}_2)$.

Proof. 1. Suppose $x \in \bigcap(\mathscr{S}_2)$ so that $x \in A$ for all $A \in \mathscr{S}_2$. Then, since $\mathscr{S}_1 \subseteq \mathscr{S}_2$, it follows that $x \in B$ for all $B \in \mathscr{S}_1$. Hence $x \in \bigcap(\mathscr{S}_1)$.
2. Suppose $x \in \bigcup(\mathscr{S}_1)$ so that there is an $A \in \mathscr{S}_1$ such that $x \in A$. Then, since $\mathscr{S}_1 \subseteq \mathscr{S}_2$, $A \in \mathscr{S}_2$ and hence $x \in \bigcup(\mathscr{S}_2)$.

Theorem 3. If $B \subseteq A$ for all $A \in \mathscr{S}$, then $B \subseteq \bigcap(\mathscr{S})$.
If $C \supseteq A$ for all $A \in \mathscr{S}$, then $C \supseteq \bigcup(\mathscr{S})$.

Proof. 1. Suppose $x \in B$ and $B \subseteq A$ for all $A \in \mathscr{S}$. Then clearly $x \in A$ for all $A \in \mathscr{S}$ and hence $x \in \bigcap(\mathscr{S})$.
2. Suppose $x \in \bigcup(\mathscr{S})$. Then there is a set $A_0 \in \mathscr{S}$ such that $x \in A_0$. Now, if $C \supseteq A$ for all $A \in \mathscr{S}$, then $C \supseteq A_0$ and clearly $x \in C$.

A set C is called a *least element* of the family of sets $\mathscr{S}$ if $C \in \mathscr{S}$ and $C \subseteq A$ for all $A \in \mathscr{S}$.

Theorem 4. A family of sets $\mathscr{S}$ contains a least element if and only if $D = \bigcap(\mathscr{S}) \in \mathscr{S}$. The set D is then a least element of $\mathscr{S}$ and $\mathscr{S}$ has no other least elements.

Proof. 1. Suppose $D = \bigcap(\mathscr{S}) \in \mathscr{S}$. By Theorem 1, $D \subseteq A$ for all $A \in \mathscr{S}$ and therefore D is a least element of $\mathscr{S}$.
2. Suppose $\mathscr{S}$ contains a least element D. Then $D = \bigcap(\mathscr{S})$, because
2.1. $D \subseteq A$ for all $A \in \mathscr{S}$, and hence $D \subseteq \bigcap(\mathscr{S})$ (Theorem 3)
and 2.2. $D \in \mathscr{S}$, and hence $\bigcap(\mathscr{S}) \subseteq D$ (Theorem 1).

The family of sets $\mathscr{S}$ in Example 1 has no least element because $\bigcap(\mathscr{S}) = \{0\} \notin \mathscr{S}$. However if, instead of all positive non-zero real numbers ρ, we take all real numbers ρ such that $\rho \geqslant \rho_0$ for some fixed $\rho_0 > 0$, then the intersection becomes $\bigcap(\mathscr{S}) = A_{\rho_0} \in \mathscr{S}$ and $\mathscr{S}$ has the least element A_{ρ_0}.

The *difference set* $A \setminus B$ of two sets A and B is defined by

$$A \setminus B = \{x; x \in A, x \notin B\}.$$

That is, $A \setminus B$ consists of those elements of A which are not contained in B. For example, if $A = \{1, 2, 5, 7\}$ and $B = \{5, 6, 7\}$, then $A \setminus B = \{1, 2\}$ and $B \setminus A = \{6\}$.

Problems

1. Prove the distributive laws

$$A \cap (B \cup C) = (A \cap B) \cup (A \cap C)$$

and $$A \cup (B \cap C) = (A \cup B) \cap (A \cup C)$$

2. Let E be a set and, for each subset A of E, let $\bar{A}$ denote the *complement* of A in E, i.e. the set $E \setminus A$. Prove the following rules.

(a) $\overline{A \cup B} = \bar{A} \cap \bar{B}$
(b) $\overline{A \cap B} = \bar{A} \cup \bar{B}$
(c) $A \subseteq B$ if and only if $A \cap \bar{B} = \varnothing$

1.3 Mappings

1.3.1 General Mappings

Let E and F be two sets. *A mapping of E into F* is a rule which assigns to each element $x \in E$ a unique element $y \in F$.

The element $y \in F$ which is assigned to the element $x \in E$ by the mapping f is called the *image* of x under f and is denoted by $f(x)$. The set E is called the *domain* of f and F is called the *range* of f.

Example 1. Suppose E is the set of all squares in a Euclidean plane and F is the set of all real numbers. A mapping of the domain E into the range F may be defined by assigning to each square the real number which is its area.

Mappings will usually be denoted by small Roman letters $f, g, \ldots$. (An exception will be made in the case of permutations, cf. 4.1.) When an element $x \in E$ goes into the element $y \in F$ under the mapping f, this will be indicated by the notation $x \rightarrow y = f(x)$.

Two mappings f and g of E into F will be considered to be equal if $f(x) = g(x)$ for all $x \in E$, and this will be written briefly as $f = g$.

Not every element of F needs to be the image of an element in E. The images of the elements of E form a subset of F which we call the *image of E under f* and denote by $f(E)$.

Thus $f(E) = \{y; y \in F$ and there is an element $x \in E$ such that $y = f(x)\}$.

If it happens that $f(E) = F$, we say that f is a mapping of E *onto* F.

More generally, if A is a subset of E, we use the notation $f(A)$ for the set of all the images of the elements of A.

That is, $f(A) = \{y;$ there is an $x \in A$ such that $f(x) = y\}$.

Now let E, F and G be three sets and let f be a mapping of E into F and g

a mapping of F into G. We can define a mapping of E into G by assigning to the element $x \in E$ the element $z = g(f(x))$ in G. This mapping is called the product of g and f and we denote it by gf (in this order!). The product mapping gf is therefore defined by

$$gf(x) = g(f(x)) \text{ for all } x \in E.$$

It is important to remember the order of the factors. The product fg may have no meaning in general and even when it has (e.g., when $E = F = G$) fg need not be equal to gf.

Example 2. Suppose $E = F = G = \{1, 2, 3\}$. Let f be defined by $f(1) = 2$; $f(2) = 3$; $f(3) = 1$ and g by $g(1) = 1$; $g(2) = 3$; $g(3) = 2$. Then, for example, $fg(2) = f(3) = 1$ and $gf(2) = g(3) = 2$ and hence $fg \neq gf$.

Now let H be a fourth set and let h be a mapping of G into H. Then both the products $h(gf)$ and $(hg)f$ have a meaning and, in the following fundamental Theorem, we prove that they are equal.

Theorem 1. Multiplication of mappings is associative.

Proof. For $x \in E$, let $y = f(x) \in F$, $z = g(y) \in G$ and $u = h(z) \in H$. Then $gf(x) = z$ and $h(gf)(x) = h(z) = u$. Similarly $hg(y) = u$ and therefore $(hg)f(x) = u$. Thus $h(gf)(x) = (hg)f(x)$ for all $x \in E$ and hence $h(gf) = (hg)f$.

In view of this theorem it is possible to omit the brackets and simply write hgf for the product of h, g and f.

If f is a mapping of E into F and K is a subset of F, we define the *inverse image* of K under f to be the subset of E given by

$$f^{-1}(K) = \{x; x \in E, f(x) \in K\}.$$

That is, $f^{-1}(K)$ is the subset of E which consists of all those elements whose images under f are in K. Note that the symbol f^{-1} does not represent a mapping in general because it is possible for many elements in E to have the same image in F. Clearly $f^{-1}(F) = f^{-1}(f(E)) = E$, but we also note that it is possible for $f^{-1}(K)$ to be the empty set and that this will happen when $K \cap f(E) = \varnothing$.

If K consists of a single element, say $K = \{y\}$, we will write $f^{-1}(y)$ for $f^{-1}(\{y\})$ and this then means the set of all $x \in E$ for which $f(x) = y$.

Example 3. As in Example 1, let E be the set of all squares in a Euclidean plane, F the set of all real numbers and f the mapping which assigns to each square its area. If $\xi \in F$ and $\xi \leqslant 0$, then $f^{-1}(\xi) = \varnothing$ but, if $\xi > 0$, then $f^{-1}(\xi)$ is the set of all squares of side $\sqrt{\xi}$.

1.3.2 One-to-One Mappings

A mapping f is called *one-to-one* (1-1) when no two different elements in the domain have the same image in the range, i.e. when

$$x_1 \neq x_2 \quad \text{implies} \quad f(x_1) \neq f(x_2)$$

or, alternatively,

$$f(x_1) = f(x_2) \quad \text{implies} \quad x_1 = x_2.$$

If E is a finite set, a mapping f of E is clearly 1-1 if and only if the image $f(E)$ contains exactly the same number of elements as E.

When f is 1-1, $f^{-1}(y)$ contains exactly one element $x \in E$ for each element $y \in f(E)$. Putting $f^{-1}(y) = x$ (instead of $\{x\}$), f^{-1} can now be thought of as a mapping from $f(E)$ *onto* E and we call it the *inverse mapping of f*. The mapping f^{-1} is also 1-1. Obviously $f^{-1}f$ is the mapping of E onto itself under which each $x \in E$ goes into itself. We call this the *identity mapping* on E. Similarly ff^{-1} is the identity mapping on $f(E)$.

Theorem 2. The product of two 1-1 mappings is 1-1.

Proof. Let f and g be 1-1 mappings and suppose that gf is defined. If $gf(x_1) = gf(x_2)$, then $g(f(x_1)) = g(f(x_2))$ and, since g is 1-1, $f(x_1) = f(x_2)$ and hence $x_1 = x_2$.

1.3.3 Mappings of a Set into Itself

It is possible for the domain and the range of a mapping to be the same set and, in this case, we will be dealing with a mapping of a set into itself. The best known examples of this are provided by the concept of a function. For example the function $y = \sin x$ assigns to each real number x the real number $\sin x$, so that the function is a mapping of the set of real numbers into itself.

The 1-1 mappings of a set E *onto* itself are of particular interest. (Note that if E is finite, a mapping of E into itself is 1-1 if and only if it is onto E so that only one of these two properties needs to be assumed.)

If we denote by $T(E)$ the set of all 1-1 mappings of the set E onto itself, then $T(E)$ has the following properties.

1. If $f, g \in T(E)$, then fg and $gf \in T(E)$.
2. If $f \in T(E)$, then $f^{-1} \in T(E)$.
3. The identity mapping on E is in $T(E)$.

Together with the associativity of the multiplication of mappings proved earlier in Theorem 1.3; 1, properties 1, 2, 3 show that $T(E)$ is a *group*. (For the concept of a group, see 2.1.1 or [25] pp. 1–3.)

Example 4. Let $E=\{1,2,3\}$. The group $T(E)$ consists of the following six mappings,

$$f_1 = \begin{pmatrix} 1 & 2 & 3 \\ 1 & 2 & 3 \end{pmatrix} \qquad f_4 = \begin{pmatrix} 1 & 2 & 3 \\ 1 & 3 & 2 \end{pmatrix}$$

$$f_2 = \begin{pmatrix} 1 & 2 & 3 \\ 2 & 3 & 1 \end{pmatrix} \qquad f_5 = \begin{pmatrix} 1 & 2 & 3 \\ 3 & 2 & 1 \end{pmatrix}$$

$$f_3 = \begin{pmatrix} 1 & 2 & 3 \\ 3 & 1 & 2 \end{pmatrix} \qquad f_6 = \begin{pmatrix} 1 & 2 & 3 \\ 2 & 1 & 3 \end{pmatrix}$$

In these, the mapping f is defined by writing the elements of E in the first row and then under each element its image under f. Thus f_1 is the identity mapping, and, for example, $f_2^{-1}=f_3$, $f_5^{-1}=f_5$. It is clear that $f_1, f_2, \ldots, f_6$ are just the permutations of the three elements $1,2,3$. Permutations of a general set of elements will be introduced later in 4.1 by using this idea.

Problems

1. Let f be a mapping of the set E into the set F and let A, B be subsets of E. Prove that $f(A \cup B)=f(A) \cup f(B)$ and $f(A \cap B) \subseteq f(A) \cap f(B)$. Give an example to show that the $\subseteq$ sign cannot be replaced by $=$ in the second of these rules. What more can be said if f is 1-1?

2. Let f be a mapping of E into F and let A, B be subsets of E. Further let $A = f^{-1}(A^*)$ and $f(B) = B^*$. Prove that $f(A \cap B) = A^* \cap B^*$.

CHAPTER 2

VECTOR SPACES

2.1 The Concept of a Vector Space, Examples

Linear algebra can be characterized as the study of vector spaces. Vector spaces, which are sometimes also called linear spaces, are a particular kind of algebraic system and accordingly they are sets in which certain operations on the elements are defined. (For the general concept of an algebraic system, see for example [24].)

2.1.1 Real Vector Spaces

Let E be a set whose elements are denoted by small Roman letters, and suppose that there is a rule which assigns to each ordered pair of elements $x, y \in E$ a further element $z \in E$ which we write as the sum of x and y, i.e. we express the relationship between z and x, y by the formula

$$z = x + y.$$

The construction of z from x and y is known as a *binary operation* on the set E which in this case is written as *addition*. We have said that z should be constructed from the *ordered* pair x, y (i.e., the pair y, x is different from the pair x, y), because initially we do not want to assume that the addition is commutative, i.e., that $x + y = y + x$ for all $x, y \in E$. We will be able to prove this as a consequence of the other conditions to be introduced later.

Now suppose that there is a further rule which assigns to each real number α and each element $x \in E$ a further element $u \in E$ which we write as

$$u = \alpha x.$$

The element u will be referred to as the product of the element $x \in E$ with the real number (or the *scalar*) α. The construction of u from α and x is a binary operation in the set of all real numbers and of elements of E which we will call *multiplication by scalars*.

Definition 1. The set E together with the operations considered above is a 'real vector space', if the operations satisfy the following seven axioms.

9

A1. Addition is *associative*

$$x+(y+z) = (x+y)+z \text{ for all } x, y, z \in E$$

A2. There is a *zero-element* $0 \in E$ such that

$$x+0 = 0+x = x \text{ for all } x \in E$$

A3. To each $x \in E$, there is an *inverse element* $(-x) \in E$ such that

$$(-x)+x = x+(-x) = 0$$

Except when there may be some risk of misunderstanding we will normally write $-x$ in place of $(-x)$.

M1. $1x = x$ for all $x \in E$

M2. Multiplication by scalars is *associative*

$$\alpha(\beta x) = (\alpha\beta)x \text{ for all scalars } \alpha, \beta \text{ and all } x \in E$$

D1. $\alpha(x+y) = \alpha x + \alpha y$ for all scalars α and all $x, y \in E$
D2. $(\alpha+\beta)x = \alpha x + \beta x$ for all scalars α, β and all $x \in E$

D1 and D2 are known as *distributive laws*.

We will prove in 2.2.5 that, as a consequence of these seven axioms, the addition is also *commutative*. Because this is so important, we will already make a note of it here in the form of a further axiom.

A4. $x+y = y+x$ for all $x, y \in E$ (see Theorem 2.2.1)

An algebraic system which has a binary operation satisfying the axioms A1, A2 and A3 is called a *group*. There are many examples of groups in which the operation is not commutative (e.g., the groups $A(E)$ and GL_n in 5.3, where the operation is written as multiplication instead of addition). Consequently, in the case of a vector space, the axioms A1, A2 and A3 will not be sufficient in themselves to prove A4, and the other axioms must also be used in the proof.

The elements of a vector space will usually be referred to as *vectors*. This name comes from the first of the following examples of vector spaces in which the elements are vectors of 3-dimensional space in the sense of elementary geometry. As before, we will use small Roman letters for vectors and small Greek letters for scalars throughout the rest of this work.

The significance of Linear Algebra in mathematics and in its applications stems from the fact that various types of vector space arise naturally in many different branches of mathematics. We will now set out some examples of these for future reference.

Example 1. We start with an (affine or Euclidean) 2-dimensional plane or a 3-dimensional space and consider vectors in the sense of elementary geometry. These can be represented by directed segments (arrows) where two arrows which are obtained from each other by a parallel shift represent the same vector. For these vectors, an addition and a multiplication by real scalars are defined in a way which is well-known and may be remembered with the help of Fig. 2.

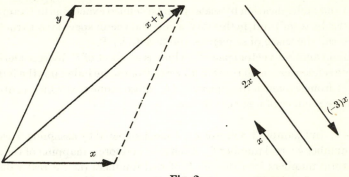

Fig. 2.

It is possible to prove by geometrical methods that these two operations satisfy the seven axioms of Definition 1. The set E of all vectors of the plane or of the 3-space is therefore a real vector space with this addition and multiplication by scalars. We will refer to these two spaces as G_2 (for the plane) and G_3 (for the space). Note that they are different from the plane and space which appear in their construction and which are of course sets of points.

This example is of particular importance for two reasons.

1. It makes possible the use of Linear Algebra in the development of analytic geometry.

2. It enables us to illustrate results in Linear Algebra geometrically in a plane or in 3-space.

Example 2. Let n be a natural number and let E be the family of all ordered sets of n real numbers (n-tuples). If $x = (\xi_1, \ldots, \xi_n) \in E$ and $y = (\eta_1, \ldots, \eta_n) \in E$ are two such ordered sets, we define their sum by

$$x + y = (\xi_1 + \eta_1, \ldots, \xi_n + \eta_n) \in E,$$

and the product of x and the scalar α by

$$\alpha x = (\alpha \xi_1, \ldots, \alpha \xi_n).$$

It is easy to verify that the seven axioms are satisfied. In particular the zero-element is $0 = (0, \ldots, 0) \in E$. (Note that here the left-hand zero denotes the zero-element or zero-vector of E, while the zeros inside the brackets denote the real number 0.) Further $-x = (-\xi_1, \ldots, -\xi_n)$. The real vector space so defined is called the space of n-tuples and we will refer to it as R_n.

Example 3. Instead of ordered sets of n real numbers, we can also consider countably infinite sequences of real numbers, i.e., $x = (\xi_1, \xi_2, \xi_3, \ldots)$. Addition and multiplication by scalars can be defined exactly as in Example 2 (viz. 'term by term') and, in this way, a new real vector space is constructed which we call the space of sequences and denote by F.

We obtain another vector space F_0 which is a subset of F, by considering only those sequences which contain all zeros from some index on (the index may differ from sequence to sequence), i.e., those sequences which contain only a finite number of non-zero terms.

Example 4. In Example 3, a vector $x \in F$ is constructed by assigning to each natural number k a real number ξ_k. Each x is therefore a mapping of the set N of natural numbers into the set R of real numbers (i.e., a real-valued function defined on N). The same is true in Example 2 when N is replaced by $\{1, 2, \ldots, n\}$.

We obtain a generalization of these examples by replacing N with an arbitrary set A. We denote by $F(A)$ the set of all mappings f of the set A into the set R of real numbers. If we now define for mappings $f, g \in F(A)$ and scalar α

$$f + g \qquad \text{by} \quad (f+g)(z) = f(z) + g(z)$$
and $\qquad\qquad\qquad\qquad\qquad\qquad\qquad\qquad\qquad\qquad$ (1)
$$\alpha f \qquad \text{by} \quad (\alpha f)(z) = \alpha f(z)$$

for all $z \in A$, then $F(A)$ becomes a real vector space.

By analogy with Example 3, we can further consider the set $F_0(A) \subseteq F(A)$ which consists of those mappings $f \in F(A)$ such that $f(z) \neq 0$ for only a finite number of elements $z \in A$. Obviously (1) defines operations in $F_0(A)$ and $F_0(A)$ is itself a vector space.

Example 5. Again let n be a natural number and let P_n be the set of all real polynomials x in a real variable τ which have degree at most n.

$$x(\tau) = \sum_{k=0}^{n} \alpha_k \tau^k \qquad (\alpha_0, \ldots, \alpha_n \text{ real}).$$

We define addition and multiplication by scalars in the usual way for functions, i.e. we put

$$z = x + y \quad \text{when} \quad z(\tau) = x(\tau) + y(\tau) \quad \text{for all } \tau \in R$$
and $\quad u = \alpha x \quad \text{when} \quad u(\tau) = \alpha x(\tau) \qquad \text{for all } \tau \in R.$

With these definitions $x+y$ and αx are polynomials of degree at most n and the seven axioms are satisfied. This real vector space of polynomials of degree at most n is denoted by P_n.

Example 6. In the same way, we can consider the set P of all real polynomials (without restriction on the degree). We then obtain the space of polynomials P.

Example 7. Let C be the set of continuous real-valued functions $x(\tau)$ on the real interval $-1 \leqslant \tau \leqslant +1$. Addition and multiplication by scalars are defined as in Example 5. Since, from $x, y \in C$, it follows that $x+y \in C$ and $\alpha x \in C$ for all scalars α, and since the axioms are satisfied, we obtain another real vector space which will be denoted by C.

Of course we can replace the interval $-1 \leqslant \tau \leqslant +1$ by an arbitrary closed interval $\rho \leqslant \tau \leqslant \sigma$ to obtain a vector space $C(\rho, \sigma)$.

Example 8. Let τ be a variable angle measured in radians. The functions $\cos k\tau$ and $\sin k\tau$ are periodic of periodicity 2π for each natural number k and so can be thought of as real-valued functions on the circumference of a circle. The same is true for all finite sums of the form

$$x(\tau) = \tfrac{1}{2}\alpha_0 + \sum_{k=1}^{\infty} \{\alpha_k \cos k\tau + \beta_k \sin k\tau\} \tag{2}$$

with real coefficients α_k and β_k, all but a finite number of which are zero. (The factor $\tfrac{1}{2}$ in the term $\tfrac{1}{2}\alpha_0$ is introduced for purely technical reasons—see Example 3.1;8.) If we again define addition and multiplication by scalars as in Example 5, we see that, if x and y are of the form (2), then so are $x+y$ and αx. Also the seven axioms are satisfied and hence we obtain the real vector space of *trigonometric polynomials* which we denote by T.

Example 9. Let $\xi_1, \ldots, \xi_n$ be real variables. A linear form in these variables is a function x which can be represented in the form

$$x(\xi_1, \ldots, \xi_n) = \alpha_1 \xi_1 + \ldots + \alpha_n \xi_n = \sum_{k=1}^{n} \alpha_k \xi_k$$

with real coefficients $\alpha_1, \ldots, \alpha_k$. It is easy to verify that the set of all linear forms in the variables $\xi_1, \ldots, \xi_n$ together with their natural addition and multiplication by scalars is a real vector space L_n.

Example 10. The sum $x+y$ of two complex numbers and the product αx of a complex number x by a real number α are again complex numbers and the seven axioms are satisfied. Hence the set of complex numbers with the usual addition and multiplication is a real vector space which we will denote by K. (In this context, only the multiplication by real numbers is relevant and the multiplication of arbitrary complex numbers is not needed.)

Example 11. In Example 10, the set of complex numbers can be replaced by the set of real numbers. The latter then becomes a real vector space which we call the vector space of real scalars and denote by S_R.

2.1.2 Complex Vector Spaces

The definition of the concept of a vector space can be straight-forwardly modified so that the complex numbers appear as scalars in place of the real numbers. We will then use the term 'complex vector space'.

Definition 2. A 'complex vector space' E is a set in which an addition $(x, y \in E \to z = x + y \in E)$ and a multiplication by complex numbers $(x \in E, \alpha \text{ complex} \to u = \alpha x \in E)$ are defined in such a way that the seven axioms of Definition 1 are satisfied.

We will again refer to the elements of a complex vector space as vectors and to the complex numbers as scalars in this context.

The theories of real and of complex vector spaces have very many results in common. Because of this, it is convenient in the following to make the convention that all results, which are not specifically stated to apply to only one type of vector space, apply to both.

Examples of Complex Vector Spaces

Example 12. In Examples 2 and 3 (R_n, F, F_0), if we use ordered sets and sequences of complex numbers (instead of real numbers) and use complex scalars, we obtain three examples of complex vector spaces. Similarly we obtain a complex version of Example 4, by replacing $F(A)$ with the set of all mappings of A into the complex numbers and at the same time use complex scalars.

Example 13. Corresponding to Examples 5 and 6 $(P_n$ and $P)$, we obtain complex vector spaces by leaving the variable τ to be real but allowing the coefficients α_k of the polynomials to be complex and using complex scalars.

Notice that the set of polynomials with complex coefficients can also be made into a real vector space by considering only the multiplication by real scalars. This set becomes a complex vector space as soon as arbitrary complex numbers are allowed as scalars.

Naturally we could also take τ to be a complex variable and, in this way, we would obtain new examples of vector spaces insofar as the elements are new. However it is well known that a polynomial in a complex variable τ is completely determined by its values on the real numbers so that in fact no essentially different example is found in this way.

Example 14. Example 7 *(C)* can be made into a complex vector space by taking the functions $x(\tau)$ to be complex-valued continuous functions on the real interval $-1 \leqslant \tau \leqslant +1$. The sum of two functions of this type and the product of one by a complex scalar are again functions of the same type.

Example 15. A complex vector space is constructed from Example 8 *(T)*, if the coefficients α_k and β_k are allowed to be complex and the scalars are taken to be complex. In view of Euler's formulae

$$e^{ik\tau} = \cos k\tau + i \sin k\tau$$

$$\cos k\tau = \frac{e^{ik\tau} + e^{-ik\tau}}{2} \qquad \sin k\tau = \frac{e^{ik\tau} - e^{-ik\tau}}{2i}$$

the complex trigonometric polynomials can be written more concisely in the form

$$x(\tau) = \sum_{k=-\infty}^{+\infty} \alpha_k e^{ik\tau} \qquad (\alpha_k = 0 \text{ for all but a finite number of } k\text{'s}).$$

Example 16. The complex version of Example 9 (L_n) is also easy to define. We merely have to take the coefficients α_k and the scalars to be complex. It is also possible to choose the variables ξ_k to be real or complex.

Example 17. Finally the set of all complex numbers is itself a complex vector space with the usual addition and multiplication. We call it the vector space of complex scalars and denote it by S_K.

Problems

1. Is the set of all integers with their usual addition and multiplication by real scalars a vector space?

2. Show that the set of all vectors $x = (\xi_1, \xi_2, \xi_3) \in R_3$ for which $2\xi_1 - \xi_2 + \xi_3 = 0$ is a vector space with the operations defined in R_3. Is this still true for the set of those vectors for which $2\xi_1 - \xi_2 + \xi_3 = 1$?

3. Show that the set of all polynomials $x(\tau) \in P$ (Example 2.1; 6) for which $\int_{-1}^{+1} x(\tau) \, d\tau = 0$ is a vector space with the operations defined in P. Is this still true if the condition is $\int_{-1}^{+1} x(\tau) \, d\tau = 1$?

15

4. Let E_1, E_2 be real vector spaces. Let F be the set of all ordered pairs (x_1, x_2), where $x_1 \in E_1$ and $x_2 \in E_2$, and define

$$(x_1, x_2) + (y_1, y_2) = (x_1 + y_1, x_2 + y_2)$$
$$\alpha(x_1, x_2) = (\alpha x_1, \alpha x_2).$$

Prove that F is a vector space. This is known as the *direct product* of E_1 and E_2. (If E_2 is spanned by a single element $u \neq 0$ (see Definition 2.4; 2), then the construction of the direct product is also referred to as adjoining the vector u to the space E_1.)

2.2 Rules for Calculation in Vector Spaces

2.2.1 Sums of Finitely Many Vectors

Addition in a vector space is initially defined only for two terms. If we are given three vectors $x, y, z \in E$, we can first form $x + y$ and then the sum of this vector with z, i.e., $(x + y) + z \in E$. Similarly we can form $x + (y + z) \in E$. The associative law A1 now says that these two expressions represent the same vector. In other words, it does not matter in which way the three terms are bracketed together (for a given order) and so the brackets may be omitted. Hence, for given x, y, z, $w = x + y + z$ is a well-defined vector in E. The same is true for an arbitrary number of vectors. For example, for four vectors, in view of A1,

$$x + (y + (z + w)) = (x + y) + (z + w) = ((x + y) + z) + w$$

so that it is possible to write $x + y + z + w$.

In order to avoid possible misunderstandings, we note here that a sum of *finitely many* vectors has a well-defined meaning but that *infinite series* of vectors are meaningless (nevertheless see 12.3).

A convenient notation for a sum of finitely many vectors is obtained by using the summation sign.

$$x = x_1 + \ldots + x_n = \sum_{k=1}^{n} x_k.$$

2.2.2 The Zero-Vector and Inverse Vectors

In connection with A2, we note that, in a vector space E, there is only *one* zero-element. If 0 and 0^* are two such elements, i.e. $0 + x = x$ and $y + 0^* = y$ for all $x, y \in E$, then in particular $0 = 0 + 0^* = 0^*$. (Put $x = 0^*$ and $y = 0$.)

Similarly we note that, in connection with A3, there is only *one* inverse vector of a given vector $x \in E$. If $(-x)$ and $(-x)^*$ are two such vectors, then

$$(-x)^* = 0 + (-x)^* = ((-x) + x) + (-x)^*$$
$$= (-x) + (x + (-x)^*) = (-x) + 0 = (-x).$$

Note that we have not used either $x+(-x)=0$ or $(-x)^*+x=0$ in proving this. Axiom A3 could therefore be stated in the following weaker form: 'To each $x \in E$ there is a left inverse element $(-x)$ such that $(-x)+x=0$ and a right inverse element $(-x)^*$ such that $x+(-x)^*=0$'. From the above argument it follows that $(-x)=(-x)^*$. A corresponding result is also true for the zero-vector.

According to A3, $-x$ also has an inverse vector which is actually equal to x again (because $(-x)+x=x+(-x)=0$). Thus

$$-(-x) = x. \tag{1}$$

Since $(x+y)+((-y)+(-x))=x+((y+(-y))+(-x))=x+(-x)=0$ and similarly $((-y)+(-x))+(x+y)=0$, the inverse of a sum of vectors x and y is given by

$$-(x+y) = (-y)+(-x). \tag{2}$$

2.2.3 Subtraction of Vectors

Corresponding to two given vectors $x, y \in E$, there is a unique vector $z \in E$ such that

$$z+x = y \tag{3.1}$$

(viz. $z=y+(-x)$).
This follows because, adding $-x$ to both sides of equation(3.1) on the right, we have

$$y+(-x) = (z+x)+(-x) = z+(x+(-x)) = z$$

and with this z we obtain

$$z+x = (y+(-x))+x = y+0 = y.$$

Instead of $y+(-x)$ we usually write $y-x$ and call this the difference of y and x (in this order). We must remember however that $y-x$ has no other meaning than $y+(-x)$. The two equations

$$z+x = y \quad \text{and} \quad z = y-x \tag{3.2}$$

therefore have the same meaning.

2.2.4 Rules for Multiplication by Scalars

If $x \in E$ and α is a scalar, then

$$\alpha x = 0 \text{ if and only if } \alpha = 0 \quad \text{or} \quad x = 0. \tag{4}$$

Proof. 1. For an arbitrary scalar α,

$$\alpha 0 = \alpha(0+0) = \alpha 0 + \alpha 0$$

17

and, adding $-(\alpha 0)$ to both sides, it follows that $\alpha 0 = 0$.

2. For an arbitrary vector $x \in E$

$$0x = (0+0)x = 0x + 0x$$

and, adding $-(0x)$ to both sides, it follows that $0x = 0$.

3. Let $\alpha x = 0$ and assume that $\alpha \neq 0$.
By 1, M1 and M2, it follows that

$$x = 1x = \left(\frac{1}{\alpha}\alpha\right)x = \frac{1}{\alpha}(\alpha x) = \frac{1}{\alpha}0 = 0.$$

Since $x+(-1)x = 1x+(-1)x = 0x = 0$, and similarly $(-1)x + x = 0$, it follows that

$$-x = (-1)x. \tag{5}$$

Multiplying both sides of (5) by an arbitrary scalar α, it also follows that

$$\alpha(-x) = (-\alpha)x = -(\alpha x). \tag{6}$$

If k is a natural number, then

$$kx = x+x+\ldots+x \ (k \text{ terms}). \tag{7}$$

Proof. The equation (7) is true for $k = 1$ (M1). We use induction on k to prove (7) in general. Assume that (7) is true for a sum of $k-1$ terms, then

$$kx = ((k-1)+1)x = (k-1)x + 1x = (x+x+\ldots+x)+x$$

where there are $k-1$ terms in the brackets. Therefore (7) is also true for k and hence, by induction, for all natural numbers.

By using (6), it also follows that

$$(-x)+(-x)+\ldots+(-x) = k(-x) = (-k)x = -(kx)$$

where there are k terms on the left-hand side.

2.2.5 Commutativity of Addition

Theorem 1. The addition in a vector space E is commutative, i.e.,

$$x+y = y+x \text{ for all } x, y \in E.$$

Proof. $\begin{aligned}
x+y &= -(-x)+(-(-y)) &&(1)\\
&= -((-y)+(-x)) &&(2)\\
&= -((-1)y+(-1)x) &&(5)\\
&= -((-1)(y+x)) &&(D1)\\
&= -(-(y+x)) &&(5)\\
&= y+x. &&(1)
\end{aligned}$$

In view of the commutativity and associativity of addition, the sum of a finite number of vectors is completely determined by these vectors and does not depend on their order or on the way in which they are bracketed.

In the later sections we will often make use of the rules of calculation which we have proved here without making any special reference to them. However, the reader is advised occasionally to see how the steps in later proofs may be reduced to the axioms and these rules for calculation.

2.3 Linear Combinations and Calculations with Subsets of a Vector Space

2.3.1 Linear Combinations

Suppose E is a vector space, $x_1,\dots,x_n \in E$, and $\alpha_1,\dots,\alpha_n$ are scalars. For each index k, we can form the vector $\alpha_k x_k$ and then the sum of these

$$y = \sum_{k=1}^{n} \alpha_k x_k = \alpha_1 x_1 + \dots + \alpha_n x_n \in E. \tag{1}$$

An expression of this form is called a *linear combination* of the vectors $x_1,\dots,x_n$.

We will often have to consider linear combinations of vectors from a subset A of E, i.e., expressions of the form (1), where $x_1,\dots,x_n \in A$. If A is finite, say $A = \{x_1,\dots,x_n\}$, all linear combinations of elements from A may be expressed in the form (1). On the other hand, if A is arbitrary, we can express all linear combinations of vectors in A in the form

$$y = \sum_{x \in A} \alpha_x x, \tag{2}$$

where it will always be understood, in an expression of the form (2), that the scalars α_x are non-zero for only a finite number of the vectors $x \in A$ and that terms with $\alpha_x = 0$ are to be left out of the summation. Accordingly (2) is to be read as follows: x runs through all the vectors of the subset A of E. To each $x \in A$ there is assigned a scalar α_x and the product $\alpha_x x$ is formed. y is the sum of the finite set of the vectors $\alpha_x x$ for which $\alpha_x \neq 0$.

2.3.2 Calculations with Subsets of a Vector Space

Let E be a vector space, let X and Y be subsets of E and let α be a scalar.

Definition 1. The sum $Z = X + Y$ of the sets $X \subseteq E$ and $Y \subseteq E$ is the subset Z of E given by $Z = \{z; z = x + y, x \in X, y \in Y\}$.

2. The product $U = \alpha X$ of the set $X \subseteq E$ with the scalar α is the subset $U \subseteq E$ given by $U = \{u; u = \alpha x, x \in X\}$.

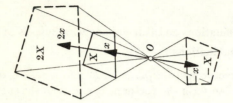

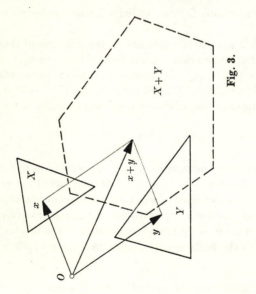

Fig. 3.

Thus the set $X + Y$ is constructed by forming all possible sums of vectors from X with vectors from Y and αX is constructed by multiplying all the vectors of X by α. We write $-X$ instead of $(-1)X$ and $Y - X$ instead of $Y + (-X)$.

Using Definition 1, we can also define linear combinations $\alpha X + \beta Y$ of the sets X and Y, and more generally of a finite number of subsets.

The operations we have defined are illustrated by Fig. 3 which refers to the vector space G_2 of vectors in a plane. (Example 2.1; 1.) In this picture the set $X \subseteq G_2$ is represented by the figure which is formed by the end points of the vectors $x \in X$ when these vectors are drawn as arrows from a fixed point O.

By using similar diagrams, we are able to demonstrate that some expected rules for calculation are not in fact true. For example $X + (-X) = \{0\}$ if and only if X contains only one vector. The relation $A + A = 2A$ is not correct for example when A consists of two distinct non-zero vectors. However, it is correct when A is a convex set (see 2.6).

Problems

1. The set of all vectors $x = (\xi_1, \xi_2, \xi_3, \xi_4) \in R_4$, which are linear combinations of the three vectors $(1, 1, -1, -1)$, $(1, 0, 1, 0)$, $(0, 1, 1, 0)$ is a vector space with the operations as defined in R_4. What is the obvious generalization of this?
2. Show that every vector $x \in R_4$ can be written in exactly one way as a linear combination of the four vectors $(1, 1, 0, 0)$, $(1, -1, 0, 0)$, $(0, 0, 1, 1)$, $(0, 0, 1, -1)$.

2.4 Subspaces of a Vector Space

2.4.1 Subspaces

Definition 1. A subset L of a vector space E is a subspace of E if L is itself a vector space with the binary operations defined in E.

Clearly for a subset L of E to be a subspace it is necessary that,

$$\text{if } x, y \in L, \quad \text{then } x + y \in L \tag{1}$$

and $\qquad$ if $x \in L$, $\quad$ then $\alpha x \in L$ for all scalars α. $\hfill (2)$

If 0^* is the zero element of the subspace L, then $0^* + 0^* = 0^*$. Adding the inverse -0^* of 0^* in E to both sides of this equation, we have $0^* = 0^* + 0^* + (-0^*) = 0^* + (-0^*) = 0$, i.e., L must have the same zero-element as E. Hence, for the set L to be a subspace, it is also necessary that

$$0 \in L. \tag{3}$$

21

Similarly, it is also necessary that,

$$\text{if } x \in L, \qquad \text{then } -x \in L. \tag{4}$$

The inverse $-x$ of x in E is also the inverse of x in L.

The conditions (1), (2), (3), (4) are in fact sufficient for a subset L of E to be a subspace. Conditions (1) and (2) mean that the binary operations defined in E are also binary operations for L. The axioms A1, M1, M2, D1, D2 are satisfied in L because they are satisfied by all vectors of E, and conditions (3) and (4) mean that A2 and A3 are also satisfied in L.

Actually (1), (2), (3) and (4) are not independent and in fact, when L is not the empty set, (3) and (4) follow from (1) and (2). From (2), it follows that if $x \in L$, then $(-1)x = -x \in L$, i.e., that (4) is satisfied. Then it follows from (1) that, for any vector $x \in L (L \neq \varnothing)$, $x + (-x) = 0 \in L$, i.e., that (3) is satisfied. Since a vector space L is never empty (0 always belongs to L), we have proved the following Theorem.

Theorem 1. A subset L of a vector space E is a subspace of E if and only if
 1. $L \neq \varnothing$,
 2. $x, y \in L$ implies $x + y \in L$,
and 3. $x \in L$ implies $\alpha x \in L$ for all scalars α.

The conditions 2. and 3. can be combined into the one condition

$$x, y \in L \text{ implies } \alpha x + \beta y \in L \text{ for all scalars } \alpha, \beta. \tag{5}$$

By repeated application of (5), it is clear that a subspace contains all linear combinations of its elements.

Every vector space E is obviously a subspace of itself and the subset $\{0\}$ is a subspace of E. Subspaces L of E which are not equal to E are referred to as *proper subspaces* of E.

Example 1. In G_2 (Example 2.1; 1), the vectors which are parallel to a fixed line form a proper subspace of G_2. In G_3 the vectors parallel to a fixed line or to a fixed plane also form proper subspaces of G_3. In both cases, these methods produce all possible proper subspaces except $\{0\}$.

Theorem 2. If L_1 is a subspace of E and L_2 is a subspace of L_1, then L_2 is a subspace of E.
The proof is easy using (5).

Theorem 3. The intersection of a family of subspaces of a vector space E is a subspace of E.

Proof. Let $\mathscr{S}$ be a family of subspaces of E and let $H = \bigcap(\mathscr{S})$. We show that (5) is satisfied in H. Suppose $x, y \in H$, i.e., $x, y \in L$ for all subspaces $L \in \mathscr{S}$.

Then by (5) $\alpha x + \beta y \in L$ for all scalars α, β and all $L \in \mathscr{S}$. Therefore $\alpha x + \beta y \in \bigcap(\mathscr{S}) = H$ for all α, β. Further $H \neq \varnothing$ because $0 \in L$ for all $L \in \mathscr{S}$ and hence $0 \in H$.

2.4.2 Linear Hulls

Now let A be a subset of a vector space E, and let $\mathscr{S}_A$ be the family of all subspaces L of E which contain A. By Theorem 3, $H_A = \bigcap(\mathscr{S}_A)$ is again a subspace of E.

Definition 2. The subspace H_A is the subspace of E 'spanned' by the subset $A \subseteq E$ (or the 'linear hull' of A) and is denoted by $L(A)$.

If A consists of only a finite number of elements, $A = \{a_1, \ldots, a_n\}$ say, we will denote the linear hull by $L(a_1, \ldots, a_n)$. Since $A \subseteq L$ for all $L \in \mathscr{S}_A$, it follows from Theorem 1.2;3 that $A \subseteq L(A)$. Hence by definition $L(A) \in \mathscr{S}_A$, and, by Theorem 1.2;4, $L(A)$ is the least element of $\mathscr{S}_A$, i.e.,

Theorem 4. The linear hull $L(A)$ of a subset $A \subseteq E$ is the smallest subspace of E which contains A.

Theorem 5. If $A \subseteq E$ is a subspace of E, then $L(A) = A$.

Proof. Because $A \subseteq A$, $A \in \mathscr{S}_A$ and hence $L(A) \subseteq A$. On the other hand, we have already shown that $A \subseteq L(A)$.

Theorem 6. If $A_1 \subseteq A_2$, then $L(A_1) \subseteq L(A_2)$.

Proof. If $L \in \mathscr{S}_{A_2}$, then $A_2 \subseteq L$ and hence $A_1 \subseteq L$. Therefore $L \in \mathscr{S}_{A_1}$. Hence $\mathscr{S}_{A_2} \subseteq \mathscr{S}_{A_1}$ and the assertion follows from Theorem 1.2;2.

From these theorems, it follows that the construction of linear hulls has the following three properties.
1. $A \subseteq L(A)$ (it extends A)
2. If $A_1 \subseteq A_2$, then $L(A_1) \subseteq L(A_2)$ (it is monotone)
3. $L(L(A)) = L(A)$ (it is idempotent)
(3. follows from Theorem 5 by substituting $L(A)$ for A.)
In general, a correspondence $A \to L(A)$ with these three properties is called a *closure operator*. Such closure operators occur in many parts of mathematics (e.g., the convex hull of a set of vectors (2.6), the topological closure of a set in a topological space or the algebraic closure of a field).

Theorem 7. If $A \subseteq L(B)$ and $B \subseteq L(A)$, then $L(A) = L(B)$.

23

Proof. In view of 2. and 3., and since $A \subseteq L(B)$

$$L(A) \subseteq L(L(B)) = L(B). \qquad \text{Similarly } L(B) \subseteq L(A).$$

Theorem 8. *If A is a non-empty set, then $L(A)$ is the set of all linear combinations of vectors in A.*

Proof. Let the set of all linear combinations of vectors in A be denoted by $L^*(A)$.

1. Since $L(A)$ is a subspace of E and $A \subseteq L(A)$, it follows from (5) that $L^*(A) \subseteq L(A)$.

2. Since $A \subseteq L^*(A)$ and A is not empty, $L^*(A)$ is also not empty and so we can use condition (5) to show that $L^*(A)$ is a subspace of E. If $x = \sum\limits_{z \in A} \alpha_z z$ and $y = \sum\limits_{z \in A} \beta_z z$ are arbitrary vectors in $L^*(A)$, then $\lambda x + \mu y = \sum\limits_{z \in A} (\lambda \alpha_z + \mu \beta_z) z$ and is in $L^*(A)$ for all scalars λ, μ. That is, condition (5) is satisfied.

3. Since $A \subseteq L^*(A)$, $L^*(A) \in \mathscr{S}_A$ and hence, by Theorem 1,2:1, $L(A) = \bigcap (\mathscr{S}_A) \subseteq L^*(A)$.

Definition 3. *If $A \subseteq E$ and $L(A) = E$, then A is called a 'spanning set' for E.*

By Theorem 8, every vector $z \in E$ can be written in the form $z = \sum\limits_{x \in A} \alpha_x x$, when A is a non-empty spanning set for E. The empty set can only be a spanning set for E when $E = \{0\}$.

Example 2. *A subset $A \subseteq G_3$ (Example 2.1:1) is a spanning set for G_3 if and only if A contains three vectors which are not parallel to the same fixed plane.*

2.4.3 Direct Sums of Subspaces

Theorem 9. *The sum of two subspaces L_1, L_2 of a vector space E is equal to the linear hull of their set union, i.e., $L_1 + L_2 = L(L_1 \cup L_2)$. In particular, this means that $L_1 + L_2$ is a subspace of E.*

Proof. 1. Suppose $x, y \in L_1 + L_2$, i.e., $x = x_1 + x_2$ and $y = y_1 + y_2$ where $x_1, y_1 \in L_1$ and $x_2, y_2 \in L_2$. Then $\alpha x + \beta y = (\alpha x_1 + \beta y_1) + (\alpha x_2 + \beta y_2) \in L_1 + L_2$ for all scalars α, β. Hence $L_1 + L_2$ is a subspace of E. Since $0 \in L_1$, $L_1 + L_2 \supseteq L_2$ and similarly $L_1 + L_2 \supseteq L_1$. Therefore $L_1 + L_2 \supseteq L_1 \cup L_2$ and hence $L_1 + L_2 \supseteq L(L_1 \cup L_2)$.

2. By Theorem 8, $L(L_1 \cup L_2)$ contains every sum of a vector from L_1 with a vector from L_2 and hence $L(L_1 \cup L_2) \supseteq L_1 + L_2$.

Definition 4. If $L_1 \cap L_2 = \{0\}$, the subspace $L_1 + L_2 = L$ is called the 'direct sum' of the subspaces L_1 and L_2 and in this case we write $L = L_1 \oplus L_2$.

Theorem 10. If $L = L_1 \oplus L_2$, then each vector $x \in L$ has a unique decomposition into a sum $x = x_1 + x_2$ where $x_1 \in L_1$ and $x_2 \in L_2$.

Proof. It is clear that x has such a decomposition because $L = L_1 + L_2$. Suppose that $x = x_1 + x_2$ and $x = y_1 + y_2$ are two such decompositions, then $x_1 - y_1 = y_2 - x_2 \in L_1 \cap L_2 = \{0\}$ and hence $x_1 = y_1$ and $x_2 = y_2$.

The extension of the concept of a direct sum to include the case of an arbitrary finite number of subspaces is straightforward, and Theorem 10 remains true in its correspondingly extended form.

Problems

1. Does the vector $x = (-4, -5, 7)$ belong to the linear hull $L(x_1, x_2)$ of the vectors $x_1 = (1, -1, 2)$ and $x_2 = (2, 1, -1)$?

2. Let L_1, L_2 be subspaces of the vector space E. Prove that, if $L_1 \cup L_2$ is also a subspace of E, then $L_1 \subseteq L_2$ or $L_2 \subseteq L_1$.

3. Show that the set of all continuous functions $x(\tau) \in C$ for which $x(\tau) = x(-\tau)$ $(-1 \leqslant \tau \leqslant +1)$ is a subspace L of C.

4. Show that the set of all $x(\tau) \in C$ for which $\int_{-1}^{+1} \tau x(\tau) \, d\tau = 0$ is a subspace of C. Is this still true when the condition is $\int_{-1}^{+1} \tau x(\tau) \, d\tau = 1$?

5. Let $\tau_1, \ldots, \tau_n$ be real numbers. Show that the set of all polynomials $x(\tau) \in P$ for which $x(\tau_k) = 0$ $(k = 1, \ldots, n)$ is a subspace of P. Is this still true when the condition is $x(\tau_k) = \tau_k$ $(k = 1, \ldots, n)$?

6. The direct product F of the vector spaces E_1 and E_2 (Problem 2.1;4) contains the subspaces $L_1 = \{(x_1, x_2); x_2 = 0\}$ and $L_2 = \{(x_1, x_2); x_1 = 0\}$. Show that $F = L_1 \oplus L_2$.

7. Let L_1 and L_2 be subspaces of the vector space E. Prove that $L_1 + L_2 = L_1 \oplus L_2$ if and only if, from $x_1 \in L_1$, $x_2 \in L_2$ and $x_1 + x_2 = 0$, it follows that $x_1 = x_2 = 0$.

2.5 Cosets and Quotient Spaces

2.5.1 Cosets

If L is a subspace of the vector space E and x is a vector in E, we can form the sum $L + \{x\}$ which we will write briefly as $L + x$. Thus $L + x$ is the set of vectors $y \in E$ which can be written in the form $y = a + x$ with $a \in L$.

Definition 1. If L is a subspace of the vector space E, then for each $x \in E$, the set $L + x$ is said to be a 'coset' of L.

The *cosets* defined here are more usually known as *linear manifolds*. We will use the shorter terminology because these cosets are in fact the additive cosets of the subspace in the group theoretic sense.

To avoid possible misunderstandings, we note straight away that the coset $L + x$ is not a subspace of E unless $x \in L$.

Example 1. Let L be the subspace of the space G_3 (Example 2.1;1) which consists of all the vectors parallel to a fixed line g. Each coset of L consists of those vectors whose endpoints (when all vectors are drawn as arrows from a fixed point O) lie on a fixed line parallel to g. All the cosets of L can be obtained in this way.

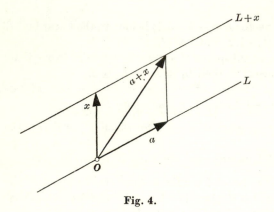

Fig. 4.

Example 2. If $L = E$, then L is the only coset of itself. If $L = \{0\}$, then each coset consists of exactly one vector and each vector forms a coset. Every subspace L of E is a coset of itself, because obviously $L = L + 0$.

Theorem 1. $L + x = L$ *if and only if* $x \in L$.

Proof. 1. Suppose $L + x = L$, then $0 = a + x$ for some $a \in L$ and $x = -a \in L$.

2. Suppose $x \in L$, then obviously $L + x \subseteq L$. Since $a = (a - x) + x$ and $a - x \in L$ for each $a \in L$, we also have $L \subseteq L + x$. Hence $L + x = L$.

If $x \notin L$, then L and $L + x$ are disjoint. Because, if $y \in L \cap (L + x)$, then $y \in L$ and $y = a + x$ where $a \in L$ and hence $x \in L$.

Theorem 2. $L + x_1 = L + x_2$ *if and only if* $x_1 - x_2 \in L$.

Proof. It is easy to verify that $(L+x)+y = L+(x+y)$.

1. If $L+x_1 = L+x_2$, then, adding $(-x_2)$, $L+(x_1-x_2) = L+0 = L$ and hence $x_1 - x_2 \in L$. (Theorem 1).

2. If $x_1 - x_2 \in L$, then $L+(x_1-x_2) = L$ and hence $L+x_1 = L+x_2$.

If $x_1 - x_2 \notin L$, then $L+(x_1-x_2)$ is disjoint from L and therefore $L+x_1$ and $L+x_2$ are also disjoint. *Two cosets of L are therefore either identical or disjoint.*

Theorem 3. *Let N be a coset of a subspace L of E and let $y \in E$. Then $N = L+y$ if and only if $y \in N$.*

Proof. Let $N = L+x$. By Theorem 2, $L+x = L+y$ if and only if $y-x \in L$, i.e., if and only if $y \in L+x = N$.

Theorem 4. *A subspace L is uniquely determined by a coset of L.*

Proof. Suppose N is a coset of two subspaces L_1 and L_2. Then, for every $y \in N$, $N = L_1 + y = L_2 + y$ and, by adding $-y$, $L_1 = L_1 + 0 = L_2 + 0 = L_2$.

2.5.2 Quotient Spaces

The cosets of a subspace L of E can be added according to Definition 2.3;1. Thus

$$(L+x)+(L+y) = L+(x+y). \tag{1}$$

$z \in (L+x)+(L+y)$ means that $z = (a_1+x)+(a_2+y)$ with $a_1, a_2 \in L$, and therefore $z = (a_1+a_2)+(x+y) \in L+(x+y)$.

$z \in L+(x+y)$ means that $z = a+(x+y)$ with $a \in L$ and therefore $z = (0+x)+(a+y) \in (L+x)+(L+y)$.

Similarly, it is possible to multiply cosets by scalars and then, for $\alpha \neq 0$,

$$\alpha(L+x) = L+\alpha x, \tag{2}$$

which can be proved in the same way as (1). (For $\alpha = 0$, the equation (2) may be used as the definition of $0(L+x)$.)

Theorem 5. *If N, N_1, N_2 are cosets of a subspace L of E, then $N_1 + N_2$ and αN are also cosets of L.*
Alternatively. Addition and multiplication by scalars are binary operations in the family $\mathcal{N}_L$ of all cosets of L.

The question now arises as to whether $\mathcal{N}_L$ is a vector space with these binary operations. In order to answer this, we must see if the axioms of Definition 2.1;1 are satisfied.

A1. By (1), $\{(L+x_1)+(L+x_2)\}+(L+x_3) = \{L+(x_1+x_2)\}+(L+x_3)$
$$= L+(x_1+x_2+x_3)$$

and the same is true for $(L+x_1)+\{(L+x_2)+(L+x_3)\}$.

A2. L is the zero element of $\mathcal{N}_L$, since

$$L+(L+x) = L+x \text{ for all } x \in E.$$

A3. The inverse element of $L+x$ is $L-x$.

M1. By (2), $1(L+x)=L+1x=L+x$.

M2. $\alpha(\beta(L+x)) = \alpha(L+\beta x)$
$$= L+\alpha(\beta x) = L+(\alpha\beta)x$$
$$= (\alpha\beta)(L+x) \text{ by (2)}.$$

D1. $\alpha\{(L+x)+(L+y)\} = \alpha(L+(x+y))$
$$= L+\alpha(x+y) = (L+\alpha x)+(L+\alpha y)$$
$$= \alpha(L+x)+\alpha(L+y) \text{ by (1) and (2)}.$$

D2. $(\alpha+\beta)(L+x) = L+(\alpha+\beta)x$
$$= L+\alpha x+\beta x$$
$$= \alpha(L+x)+\beta(L+x) \text{ by (1) and (2)}.$$

Our question can therefore be answered positively.

Definition 2. The vector space whose elements are the cosets of the subspace L of E and whose binary operations are defined by (1) and (2) is called the 'quotient space' of E by L and is denoted by E/L.

Theorem 6. Let L be a subspace of the vector space E and let D be a subspace of L. Then L/D is a subspace of E/D, and a coset $D+x \in L/D$ if and only if $x \in L$.

Proof. The space L/D consists of those cosets $D+x$ for which $x \in L$ and is therefore a subspace of E/D. Now, if $D+x_1=D+x$ where $x \in L$, then by Theorem 2, $x_1-x \in D \subseteq L$ and hence $x_1 \in L$.

In view of Theorem 6 we can now associate with each subspace L lying between D and E ($D \subseteq L \subseteq E$) the subspace L/D of E/D, i.e., we have a mapping

$$L \to L/D. \tag{3}$$

Theorem 7. The correspondence (3) defines a 1-1 mapping of the set of all subspaces of E which contain D onto the set of all subspaces of E/D. If $D \subseteq L_1 \subseteq L_2$, then $L_1/D \subseteq L_2/D$ and conversely.

Proof. 1. An arbitrary subspace $\bar{L}$ of E/D is the image of the subspace L of E which is the union of all the cosets of D which are in $\bar{L}$. Hence the mapping is onto E/D.

2. Suppose $L_1/D \subseteq L_2/D$. If $x_1 \in L_1$, then $D+x_1 \in L_1/D \subseteq L_2/D$ and, by Theorem 6, $x_1 \in L_2$. Hence $L_1 \subseteq L_2$. It is easy to see the converse, that if $L_1 \subseteq L_2$ then $L_1/D \subseteq L_2/D$.

3. Suppose $L_1/D = L_2/D$, then, by 2., $L_1 \subseteq L_2$ and $L_2 \subseteq L_1$ so that $L_1 = L_2$ and the mapping is 1-1.

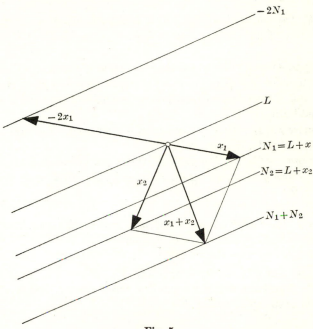

Fig. 5.

Theorem 8. If E is the sum of two subspaces, say, $E = L_1 + L_2$ and $L_1 \cap L_2 = D$ then $E/D = L_1/D \oplus L_2/D$ (see Definition 2.4;4).

Proof. 1. For each coset $D+x \in E/D$ there is a decomposition into

$$D+x = D+(x_1+x_2) = (D+x_1)+(D+x_2),$$

where $x_1 \in L_1$ and $x_2 \in L_2$. Hence $D+x \in L_1/D + L_2/D$.

2. If $D+x \in L_1/D \cap L_2/D$, then $x \in L_1 \cap L_2 = D$ and therefore $D+x = D$. Hence $L_1/D \cap L_2/D = \{D\}$ consists only of the zero-element of E/D.

29

2.5.3 The Generation of Cosets

In this section we investigate the possibility of finding a smallest coset $N(A)$ which contains a given set A of vectors from E. The procedure is similar to the one used in 2.4, where the problem was to find the smallest subspace $L(A)$ containing A.

Let $\mathcal{N}$ be a family of cosets in E and let $\mathcal{L}$ be the family of subspaces which correspond to these cosets. (Different cosets in $\mathcal{N}$ may correspond to different subspaces of E.)

Theorem 9. The intersection $\bigcap(\mathcal{N}) \subseteq E$ is either empty or it is a coset in E. In the latter case $\bigcap(\mathcal{L})$ is the subspace which corresponds to $\bigcap(\mathcal{N})$.

Proof. Suppose that $\bigcap(\mathcal{N}) \neq \varnothing$ so that there is an element $x_0 \in \bigcap(\mathcal{N})$. It is sufficient to show that $\bigcap(\mathcal{N}) = x_0 + \bigcap(\mathcal{L})$, because $\bigcap(\mathcal{L})$ is certainly a subspace of E. The subspace of E which corresponds to the coset $N \in \mathcal{N}$ will be denoted by L_N.

1. Suppose $x \in x_0 + \bigcap(\mathcal{L})$. Then $x = x_0 + y$ where $y \in L_N$, and therefore $x \in N$ for all $N \in \mathcal{N}$. Hence $x \in \bigcap(\mathcal{N})$.

2. Suppose $x \in \bigcap(\mathcal{N})$. For each N, there is a decomposition of x into $x = x_0 + y$ where $y \in L_N$. Therefore $x - x_0 \in L_N$ for all $N \in \mathcal{N}$ and hence $x \in x_0 + \bigcap(\mathcal{L})$.

Definition 3. The coset $N(A)$ generated by a non-empty subset A of E is the intersection of the family of cosets in E which contain A.

Since A is assumed to be non-empty and $A \subseteq N(A)$, it follows that $N(A)$ is a coset in E and indeed it is the smallest coset which contains A. $N(A)$ is also known as the *affine hull* of A.

Theorem 10. The coset $N(A)$ generated by $A \subseteq E$ consists of those linear combinations of elements of A in which the sum of the coefficients is 1. The subspace corresponding to $N(A)$ consists of those linear combinations of elements of A in which the sum of the coefficients is 0.

Proof. 1. Using Theorem 9, it is easy to see that the subspace corresponding to $N(A)$ is $L_{N(A)} = L(A - x_0)$ where x_0 is an arbitrary element of A. Then $N(A) = x_0 + L(A - x_0)$.

2. By Theorem 2.4;8, $L(A - x_0)$ consists of all linear combinations $z = \sum_{x \in A} \lambda_x (x - x_0)$. But these are exactly those linear combinations of elements of A in which the sum of coefficients is 0.

3. Therefore $N(A)$ consists of the elements of the form $z = x_0 + \sum_{x \in A} \lambda_x (x - x_0)$, i.e., linear combinations of elements of A in which the sum of the coefficients is 1.

It is easy to show that the correspondence $A \to N(A)$ is a *closure operator* (see 2.4).

Problems

1. Show that the set of all vectors $x = (\xi_1, \xi_2, \xi_3, \xi_4) \in R_4$ for which $\xi_1 + 2\xi_2 - \xi_3 + \xi_4 = 1$ and $\xi_1 - \xi_2 + 2\xi_3 - 2\xi_4 = 2$ is a coset N_1. Find the equations which describe the vectors in the corresponding subspace L. Replacing the right-hand sides of the above equations by -2 and 3, produces a second coset N_2 of the same subspace L. Find the equations which describe the vectors in the cosets $N_1 + N_2$ and $2N_1 + 3N_2$.

2. Let L_1 be the subspace of the direct product of E_1 and E_2 which was considered in Problem 2.4;6. Let N be a coset of L_1. What property do all the elements $(x_1, x_2) \in N$ have in common?

3. Answer the same question as in Problem 2 for the cosets of the subspace L of C which was considered in Problem 2.4;4.

4. Answer the same question again for the cosets of the subspace L of P which was considered in Problem 2.4;5. How many polynomials of degree at most $n-1$ does a coset of L contain, when $\tau_1, \ldots, \tau_n$ are all distinct?

5. Let L_1 and L_2 be subspaces of the space E such that $L_1 \subseteq L_2$. Show that, if N_1, N_2 are cosets of L_1, L_2, then either $N_1 \cap N_2 = \varnothing$ or $N_1 \subseteq N_2$. Conversely, if $N_1 \subseteq N_2$, then $L_1 \subseteq L_2$.

2.6 Convex Sets

In this section we restrict the discussion to *real* vector spaces.

2.6.1 Convex Sets

Definition 1. A subset C of a real vector space E is said to be 'convex' if, together with any pair of vectors $x_1, x_2 \in C$, C also contains all linear combinations $y = \lambda_1 x_1 + \lambda_2 x_2$ for which $\lambda_1, \lambda_2 \geqslant 0$ and $\lambda_1 + \lambda_2 = 1$.

The condition $\lambda_1, \lambda_2 \geqslant 0$, which is meaningless for complex numbers, is the reason for the restriction to real vector spaces.

Example 1. Which are the convex sets in the vector spaces G_2 and G_3 (Example 2.1;1)? If $\lambda_1 + \lambda_2 = 1$, then $y = \lambda_1 x_1 + \lambda_2 x_2$ can be written in the form $y = x_1 + \lambda_2 (x_2 - x_1)$. Further, if $\lambda_1, \lambda_2 \geqslant 0$, then $0 \leqslant \lambda_2 \leqslant 1$ so that, as we can see from the following diagram, the end point of y lies on the segment

31

joining the end points of x_1 and x_2, when all the vectors are drawn from a fixed point O.

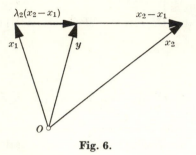

Fig. 6.

Thus, if we draw the vectors of a convex set C of G_2 or G_3 from a point O in the plane or in 3-space, we will obtain a figure C^* with the following property,

C^* contains all the points on the segment joining any two of its points. (1)

Conversely every figure C^* with property (1) gives a convex set of vectors C in G_2 or G_3 consisting of all the arrows drawn from a fixed point O to the points of C^*.

The figures C^* with property (1) are convex bodies in the usual sense of geometry. Thus we have established the connection between convex sets in a vector space and convex bodies in a plane or in 3-space.

Example 2. The set of all functions $x(\tau)$ for which $x(0)=1$ and $x(1)\geqslant 0$ is a convex set in the vector space C (Example 2.1;7). It follows from $x_1(0)=x_2(0)=1$, $x_1(1)\geqslant 0$, $x_2(1)\geqslant 0$, $\lambda_1,\lambda_2\geqslant 0$ and $\lambda_1+\lambda_2=1$, that $\lambda_1 x_1(0)+\lambda_2 x_2(0)=1$ and $\lambda_1 x_1(1)+\lambda_2 x_2(1)\geqslant 0$.

Theorem 1. If C_1 and C_2 are convex sets in a real vector space, then $C=\mu_1 C_1+\mu_2 C_2$ is also convex for all real μ_1,μ_2.

Proof. Suppose $x_1,x_2 \in C$, so that $x_1=\mu_1 c_{11}+\mu_2 c_{21}$ and $x_2=\mu_1 c_{12}+\mu_2 c_{22}$ where $c_{11},c_{12} \in C_1$ and $c_{21},c_{22} \in C_2$. If $\lambda_1,\lambda_2\geqslant 0$ and $\lambda_1+\lambda_2=1$, then

$$y = \lambda_1 x_1 + \lambda_2 x_2$$
$$= \mu_1(\lambda_1 c_{11}+\lambda_2 c_{12})+\mu_2(\lambda_1 c_{21}+\lambda_2 c_{22})$$
$$= \mu_1 c_1 + \mu_2 c_2 \quad \text{where} \quad c_1 \in C_1 \text{ and } c_2 \in C_2.$$

Hence $y \in C$.

Definition 2. Let E be a real vector space and let A be a subset of E. A 'convex linear combination' of the elements of A is a linear combination $y = \sum\limits_{x \in A} \lambda_x x$ where $\lambda_x \geqslant 0$ for all $x \in A$ and $\sum\limits_{x \in A} \lambda_x = 1$.

Theorem 2. A convex set C contains all convex linear combinations of its elements.

Proof. The proof is by induction on the number n of vectors $x_1, \ldots, x_n \in C$ in the convex linear combination $y = \lambda_1 x_1 + \ldots + \lambda_n x_n$. The theorem is true for $n = 2$. Now assume that it is true for all convex linear combinations of at most $n - 1$ vectors.

Since at least one of the coefficients, say λ_n, in the convex linear combination $y = \lambda_1 x_1 + \ldots + \lambda_n x_n$ is strictly positive, we can write y in the form

$$y = \lambda_1 x_1 + \mu \left(\frac{\lambda_2}{\mu} x_2 + \ldots + \frac{\lambda_n}{\mu} x_n \right)$$

where $\mu = \lambda_2 + \ldots + \lambda_n (\neq 0)$. The vector in the brackets is a convex linear combination of $x_2, \ldots, x_n$ and hence, by the induction hypothesis, it is in C. Since also $x_1 \in C$, $\lambda_1 + \mu = \lambda_1 + \lambda_2 + \ldots + \lambda_n = 1$ and $\lambda_1, \mu \geqslant 0$, it follows that $y \in C$. The theorem is therefore true by induction for all n.

For an arbitrary subset A of E, we will denote the set of all convex linear combinations of the vectors of A by $K^*(A)$.

Theorem 3. $K^*(A)$ is a convex set containing A.

Proof. 1. Every vector $x \in A$ is a convex linear combination $x = 1x$ of itself and therefore $K^*(A) \supseteq A$.

2. Suppose $x, y \in K^*(A)$ so that

$$x = \sum_{z \in A} \alpha_z z \quad \text{where} \quad \sum_{z \in A} \alpha_z = 1 \quad \text{and} \quad \alpha_z \geqslant 0 \text{ for all } z \in A,$$

and $\quad y = \sum\limits_{z \in A} \beta_z z \quad \text{where} \quad \sum\limits_{z \in A} \beta_z = 1 \quad \text{and} \quad \beta_z \geqslant 0 \text{ for all } z \in A.$

Now, if $\lambda + \mu = 1$ and $\lambda, \mu \geqslant 0$,

$$\lambda x + \mu y = \sum_{z \in A} (\lambda \alpha_z + \mu \beta_z) z$$

and in this $\sum\limits_{z \in A} (\lambda \alpha_z + \mu \beta_z) = 1$ and $\lambda \alpha_z + \mu \beta_z \geqslant 0$ for all $z \in A$. Hence $\lambda x + \mu y \in K^*(A)$ and $K^*(A)$ is convex.

33

Theorem 4. Let $\mathscr{S}$ be a family of convex sets in a real vector space E, then $\bigcap(\mathscr{S})$ is also convex.

Proof. Suppose $x_1, x_2 \in \bigcap(\mathscr{S})$, so that $x_1, x_2 \in C$ for all $C \in \mathscr{S}$. Now, if $\lambda_1, \lambda_2 \geqslant 0$ and $\lambda_1 + \lambda_2 = 1$, then $\lambda_1 x_1 + \lambda_2 x_2 \in C$ for all $C \in \mathscr{S}$. That is, $\lambda_1 x_1 + \lambda_2 x_2 \in \bigcap(\mathscr{S})$.

Definition 3. Let E be a real vector space. The 'convex hull' $K(A)$ of a subset A of E is the intersection of the family of convex sets which contain A.

If A is finite, say $A = \{a_1, \dots, a_n\}$, we will write $K(A) = K(a_1, \dots, a_n)$. It follows immediately from Theorem 4 that $K(A)$ is convex.

Theorem 5. The correspondence $A \to K(A)$ is a closure operator, i.e.,
1. $K(A) \supseteq A$
2. If $A_1 \subseteq A_2$, then $K(A_1) \subseteq K(A_2)$
3. $K(K(A)) = K(A)$.

Proof. 1. $K(A)$ is the intersection of sets which contain A and therefore $K(A)$ also contains A (Theorem 1.2;3).

2. Let $\mathscr{S}_i$ be the family of convex sets containing $A_i (i = 1, 2)$. Then, if $A_1 \subseteq A_2$, $\mathscr{S}_1 \supseteq \mathscr{S}_2$ and hence $K(A_1) = \bigcap(\mathscr{S}_1) \subseteq \bigcap(\mathscr{S}_2) = K(A_2)$ (Theorem 1.2;2).

3. More generally, we show that if C is convex then $K(C) = C$. In this case, C is a convex set containing itself and therefore, by Theorem 1.2;1, $K(C) \subseteq C$, but from 1. $K(C) \supseteq C$ and hence $K(C) = C$.

Since $K(A)$ is both convex and the intersection of all convex sets which contain A, $K(A)$ could also be defined as the least convex set containing A (Theorem 1.2;4).

Theorem 6. If $A \subseteq K(B)$ and $B \subseteq K(A)$, then $K(A) = K(B)$.

Proof. By Theorem 5, if $A \subseteq K(B)$, then $K(A) \subseteq K(B)$ and similarly $K(B) \subseteq K(A)$.

Theorem 7. $K(A)$ consists of the convex linear combinations of the vectors of A, i.e., $K(A) = K^*(A)$.

Proof. 1. $K(A)$ is convex and contains A and therefore, by Theorem 2, $K(A) \supseteq K^*(A)$.

2. By Theorem 3, $K^*(A)$ is a convex set containing A and therefore $K(A) \subseteq K^*(A)$ (Theorem 1.2;1).

2.6.2 Convex Cones

Let E again be a real vector space.

*Definition 4. If A is a subset of E, a vector of the form $y = \sum\limits_{x \in A} \lambda_x x$ where $\lambda_x \geqslant 0$
for all $x \in A$ is called a 'positive linear combination' of the elements of A. The
vector y is called a 'strictly positive linear combination' if not all $\lambda_x = 0$.*

*Definition 5. A subset P of E is a 'convex cone' if P contains all positive linear
combinations of any pair of its vectors.*

A convex cone is obviously convex in the sense of Definition 1. Convex
cones could also be characterized as those convex sets C which contain all
positive scalar multiples $\lambda x (\lambda \geqslant 0)$ of all vectors $x \in C$.

The theory of convex cones is developed in a way directly analogous to
that of convex sets, and we will therefore only set out the most important
results and in the main without proofs.

If K is a convex cone and $\alpha > 0$, then $\alpha K = K$. If $\alpha < 0$, $\alpha K = -K$ which is
again a convex cone. If K_1 and K_2 are convex cones, then so is $K_1 + K_2$ (see
Theorem 1).

A convex cone contains all positive linear combinations of its elements
(see Theorem 2). If A is a subset of E, we will denote the set of all positive
linear combinations of the elements of A by $P^*(A)$. $P^*(A)$ is a convex cone
which contains A. We refer to $P^*(A)$ as the *positive hull* of A (Theorem 3).

The intersection of a family of convex cones is again a convex cone
(Theorem 4), so that, if $\mathscr{S}$ is the family of all convex cones which contain the
set A, then $P(A) = \bigcap(\mathscr{S})$ is the least convex cone which contains A, and
$P(A) = P^*(A)$ (Theorem 7). The correspondence $A \to P(A)$ is a closure operator (Theorem 5).

*Definition 6. A convex cone K is said to be 'acute' if $K \cap (-K) = \{0\}$; i.e., if
$x \in K$ and $-x \in K$ implies $x = 0$.*

An acute convex cone therefore contains no inverses of any of its vectors
except the zero-vector. A convex cone K is acute if and only if $0 \in K$ is not a
strictly positive linear combination of the other vectors in K.

For every convex cone K, the set $L = K \cap (-K)$ is a subspace of E and K
is acute only when $L = \{0\}$. If $x \in K$, then the coset $L + x$ is contained in K.
The set of all cosets of L which are contained in K will be denoted by $\bar{K}$ and
clearly $\bar{K} \subseteq E/L$.

Theorem 8. If K is a convex cone in E, then $\bar{K}$ is an acute convex cone in E/L.

35

Proof. 1. Suppose $L+x_1$ and $L+x_2 \in \bar{K}$ and let $\lambda_1, \lambda_2 \geqslant 0$. Then $x_1, x_2 \in K$ and therefore $\lambda_1 x_1 + \lambda_2 x_2 \in K$ and hence

$$L + (\lambda_1 x_1 + \lambda_2 x_2) = \lambda_1(L+x_1) + \lambda_2(L+x_2) \in \bar{K}.$$

2. If $L+x \in \bar{K}$ and $-(L+x) = L-x \in \bar{K}$, then $x \in K$ and $-x \in K$ and hence $x \in K \cap (-K) = L$, i.e., $L+x = L$.

Theorem 9. If $P, Q \subseteq E$ are convex cones such that $P + Q = E$ and $P \cap (-Q) = \{0\}$, then $P = -P$ and $Q = -Q$, i.e., P and Q are subspaces of E.

Proof. Suppose $p_1 \in P$. From the conditions of the theorem, there are elements $p \in P$ and $q \in Q$ such that $-p_1 = p+q$ and hence $-q = p_1 + p$. In other words $-q \in P \cap (-Q) = \{0\}$. Thus $q = 0$ and $p_1 = -p \in -P$.

From this it follows that $P \subseteq -P$ and, multiplying by -1, that $-P \subseteq P$. Hence $P = -P$.

Similarly $Q = -Q$.

Example 3. Every subspace of a real vector space is a convex cone. A coset is a convex cone if and only if it is a subspace.

Example 4. In the vector space C (Example 2.1;7) the functions $x(\tau)$ for which $x(\tau) \geqslant 0$ for $-\frac{1}{2} \leqslant \tau \leqslant \frac{1}{2}$ form a convex cone. If $x_1(\tau) \geqslant 0$ and $x_2(\tau) \geqslant 0$ for $-\frac{1}{2} \leqslant \tau \leqslant \frac{1}{2}$ and $\lambda_1, \lambda_2 \geqslant 0$, then $\lambda_1 x_1(\tau) + \lambda_2 x_2(\tau) \geqslant 0$ for $-\frac{1}{2} \leqslant \tau \leqslant \frac{1}{2}$.

Example 5. Examples of convex cones in the vector space G_2 (Example 2.1;1) are indicated in the following diagram.

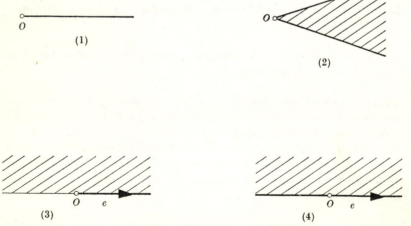

Fig. 7.

In (3) (halfplane) all positive multiples of the vector e belong to the set, but not the strictly negative multiples. In (4), all multiples of e belong to the set. (1), (2) and (3) are acute but (4) is not.

In G_3, examples of convex cones are 3-sided pyramids and circular cones (infinite in only one direction) with vertices at the zero-vector.

Problems

1. Show that the set of all vectors $x = (\xi_1, \xi_2, \xi_3, \xi_4) \in R_4$ which satisfy the homogeneous linear inequalities $\xi_1 + \xi_2 - \xi_3 - 2\xi_4 \geqslant 0$ and $2\xi_1 + \xi_2 + \xi_3 + 4\xi_4 \geqslant 0$ is a convex cone.

2. If the right-hand sides of the inequalities in Problem 1 are replaced by 2 and 3 respectively, is the set still a convex cone? Is it a convex set?

3. Show that every subspace of a real vector space is convex. Is this the case for the cosets?

4. Find conditions for a coset to be a convex cone.

5. Let E be a real vector space and let $x \in E$. Is the difference set $E \setminus \{x\}$ convex?

6. Show that, if A, B are subspaces of a vector space, then

$$K(A \cup B) = K(K(A) \cup K(B)) \quad \text{and} \quad K(A + B) = K(A) + K(B).$$

CHAPTER 3

BASES OF A VECTOR SPACE, FINITE-DIMENSIONAL VECTOR SPACES

3.1 Bases of a Vector Space

3.1.1 Linear Dependence and Independence of Vectors

Definition 1. A subset S of a vector space E is said to be 'linearly independent' if there is no linear combination $\sum_{x \in S} \alpha_x x$ which is equal to the zero-vector except the one in which $\alpha_x = 0$ for all $x \in S$.

A set which is not linearly independent is said to be *linearly dependent*.

Instead of saying that the 'set' S is linearly dependent or independent, we may say that the vectors of S are linearly dependent or independent. It is clear that a set S is linearly dependent, if there is a linear combination $y = \sum_{x \in S} \alpha_x x$ of the vectors of S which is equal to the zero vector but has some coefficients not equal to zero.

Theorem 1. A set S is linearly dependent if and only if there is a vector x_0 in S which is a linear combination of the other vectors in S.

Proof. 1. Suppose $x_0 \in S$ and $x_0 = \sum_{x \in S \setminus \{x_0\}} \alpha_x x$. Then $1x_0 - \sum_{x \in S \setminus \{x_0\}} \alpha_x x = 0$ is a linear combination of the vectors of S in which at least one coefficient is not zero.

2. Suppose S is linearly dependent, so that $\sum_{x \in S} \alpha_x x = 0$ and $\alpha_{x_0} \neq 0$ for some $x_0 \in S$. Then

$$x_0 = -\frac{1}{\alpha_{x_0}} \sum_{x \in S \setminus \{x_0\}} \alpha_x x.$$

Example 1. A subset S of the vector space G_2 (Example 2.1;1) is linearly independent only when $S = \{a\}$ and $a \neq 0$ or $S = \{a_1, a_2\}$ and a_1, a_2 are not parallel to the same line. In G_3, a subset S is linearly independent only in the following three cases.

1. $S = \{a\}$; $a \neq 0$.
2. $S = \{a_1, a_2\}$; a_1, a_2 not parallel to the same line.
3. $S = \{a_1, a_2, a_3\}$; a_1, a_2, a_3 not parallel to the same plane.

3.1.2   The Concept of a Basis

Definition 2. A 'basis' B of a vector space E is a linearly independent spanning set for E. (Definition 2.4;3).

Example 2. In $G_2(G_3)$, a linearly independent set is a basis if and only if it contains two (three) vectors.

The elements of a basis (basis vectors) will normally be denoted by the letter e.

Theorem 2. If B is a basis of E and $B \neq \varnothing$, then every vector $x \in E$ can be written in exactly one way as a linear combination

$$x = \sum_{e \in B} \xi_e e \tag{1}$$

of the elements of B. ($B = \varnothing$ is a basis if and only if $E = \{0\}$.)

Proof. 1. The fact that every vector $x \in E$ can be written in the form (1) is nothing more than the assumption that B is a spanning set.

2. Suppose $x = \sum_{e \in B} \xi_e e = \sum_{e \in B} \eta_e e$. Then $\sum_{e \in B} (\xi_e - \eta_e) e = 0$ and, because B is linearly independent, $\xi_e = \eta_e$ for all $e \in B$.

Definition 3. The scalars ξ_e which are uniquely determined when $x \in E$ is written in the form (1) are known as the 'components' of x with respect to the basis B.

Thus, for a given vector $x \in E$, there is a unique component ξ_e corresponding to each basis vector $e \in B$. We note, however, that only a finite number of these components are not equal to zero. If the basis B is finite or countably infinite, i.e., $B = \{e_1, \ldots, e_n\}$ or $B = \{e_1, e_2, e_3, \ldots\}$, we will denote the components by ξ_k instead of ξ_{e_k} and write, in place of (1),

$$x = \sum_{k=1}^{n} \xi_k e_k \quad \text{and} \quad x = \sum_{k=1}^{\infty} \xi_k e_k.$$

Example 3. In Fig. 8, which refers to the vector space G_3, $x = e_1 + e_2 + 2e_3$. Hence the components of x with respect to the basis $B = \{e_1, e_2, e_3\}$ are $\xi_1 = 1$, $\xi_2 = 1$, $\xi_3 = 2$.

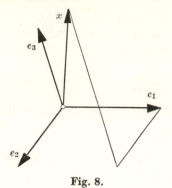

Fig. 8.

Theorem 3. The components of a linear combination of vectors are the corresponding linear combinations of the components of the vectors.

Proof. Suppose $y = \sum\limits_{k=1}^{n} \alpha_k x_k$ and $x_k = \sum\limits_{e \in B} \xi_{ke} e$ $(k = 1, \ldots, n)$. Then $y = \sum\limits_{e \in B} \eta_e e$ where $\eta_e = \sum\limits_{k=1}^{n} \alpha_k \xi_{ke}$ for all $e \in B$.

Example 4. Let $B = \{e_1, e_2, e_3\}$ be a basis of the vector space G_3 and let $x_1 = e_1 + e_2 - 2e_3$, $x_2 = 2e_1 - e_2 + e_3$. Then $y = 2x_1 - 3x_2 = -4e_1 + 5e_2 - 7e_3$. Hence the components of y are $\eta_1 = -4 = 2.1 - 3.2$, $\eta_2 = 5 = 2.1 - 3(-1)$, $\eta_3 = -7 = 2(-2) - 3.1$.

For a given vector $x \in E$, there is a component ξ_e assigned to each basis vector $e \in B$. This means that, corresponding to each $x \in E$, there is a unique element of the vector space $F_0(B)$ (Example 2.1;4). Hence a mapping f of E onto $F_0(B)$ is given by $x \rightarrow f(x) \in F_0(B)$ where $f(x)$ is the function on B which takes the value ξ_e at $e \in B$. (f is real- or complex-valued according as E is real or complex.) Theorem 3 can now be expressed by saying that the mapping f has the property

$$f\left(\sum_{k=1}^{n} \alpha_k x_k \right) = \sum_{k=1}^{n} \alpha_k f(x_k),$$

i.e., the image of a linear combination of vectors is the corresponding linear combination of the image vectors.

A mapping of one vector space into another vector space which has this property is known as a *linear mapping* (see Definition 5.1;1).

Further, f is 1-1 because, by (1), vectors which have the same components are equal.

If the basis $B = \{e_1, \ldots, e_n\}$ is finite, $F_0(B)$ can be replaced by R_n (cf. the remarks on Example 2.1;4). In this case $f(x) = (\xi_1, \ldots, \xi_n)$, i.e., to each $x \in E$ we assign its row of components.

Theorem 4. A subset S of E is linearly dependent if and only if the image set $f(S) \subseteq F_0(B)$ is linearly dependent.

Proof. Since f is a linear 1-1 mapping of E onto $F_0(B)$, Theorem 4 is a direct consequence of Theorem 5.1;5, which we may use here in advance.

Example 5. Let $B = \{e_1, e_2, e_3\}$ be a basis of G_3. Further let

$$x_1 = e_1 + 2e_2 - e_3 \in G_3 \quad \text{so that } f(x_1) = (1, 2, -1) \in R_3$$
$$x_2 = -e_1 + e_2 + 2e_3 \in G_3 \quad \text{so that } f(x_2) = (-1, 1, 2) \in R_3$$

and $x_3 = -e_1 + 4e_2 + 3e_3 \in G_3 \quad \text{so that } f(x_3) = (-1, 4, 3) \in R_3.$

Now, in R_3,

$$f(x_1) + 2f(x_2) - f(x_3) = (1, 2, -1) + 2(-1, 1, 2) - (-1, 4, 3)$$
$$= (0, 0, 0) = 0 \in R_3,$$

i.e., the rows of components are linearly dependent in R_3. Hence the same is true for x_1, x_2, x_3 in the vector space G_3.

The next two theorems show that the concept of a basis can be defined in two ways other than that in Definition 2.

Theorem 5. A 'linearly independent' set $B \subseteq E$ is a basis if and only if it is 'maximal', i.e., if every set which properly contains B is linearly dependent.

Proof. 1. Suppose B is a maximal linearly independent set and let $z \in E \setminus B$. Then $B \cup \{z\}$ is linearly dependent and there is an equation

$$\sum_{e \in B} \alpha_e e + \beta z = 0$$

in which not all the coefficients α_e and β are zero. In this case $\beta \neq 0$ because otherwise B would be linearly dependent. Hence we can solve the equation for z and $z = (-1/\beta) \sum_{e \in B} \alpha_e e$ means that $z \in L(B)$. Thus B is a spanning set for E and hence a basis.

2. Suppose B is a basis of E, that A is a subset of E properly containing B, and that $z \in A \setminus B$. Now z is a linear combination of the elements

of B (Theorem 2), so that A is linearly dependent (Theorem 1). Hence B is a maximal linearly independent set.

Theorem 6. A 'spanning set' B of E is a basis if and only if it is 'minimal', i.e., if every proper subset of B does not span E.

Proof. 1. Suppose B is a basis, A is a proper subset of B and $z \in B \setminus A$. Since B is linearly independent, $z \notin L(A)$ and therefore A is not a spanning set for E. Hence B is a minimal spanning set for E.

2. Let B be a spanning set for E which is not a basis, i.e., which is not linearly independent. Then there is an element $z \in B$ which is a linear combination of the other elements in B, i.e., $z \in L(B \setminus \{z\})$. But then $B \subseteq L(B \setminus \{z\})$ and, by Theorem 2.4;7, it follows that $L(B \setminus \{z\}) = L(B) = E$. Hence the proper subset $B \setminus \{z\} \subseteq B$ is a spanning set for E and B is not minimal.

Examples of Bases in Vector Spaces

We will describe bases for some of the examples of vector spaces introduced in 2.1, but it must be noted that, in each case, in addition to the given bases, there are infinitely many others. We note also that the following discussion holds both for real and complex vector spaces.

Example 1. G_2, G_3. We have already described all the bases of G_2 and G_3 in Example 3.1;2.

Example 2. R_n. The set of the following n vectors is a basis of R_n

$$e_1 = (1, 0, 0, \ldots, 0, 0), \quad e_2 = (0, 1, 0, \ldots, 0, 0), \ldots, \quad e_n = (0, 0, 0, \ldots, 0, 1).$$

At this point it is convenient to introduce the function which is known as the Kronecker delta δ_{ik}. The symbol δ_{ik} takes the value 1 when $i = k$ and 0 when $i \neq k$. Usually i and k will be natural numbers, but they may also be elements of an arbitrary set (Example 4). With the help of this symbol, we can describe the given basis for R_n in the form

$$e_i = (\delta_{i1}, \delta_{i2}, \ldots, \delta_{in}) \qquad (i = 1, 2, \ldots, n).$$

We show that $\{e_1, \ldots, e_n\}$ is a basis as follows. If $x = (\xi_1, \ldots, \xi_n) \in R_n$, then $x = \sum_{k=1}^{n} \xi_k e_k$ so that $e_1, \ldots, e_n$ span the whole of R_n. Also, if $\sum_{k=1}^{n} \xi_k e_k = 0$, then $(\xi_1, \ldots, \xi_n) = 0$ and $\xi_1 = \ldots = \xi_n = 0$. Hence $e_1, \ldots, e_n$ are linearly independent. At the same time, we see that, if $x = (\xi_1, \ldots, \xi_n)$, then the scalars $\xi_1, \ldots, \xi_n$ are the components of x with respect to the given basis.

Example 3. The discussion of a basis for the space of series F is outside the scope of this book because it requires transfinite methods. The countably infinite set of vectors

$$e_i = (\delta_{i1}, \delta_{i2}, \delta_{i3}, \ldots) \qquad (i = 1, 2, 3, \ldots)$$

is a basis of the vector space F_0.

Example 4. The discussion of a basis for the vector space $F(A)$ is also outside the scope of this book except when A is finite. In the latter case, the situation is analogous to the vector space R_n. The functions $e_x(x \in A)$ given by

$$e_x(y) = \delta_{xy} \qquad (y \in A)$$

form a basis B of the space $F_0(A)$, where $\delta_{xy} = 1$ if $x = y$ and $\delta_{xy} = 0$ if $x \neq y$.

Proof. 1. $\sum\limits_{x \in A} \alpha_x e_x = 0$ means $\sum\limits_{x \in A} \alpha_x e_x(y) = \alpha_y = 0$ for all $y \in A$. Therefore B is linearly independent.

  2. An arbitrary mapping $f \in F_0(A)$ can be written in the form $f = \sum\limits_{x \in A} f(x) e_x$, because $\sum\limits_{x \in A} f(x) e_x(y) = f(y)$ for all $y \in A$.

Example 5. The vector space P_n is obviously spanned by the polynomials e_i which are given by $e_i(\tau) = \tau^i$ $(i = 0, \ldots, n)$. Now suppose $\sum\limits_{i=0}^{n} \lambda_i e_i = 0$, i.e., $\sum\limits_{i=0} \lambda_i \tau^i = 0$ for all τ. From the well-known fact that a polynomial only vanishes identically if all its coefficients are zero, it follows that $\lambda_i = 0$ for $i = 0, 1, \ldots, n$. Hence $e_0, \ldots, e_n$ are linearly independent and therefore form a basis.

Example 6. The countably infinite set of polynomials e_i, where $e_i(\tau) = \tau^i$ $(i = 0, 1, 2, \ldots)$ is a basis of the space of polynomials P.

  The discussion of a basis for Example 7 (the space of continuous functions) is again outside the scope of this book.

Example 8. The countably infinite set of functions e_0, e_k and f_k $(k = 1, 2, 3, \ldots)$, where $e_0(\tau) = \frac{1}{2}$, $e_k(\tau) = \cos k\tau$ and $f_k(\tau) = \sin k\tau$, is a basis of the vector space T of trigonometric polynomials. It follows immediately from the definition of T that these functions span T. The fact that they are also linearly independent follows from the orthogonality relations which are proved in most standard texts on integral calculus.

That is, $\displaystyle\int_0^{2\pi} \cos j\tau \cos k\tau \, d\tau = 0 \quad \text{if } j \neq k$

$$\int_0^{2\pi} \sin j\tau \sin k\tau \, d\tau = 0 \quad \text{if } j \neq k$$

$$\int_0^{2\pi} \cos j\tau \sin k\tau \, d\tau = 0$$

$$\int_0^{2\pi} (\cos k\tau)^2 \, d\tau = \int_0^{2\pi} (\sin k\tau)^2 \, d\tau = \pi (k > 0)$$

Hence, if $\alpha_0 e_0 + \sum_{k=1}^{\infty} (\alpha_k e_k + \beta_k f_k) = 0$,

i.e., if $\frac{1}{2}\alpha_0 + \sum_{k=1}^{\infty} (\alpha_k \cos k\tau + \beta_k \sin k\tau) = 0$, for all τ, and we want to show that $\beta_j = 0$, say, then we multiply both sides of the equation by $\sin j\tau$ and integrate from 0 to 2π. (It is possible to do this term by term because only a finite number of the coefficients are not zero.) In view of the orthogonality relations it follows that

$$\int_0^{2\pi} \beta_j (\sin j\tau)^2 \, d\tau = \pi\beta_j = 0 \quad \text{and therefore } \beta_j = 0.$$

The proof that $\alpha_j = 0$ is just the same except that we multiply by $\cos j\tau$.

In the complex case (Example 2.1;15) another basis is the set of vectors e_k given by $e_k(\tau) = e^{ik\tau}$ $(k = 0, \pm 1, \pm 2, \ldots)$. It has already been shown that these span the vector space. Their linear independence is proved as follows.

$$\sum_{k=-\infty}^{+\infty} \alpha_k e_k = 0 \quad \text{means} \quad \sum_{k=-\infty}^{+\infty} \alpha_k e^{ik\tau} = 0 \text{ for all } \tau.$$

Multiplying by $e^{-ij\tau}$ and integrating from $-\pi$ to $+\pi$ (again only a finite number of the terms in the series are not zero), it follows that

$$\sum_{k=-\infty}^{+\infty} \alpha_k \int_{-\pi}^{+\pi} e^{i(k-j)\tau} \, d\tau = 2\pi\alpha_j = 0 \quad \text{and therefore} \quad \alpha_j = 0,$$

for $j = 0, \pm 1, \pm 2, \ldots$
(In these formulae i is the imaginary unit $\sqrt{-1}$ and j, k are integers.)

Example 9. The linear forms $e_1, \ldots, e_n$ given by

$$e_k(\xi_1, \ldots, \xi_n) = \xi_k \qquad (k = 1, \ldots, n)$$

are a basis of the vector space L_n.

Example 10. The real vector space of complex numbers has the set $\{1, i\}$ as a basis, where i is the imaginary unit.

Example 11. The vector space S_R of real scalars has the set $B = \{1\}$ as a basis, and the same is true for the vector space S_K of complex scalars (Example 17).

3.1.3 The Existence of Bases in a Vector Space

We will now show that every vector space has a basis.

The general proof is not completely elementary because it uses results which depend on the axiom of choice or equivalently on the theory of well-ordered sets. We will base our proof on a theorem due to Zorn (usually referred to as Zorn's Lemma), which we will assume without further explanation. For those readers who are not familiar with this, we will also give an elementary proof which is only valid however for those vector spaces which have a finite or countably infinite spanning set.

Theorem 7 (Zorn). *A non-empty partially ordered set, in which every non-empty chain (i.e., totally ordered subset) has an upper bound, has a maximal element.*

A proof of Zorn's Lemma can be found for example in [23] p. 197.

Theorem 8. Every vector space has a basis.

Proof. Remembering Theorem 5, we show that there exists a maximal linearly independent subset B of E. Let $\mathscr{A}$ be the family of all linearly independent subsets A of E. $\mathscr{A}$ is partially ordered by the relation $A_1 \subseteq A_2$ (for $A_1, A_2 \in \mathscr{A}$). A chain in $\mathscr{A}$ is a subset $\mathscr{B}$ of $\mathscr{A}$ with the property that, if $A_1, A_2 \in \mathscr{B}$, then either $A_1 \subseteq A_2$ or $A_2 \subseteq A_1$. For each non-empty chain, we form the union $S(\mathscr{B}) = \bigcup(\mathscr{B}) \subseteq E$. Then

1. $S(\mathscr{B})$ is linearly independent.

Suppose $\sum\limits_{x \in S(\mathscr{B})} \alpha_x x = 0$, where only finitely many of the coefficients are not zero. We denote the corresponding vectors by $x_1, \ldots, x_n$ and their coefficients by $\alpha_1, \ldots, \alpha_n$. To each vector x_k, there exists a set $A_k \in \mathscr{B}$, such that $x_k \in A_k$ $(k = 1, \ldots, n)$. Since $\mathscr{B}$ is totally ordered, one of these sets A_k will be the largest. If we denote this by A_{k_0}, then $A_k \subseteq A_{k_0}$ for all $k = 1, \ldots, n$. This means however that $x_1, \ldots, x_n \in A_{k_0}$ and it follows, from the linear independence of A_{k_0}, that $\alpha_k = 0$ for $k = 1, \ldots, n$. Hence $S(\mathscr{B})$ is linearly independent.

2. $S(\mathscr{B}) \supseteq A$ for all $A \in \mathscr{B}$. This follows immediately from Theorem 1.2;1.

Now 1. means that $S(\mathscr{B}) \in \mathscr{A}$ and 2. means that $S(\mathscr{B})$ is an upper bound for $\mathscr{B}$ in $\mathscr{A}$. Thus the partially ordered set $\mathscr{A}$ satisfies the conditions of Zorn's Lemma because $\mathscr{A}$ is not empty, since $\varnothing \in \mathscr{A}$. Therefore $\mathscr{A}$ has a maximal element B and this set B is a basis of E.

After this general proof, we will now show in a more elementary way that *a vector space E, which has a finite or a countably infinite spanning set, has a basis which is either finite or countably infinite.*

Suppose that $A \subseteq E$ is a spanning set for E which consists of the finite or countably infinite number of vectors $e_k (k = 1, 2, 3, \ldots)$ where $e_1 \neq 0$. (If $E = \{0\}$, then $B = \varnothing$ is a basis.) We construct a subset A^* of A by putting $e_1 \in A^*$ and, for $k \geqslant 2$, $e_k \in A^*$ if, and only if, e_k is not a linear combination of $e_1, \ldots, e_{k-1}$, i.e., if $e_k \notin L(e_1, \ldots, e_{k-1})$. Then A^* is a basis of E, because

1. A^* is linearly independent. If A^* were linearly dependent, then there would be an equation $\sum\limits_{k=1}^{n} \alpha_k e_k = 0$ with $n > 0$, $e_n \in A^*$ and $\alpha_n \neq 0$. This could be solved for e_n as a linear combination of $e_1, \ldots, e_{n-1}$ in contradiction of the definition of A^*.

2. A^* spans E. We only need to show that A is contained in $L(A^*)$ because then $E = L(A) \subseteq L(L(A^*)) = L(A^*)$ (Theorems 2.4;5 and 6) and hence $E = L(A^*)$.

If A were not contained in $L(A^*)$, then there would be an element $e_k \in A$ which was not in $L(A^*)$. Suppose that e_{k_0} is the first element of A with this property. Certainly $e_{k_0} \notin A^*$, and therefore $e_{k_0} \in L(e_1, \ldots, e_{k_0-1})$. (Note that, since $e_1 \in A^*$, $k_0 \geqslant 2$.) But $e_1, \ldots, e_{k_0-1} \in L(A^*)$ and hence $e_{k_0} \in L(L(A^*)) = L(A^*)$ and this is a contradiction.

We note that this elementary proof is still valid for an arbitrary spanning set A of E, if A can be so ordered that every non-empty subset of A has a first element. That this is possible for all sets (even when they are not countable) is the content of the *well-ordering theorem* (see [23] p. 198).

Theorem 9. In a vector space E, every linearly independent set C can be completed to a basis of E, i.e., there is a basis B of E such that $B \supseteq C$.

Proof. The proof is almost the same as that of Theorem 8, the only difference being that the family $\mathscr{A}$ consists of those linearly independent sets which contain C.

Theorem 10. Every spanning set C of a vector space E contains a basis, i.e., there is a basis B of E such that $B \subseteq C$.

Proof. Again the proof follows that of Theorem 8 except that here $\mathscr{A}$ is the family of linearly independent sets A which are contained in C. This shows that there is a linearly independent set B which is maximal in C, i.e., a linearly independent subset B of C such that every subset of C which properly contains B is linearly dependent. Then $L(B) \supseteq C$ and, because $L(C) = E$, $L(B) = L(L(B)) \supseteq L(C) = E$. Hence B is a basis of E.

Problems

1. Show that the vectors $x_1, \ldots, x_4$ in Problem 2.3;2 form a basis of R_4.
2. Let E be a vector space. Prove that

    (a) If S is a linearly independent subset of E and R is a subset of S, then R is linearly independent.

    (b) If S is linearly dependent and $R \supseteq S$, then R is linearly dependent.

    (c) If $0 \in S$, then S is linearly dependent.

    (d) If $x \in E$, then $\{x\}$ is linearly dependent if and only if $x = 0$.

3. Prove that, if x, y are linearly independent elements of the vector space E, then $x + y$ and $x - y$ are also linearly independent.

4. Let $\tau_0, \ldots, \tau_n$ be distinct real numbers. Show that the polynomials $x_i(\tau)$ of degree n which are given by $x_i(\tau_k) = \delta_{ik}$ $(i, k = 0, \ldots, n)$ form a basis of the vector space P_n (Example 2.1;5).

5. Let B_1 and B_2 be bases of the vector spaces E_1 and E_2 respectively. Show that the pairs $(e_1, 0)$ where $e_1 \in B_1$, and $(0, e_2)$, where $e_2 \in A_2$, form a basis of the direct product of E_1 and E_2 (see Problem 2.1;4).

6. Prove Theorems 9 and 10 in the case when E has a finite or countably infinite spanning set. (Use the proof of the existence of a basis in this special case.)

3.2 Finite-Dimensional Vector spaces

Definition 1. A vector space is said to be 'finite-dimensional' if it has a finite spanning set.

Examples. The Examples $G_2, G_3, R_n, P_n, L_n, K, S_R$ of 2.1.1 are all finite-dimensional because they all have finite bases.

In view of Theorem 3.1;10, every finite-dimensional vector space $E \neq \{0\}$ has a finite basis. The most important property of this type of vector space is that all bases contain the same number of elements (see Theorem 2). The proof of this depends on the following *Exchange Theorem* which is due to Steinitz.

Theorem 1. Let E be a vector space, $a_1,\ldots,a_n \in E$ and $L = L(a_1,\ldots,a_n)$. If the subset C of L is linearly independent, then

1. C *is finite and the number* m *of elements in* C *is* $m \leqslant n$. (*We will therefore denote the elements of* C *by* $e_1,\ldots,e_m$.)

2. *For each integer* $r, 0 \leqslant r \leqslant m$, *there are* r *of the vectors* $a_1,\ldots,a_n$ (*by renumbering if necessary, we may assume them to be* $a_1,\ldots,a_r$) *such that* $L(e_1,\ldots,e_r,a_{r+1},\ldots,a_n) = L$.

The name of the theorem comes from the fact that $a_1,\ldots,a_r$ can be exchanged with $e_1,\ldots,e_r$ without changing the linear hull L.

Proof. 1. We first prove the second part of the theorem by using induction on r. The assertion is true for $r = 0$. Assume that it is also true for $r-1$, i.e., $L = L(e_1,\ldots,e_{r-1},a_r,a_{r+1},\ldots,a_n)$. Since $e_r \in L$, e_r is a linear combination of $e_1,\ldots,e_{r-1},a_r,\ldots,a_n$. In this linear combination, at least one of the vectors $a_r,\ldots,a_n$ must have a non-zero coefficient because the e_k are linearly independent. By renumbering the a's if necessary, we may assume that this is a_r. Thus we can solve the equation for a_r, so that $a_r \in L(e_1,\ldots,e_r,a_{r+1},\ldots,a_n)$. Hence, each of the sets $\{e_1,\ldots,e_{r-1},a_r,\ldots,a_n\}$ and $\{e_1,\ldots,e_r,a_{r+1},\ldots,a_n\}$ is contained in the linear hull of the other. By Theorem 2.4;7, the linear hulls are equal and hence $L = L(e_1,\ldots,e_r,a_{r+1},\ldots,a_n)$. The second part of the theorem is therefore true for r and therefore in general.

2. If C contained more than n vectors, then, after the exchange of $a_1,\ldots,a_n$ with $e_1,\ldots,e_n$, every other vector $e \in C$ would be in $L(e_1,\ldots,e_n)$ which contradicts the linear independence of C.

Theorem 2. All bases of a finite-dimensional vector space contain the same finite number of elements.

Proof. We already know that a finite-dimensional vector space E has a finite basis. Suppose that $\{a_1,\ldots,a_n\}$ is one of these. If B is a further basis of E, the conditions of Theorem 1 are satisfied by B in the place of C. Hence the number m of elements in B is finite and $m \leqslant n$. However this means that the conditions of Theorem 1 are satisfied again but with the roles reversed so that $n \leqslant m$.

Definition 2. The common number of elements in the bases of a finite-dimensional vector space E is known as the 'dimension' of E and denoted by $\dim E$.

Example 1. The finite-dimensional vector spaces given at the beginning of this section have the following dimensions.

$$\dim G_2 = 2, \quad \dim G_3 = 3, \quad \dim R_n = n, \quad \dim P_n = n+1$$
$$\dim L_n = n, \quad \dim K = 2, \quad \dim S_R = 1.$$

Also the complex vector space S_K has dimension 1. These numbers follow from the bases given in 3.1. If $E = \{0\}$, then $\dim E = 0$ because $\varnothing$ is a basis of E.

Theorem 3. If $\dim E = n$, *then any set of* n *linearly independent vectors* $e_1, \ldots, e_n \in E$ *is a basis of* E.

Proof. By Theorem 3.1;9, there is a basis B of E such that $e_1, \ldots, e_n \in B$. But the number of elements in B is $\dim E = n$ and hence $B = \{e_1, \ldots, e_n\}$.

Theorem 4. If L *is a proper subspace of* E, *then* $\dim L < \dim E$.

Proof. A basis B^* of L is a linearly independent set in E and is therefore a subset of a basis B of E (Theorem 3.1;9). It is not possible that $B^* = B$ because L would then be equal to E.

Theorem 5. The dimension of a quotient space E/L *is given by* $\dim(E/L) = \dim E - \dim L$.

Proof. Suppose $\dim E = n$ and $\dim L = m \leqslant n$. We start with a basis of $e_1, \ldots, e_m$ of L and complete it to a basis of E (Theorem 3.1;9) by introducing $n - m$ further vectors $e_{m+1}, \ldots, e_n$. We show that the cosets $L + e_{m+1}, \ldots, L + e_n$ form a basis of E/L.

1. They are linearly independent in E/L. If $\sum\limits_{k=m+1}^{n} \alpha_k (L + e_k) = L$, then,

by 2.5;(1)/(2) and Theorem 2.5;1, $\sum\limits_{k=m+1}^{n} \alpha_k e_k \in L$. That is, $\sum\limits_{k=m+1}^{n} \alpha_k e_k = \sum\limits_{k=1}^{m} \beta_k e_k$ for some coefficients β_k. Since $e_1, \ldots, e_n$ are linearly independent, it follows that $\alpha_{m+1} = \ldots = \alpha_n = 0$.

2. They span E/L. Suppose $L + x \in E/L$. Then there is a representation of x in the form $x = \sum\limits_{k=1}^{n} \xi_k e_k$, and therefore

$$L + x = L + \sum_{k=m+1}^{n} \xi_k e_k = \sum_{k=m+1}^{n} \xi_k (L + e_k)$$

because $e_1, \ldots, e_m \in L$.

Example 2. Let $E = G_3$ and L be a 1-dimensional subspace of G_3. Then $\dim E/L = 2$ (see Fig. 9, in which $\{e_1\}$ is a basis of L, $\{e_1, e_2, e_3\}$ is a basis of E and $\{L + e_2, L + e_3\}$ is a basis of E/L).

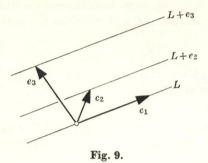

Fig. 9.

Theorem 6. If $E = L_1 \oplus L_2$, then $\dim E = \dim L_1 + \dim L_2$.

Proof. Let $\{e_1, \ldots, e_p\}$ be a basis of L_1 and $\{f_1, \ldots, f_q\}$ a basis of L_2. The union of these two bases must span E and we further show that it is linearly independent. If $\sum\limits_{i=1}^{p} \alpha_i e_i + \sum\limits_{k=1}^{q} \beta_k f_k = 0$, then by Theorem 2.4;10, both $\sum\limits_{i=1}^{p} \alpha_i e_i = 0$ and $\sum\limits_{k=1}^{q} \beta_k f_k = 0$. It follows that all the coefficients α_i and β_k vanish. Hence $\{e_1, \ldots, e_p, f_1, \ldots, f_q\}$ is a basis of E and

$$\dim E = p + q = \dim L_1 + \dim L_2.$$

Theorem 7. If $E = L_1 + L_2$ and $D = L_1 \cap L_2$, then

$$\dim E + \dim D = \dim L_1 + \dim L_2.$$

Proof. By Theorem 2.5;8, $E/D = L_1/D \oplus L_2/D$ and the assertion follows from Theorems 5 and 6.

It follows immediately from this theorem that, for any two subspaces L_1 and L_2,

$$\dim L_1 + \dim L_2 = \dim (L_1 + L_2) + \dim (L_1 \cap L_2).$$

We only need to replace E by $L_1 + L_2$.

Theorem 8. If L_1 and L_2 are subspaces of E such that $\dim L_1 + \dim L_2 > \dim E$, then there is a non-zero vector $x \in L_1 \cap L_2$.

Proof. By Theorem 7, $\dim (L_1 \cap L_2) > 0$ and hence $L_1 \cap L_2 \neq \{0\}$.

Problems

1. Prove that, in a finite-dimensional vector space E, the linear hull $L(A)$ of a subset A of E is the subspace of the least dimension which contains A.

2. Show that r arbitrary linear combinations of r linearly dependent vectors of a vector space are also linearly dependent.

3. What is the dimension of the direct product of two finite-dimensional vector spaces? (See Problems 2.1;4 and 3.1;5.)

4. With reference to Problem 2.4;6, calculate $\dim(F/L_1)$ and $\dim(F/L_2)$.

5. Similarly $\dim(C/L)$ (Problem 2.4;4).

6. Similarly $\dim(P/L)$ (Problem 2.4;5).

7. The dimension of a coset is defined to be the dimension of the corresponding subspace. Prove that, if N_1 and N_2 are cosets in a finite-dimensional vector space and if $N_1 \cap N_2 \neq \varnothing$ and $N_1 + N_2 = E$, then

$$\dim N_1 + \dim N_2 \;=\; \dim E + \dim(N_1 \cap N_2)$$

(see Theorem 2.5;9).

8. Suppose that the coset N in the vector space E is generated by a finite set of vectors, i.e., $N = N(x_1, \ldots, x_r)$. What can be said about the dimension of N?

9. Let N_1 and N_2 be cosets of dimensions r_1 and r_2 in the vector space E. Let N be the smallest coset which contains both N_1 and N_2. What can be said about $\dim N$? (Hint: the subspace corresponding to N is $L = L(L_1, L_2, x_1 - x_2)$, where $N_1 = L_1 + x_1$, $N_2 = L_2 + x_2$.)

3.3 The Exchange Method

There are many methods of numerical calculation in Linear Algebra which depend on the Exchange Theorem 3.2;1. For example, the solution of sets of linear equations by the Gaussian Algorithm, the inversion of matrices, the calculation of determinants, the Simplex Method of Linear Programming, etc. (Cf. the methods of calculation described in later chapters.) It is therefore necessary to develop the Exchange Theorem into a technique for numerical calculation.

3.3.1 The Normal Exchange Method

In the Exchange Theorem, it is assumed that the vectors $e_1, \ldots, e_m$ are linear combinations of $a_1, \ldots, a_n$. We will now suppose that these linear combinations are given, i.e., that the quantities α_{ik} in

$$e_i = \sum_{k=1}^{n} \alpha_{ik} a_k \qquad (i = 1, \ldots, m) \tag{1}$$

51

are known. With a view to producing a concise rule for calculation, we will express the relationships (1) in the form

$$
\begin{array}{c|ccc}
 & a_1 \dots\dots\dots\dots a_n \\
\hline
e_1 = & \alpha_{11} \dots\dots\dots\dots \alpha_{1n} \\
\vdots & \vdots \qquad\qquad\qquad \vdots \\
e_m = & \alpha_{m1} \dots\dots\dots\dots \alpha_{mn}
\end{array}
\tag{2}
$$

From now on we will refer to an array of this form as an *exchange tableau* or briefly as a *tableau*. (In analogy with the usual notation of the Simplex Method.)

Example 1. The tableau

$$
\begin{array}{c|ccc}
 & a_1 & a_2 & a_3 \\
\hline
e_1 = & 2 & 1 & -1 \\
e_2 = & -1 & 2 & 3
\end{array}
$$

expresses the linear relations

$$e_1 = 2a_1 + a_2 - a_3 \quad \text{and} \quad e_2 = -a_1 + 2a_2 + 3a_3.$$

The second part of the Exchange Theorem now states that, for each $r \leqslant m$, there are r of the vectors $a_1, \ldots, a_n$ which can be exchanged with $e_1, \ldots, e_r$ without changing the linear hull $L = L(a_1, \ldots, a_n)$. Thus every vector x which is a linear combination of $a_1, \ldots, a_n$ can also be expressed as a linear combination of $e_1, \ldots, e_r, a_{r+1}, \ldots, a_n$ (if $a_1, \ldots, a_r$ are the exchanged vectors). In particular, $a_1, \ldots, a_r$ can be expressed as linear combinations of these vectors. Our aim now is to find the tableau corresponding to these new linear combinations. We will start by considering the case $m = n = 2$.

We then have a tableau

$$
\begin{array}{c|cc}
 & a_1 & a_2 \\
\hline
e_1 = & \alpha_{11} & \alpha_{12} \\
e_2 = & \alpha_{21} & \alpha_{22}
\end{array}
\tag{3}
$$

which expresses the relations

$$e_1 = \alpha_{11} a_1 + \alpha_{12} a_2$$
$$e_2 = \alpha_{21} a_1 + \alpha_{22} a_2 \tag{4}$$

We solve the first equation for a_1, assuming that $\alpha_{11} \neq 0$, and substitute the result in the second, to obtain

$$a_1 = \frac{1}{\alpha_{11}} e_1 - \frac{\alpha_{12}}{\alpha_{11}} a_2$$

$$e_2 = \frac{\alpha_{21}}{\alpha_{11}} e_1 + \left(\alpha_{22} - \alpha_{21} \frac{\alpha_{12}}{\alpha_{11}} \right) a_2 \tag{5}$$

Thus we have exchanged a_1 and e_1. The relations of (5) can now be written in the form of a tableau again.

	e_1	a_2	
$a_1 =$	$\dfrac{1}{\alpha_{11}}$	$-\dfrac{\alpha_{12}}{\alpha_{11}}$	
$e_2 =$	$\dfrac{\alpha_{21}}{\alpha_{11}}$	$\alpha_{22} - \alpha_{21} \dfrac{\alpha_{12}}{\alpha_{11}}$	(6)

Tableau (6) is therefore a *consequence* of tableau (3) and the transfer from (3) to (6) simply means the exchange of a_1 and e_1. It is possible if and only if $\alpha_{11} \neq 0$. A transfer from one tableau to another which corresponds to an exchange of two vectors in this way will be referred to as an *exchange step*.

If $\alpha_{12} \neq 0$, we could equally well exchange a_2 and e_1 to obtain the following tableau

	a_1	e_1	
$a_2 =$	$-\dfrac{\alpha_{11}}{\alpha_{12}}$	$\dfrac{1}{\alpha_{12}}$	
$e_2 =$	$\alpha_{21} - \alpha_{22} \dfrac{\alpha_{11}}{\alpha_{12}}$	$\dfrac{\alpha_{22}}{\alpha_{12}}$	(7)

which is also a consequence of (3). Similarly, we can exchange a_1 or a_2 with e_2 when $\alpha_{21} \neq 0$ or $\alpha_{22} \neq 0$.

It is easy to see that this exchange method can also be carried through in the general case when m and n need not be 2. In order to be able to describe it in concise terms, we introduce the following useful notation.

If, in tableau (2), e_i and a_k are exchanged the coefficient $\alpha_{ik} \neq 0$ will be referred to as the *pivot*. Thus the pivot is the coefficient which lies to the right of e_i and underneath a_k in the tableau. In the transfer from (3) to (6), α_{11} is the pivot and, in the transfer from (3) to (7), α_{12} is the pivot. The row of coefficients which contains the pivot will be referred to as the *pivotal row* and analogously the column containing the pivot as the *pivotal column*. Thus, for instance, in the transfer from (3) to (7), the pivotal row consists of α_{11}, α_{12} and the pivotal column of α_{12}, α_{22}.

It is now easy to verify in general that an exchange step can be performed by the following rules.

Rules for an Exchange Step

1. The initial tableau is extended by an extra row—called the *cellar row*. This has an empty space under the pivotal column and its other coefficients are the corresponding ones of the pivotal row divided by the pivot and multiplied by -1.

2. In the new tableau, the reciprocal of the pivot appears in place of the pivot.

3. The rest of the pivotal row is replaced by the cellar row.

4. The rest of the pivotal column is divided by the pivot.

5. To every other coefficient is added the product of the corresponding coefficient of the cellar row and the corresponding coefficient of the pivotal column.

Example 2. We will illustrate these rules with a numerical example by exchanging a_1 and e_1 in Example 1, which is possible because $\alpha_{11} = 2 \neq 0$.

The tableau extended by the cellar row is (the pivot is indicated by *)

	a_1	a_2	a_3
$e_1 =$	2*	1	-1
$e_2 =$	-1	2	3
	*	$-\frac{1}{2}$	$\frac{1}{2}$

(8)

The new tableau becomes

	e_1	a_2	a_3
$a_1 =$	$\frac{1}{2}$	$-\frac{1}{2}$	$\frac{1}{2}$
$e_2 =$	$-\frac{1}{2}$	$\frac{5}{2}$	$\frac{5}{2}*$
	$\frac{1}{5}$	-1	$*$

$$(9)$$

i.e., we have the relations

$$a_1 = \tfrac{1}{2}e_1 - \tfrac{1}{2}a_2 + \tfrac{1}{2}a_3, \qquad e_2 = -\tfrac{1}{2}e_1 + \tfrac{5}{2}a_2 + \tfrac{5}{2}a_3.$$

In view of the last tableau, we can now exchange a further a_k with e_2 and in fact both a_2 and a_3 are possible because both corresponding coefficients are $\frac{5}{2} \neq 0$. We choose a_3 and have already added the corresponding cellar row. The next tableau becomes

	e_1	a_2	e_2
$a_1 =$	$\frac{3}{5}$	-1	$\frac{1}{5}$
$a_3 =$	$\frac{1}{5}$	-1	$\frac{2}{5}$

$$(10)$$

i.e., $a_1 = \tfrac{3}{5}e_1 - a_2 + \tfrac{1}{5}e_2$, $a_3 = \tfrac{1}{5}e_1 - a_2 + \tfrac{2}{5}e_2$. Hence a_1 and a_3 are expressed as linear combinations of e_1, a_2, e_2 and it is now possible to express any linear combination of a_1, a_2, a_3, as a linear combination of e_1, a_2, e_2. (Cf. Example 4.)

Since (9) is a consequence of (8) and (10) is a consequence of (9), it follows that (10) is a consequence of (8). In general, a sequence of tableaux is said to be *coherent* when each is obtained from its predecessor by an exchange step. Thus, in a coherent sequence of tableaux, the last is a consequence of the first. It is also true that the first tableau of a coherent sequence is a consequence of the last. To prove this it is clearly sufficient to show that, if the second of two tableaux is obtained from the first by an exchange step, then the first is a consequence of the second. This is not difficult to verify. For instance, in (6), if we exchange a_1 and e_1 using the coefficient $1/\alpha_{11}$ as a pivot, we return to tableau (3) so that (3) is a consequence of (6). The same method applies in the general case and this property will be referred to as the *reversibility* of the exchange method. Thus, in a coherent sequence of tableaux all the tableaux will be correct if any one of them is correct. (A tableau

55

is said to be correct when the linear relations which it represents are satisfied.)

Finally we remark that, in the exchange method, we have made no use of the linear independence of the vectors e_i. Thus the method is also applicable when e_i are linearly dependent. The dependence will simply mean that eventually not every e_i will be exchangeable for an a_k. This situation will be reached when all the possible pivotal coefficients are zero.

3.3.2   The Transposed Exchange Method

So far we have interpreted tableau (2) as representing the linear relations (1). From now on, we will refer to this interpretation as *normal* or *horizontal* and correspondingly there will also be an interpretation which we will refer to as *transposed* or *vertical*. In this second interpretation, tableau (2) expresses the vectors a_k (in the top row) as linear combinations of the vectors e_i (on the left-hand side of the tableau) where the coefficients for a_k are written in the kth column, i.e., tableau (2) now represents the linear relations

$$a_k = \sum_{i=1}^{m} \alpha_{ik} e_i \qquad (k = 1, \ldots, n). \tag{11}$$

Clearly, it is still possible to perform the exchange method in this interpretation. For example, starting with tableau (3), which now represents the relations

$$\begin{aligned} a_1 &= \alpha_{11} e_1 + \alpha_{21} e_2 \\ a_2 &= \alpha_{12} e_1 + \alpha_{22} e_2, \end{aligned} \tag{12}$$

and exchanging a_1 and e_1, we obtain the relations

$$e_1 = \frac{1}{\alpha_{11}} a_1 - \frac{\alpha_{21}}{\alpha_{11}} e_2$$

$$a_2 = \frac{\alpha_{12}}{\alpha_{11}} a_1 + \left(\alpha_{22} - \alpha_{12} \frac{\alpha_{21}}{\alpha_{11}} \right) e_2 \tag{13}$$

which are represented in transposed interpretation by the tableau

	$e_1 =$	$a_2 =$	
a_1	$\dfrac{1}{\alpha_{11}}$	$\dfrac{\alpha_{12}}{\alpha_{11}}$	(14)
e_2	$-\dfrac{\alpha_{21}}{\alpha_{11}}$	$\alpha_{22} - \alpha_{12}\dfrac{\alpha_{21}}{\alpha_{11}}$	

Now, if we replace a_1 by $-a_1$ on the left-hand side, which means that we must also change the signs of all the coefficients in the first row, and if we similarly replace e_1 by $-e_1$ in the top row, we obtain the following tableau which represents the same relations as (14) in the transposed interpretation.

$$
\begin{array}{c|cc}
 & -e_1 = & a_2 = \\
\hline
-a_1 & \dfrac{1}{\alpha_{11}} & -\dfrac{\alpha_{12}}{\alpha_{11}} \\[2ex]
e_2 & \dfrac{\alpha_{21}}{\alpha_{11}} & \alpha_{22} - \alpha_{12}\dfrac{\alpha_{21}}{\alpha_{11}}
\end{array}
\qquad (15)
$$

This tableau now contains the same coefficients as tableau (6). Thus we have found the following rule which is clearly true in general.

In a tableau with vertical interpretation, an exchange step can be carried out as follows.

A. The rules $1, \ldots, 5$ of the normal exchange method are applied to the coefficients of the tableau.

B. Both of the exchanged vectors are multiplied by -1.

This close relationship between the exchange methods for the two interpretations is known as the *Duality Law of the Exchange Method*. It is very closely connected with many other duality laws in Linear Algebra (see, e.g., 8.6).

In practice, the operation of multiplying the exchanged vectors by -1 is most easily carried out if, in all the tableaux, we imagine the left-hand vectors to have negative signs and the top row of vectors to have positive signs.

The tableau for the relations (12) then appears as follows.

$$
\begin{array}{c|cc}
 & a_1 = & a_2 = \\
\hline
-e_1 & -\overset{*}{\alpha_{11}} & -\alpha_{12} \\
-e_2 & -\alpha_{21} & -\alpha_{22} \\
\hline
 & * & -\dfrac{\alpha_{12}}{\alpha_{11}}
\end{array}
\qquad (16)
$$

where the cellar row for the exchange of a_1 and e_1 has already been adjoined. After this exchange we obtain the tableau

	$e_1 =$	$a_2 =$
$-a_1$	$-\dfrac{1}{\alpha_{11}}$	$-\dfrac{\alpha_{12}}{\alpha_{11}}$
$-e_2$	$\dfrac{\alpha_{21}}{\alpha_{11}}$	$-\alpha_{22} + \alpha_{21}\dfrac{\alpha_{12}}{\alpha_{11}}$

(17)

and, in vertical interpretation, this expresses the relations (13)—as indeed it should.

Example 3. Suppose that we are given the relations

$$e_1 = a_1 + a_2 - a_3$$
$$e_2 = a_1 - 2a_2 + a_3$$

(18)

and we wish to express a_1 and a_2 as linear combinations of e_1, e_2, a_3 by applying the transposed exchange method.

The initial tableau is

	$e_1 =$	$e_2 =$
$-a_1$	-1^*	-1
$-a_2$	-1	2
$-a_3$	1	-1
	$*$	-1

(19)

2nd Tableau

	$a_1 =$	$e_2 =$
$-e_1$	-1	-1
$-a_2$	1	3^*
$-a_3$	-1	-2
	$-\frac{1}{3}$	$*$

(20)

58

3rd Tableau

	$a_1 =$	$a_2 =$
$-e_1$	$-\frac{2}{3}$	$-\frac{1}{3}$
$-e_2$	$-\frac{1}{3}$	$\frac{1}{3}$
$-a_3$	$-\frac{1}{3}$	$-\frac{2}{3}$

(21)

This last tableau (21) states that

$$a_1 = \tfrac{2}{3}e_1 + \tfrac{1}{3}e_2 + \tfrac{1}{3}a_3$$
$$a_2 = \tfrac{1}{3}e_1 - \tfrac{1}{3}e_2 + \tfrac{2}{3}a_3$$

(22)

Example 4. Suppose

$$f_1 = e_1 + e_2$$
$$f_2 = e_1 - 2e_2$$
$$x = 3e_1 + 4e_2$$

(23)

The problem is to express x as a linear combination of f_1 and f_2. We solve this by constructing the tableau for the relations (23) and then exchanging e_1 and e_2 with f_1 and f_2. As an illustration we will carry out both the normal and transposed methods, writing the tableaux for horizontal interpretation on the left and those for the vertical interpretation on the right.

1st Tableau

	e_1	e_2
$f_1 =$	1	1^*
$f_2 =$	1	-2
$x =$	3	4
	-1	$*$

	$f_1 =$	$f_2 =$	$x =$
$-e_1$	-1	-1	-3
$-e_2$	-1^*	2	-4
	$*$	2	-4

2nd Tableau

	e_1	f_1
$e_2 =$	-1	1
$f_2 =$	3^*	-2
$x =$	-1	4
	$*$	$\frac{2}{3}$

	$e_2 =$	$f_2 =$	$x =$
$-e_1$	1	-3^*	1
$-f_1$	-1	2	-4
	$\frac{1}{3}$	$*$	$\frac{1}{3}$

3rd Tableau

	f_2	f_1
$e_2 =$	$-\frac{1}{3}$	$\frac{1}{3}$
$e_1 =$	$\frac{1}{3}$	$\frac{2}{3}$
$x =$	$-\frac{1}{3}$	$\frac{10}{3}$

	$e_2 =$	$e_1 =$	$x =$
$-f_2$	$\frac{1}{3}$	$-\frac{1}{3}$	$\frac{1}{3}$
$-f_1$	$-\frac{1}{3}$	$-\frac{2}{3}$	$-\frac{10}{3}$

From each of these tableaux, we obtain the solution $x = \frac{10}{3} f_1 - \frac{1}{3} f_2$.

Note that in this example, it was not really necessary to evaluate the first row and respectively the first column of the second tableau and similarly only the last row and respectively the last column of the third tableau were actually needed.

Exercises

1. Given the relations

$$e_1 = a_1 + 2a_2 - 2a_3$$
$$e_2 = 3a_1 - a_2 - a_3$$
$$e_3 = 5a_1 + 3a_2 - 2a_3$$
$$\overline{}$$
$$x = a_1 + 2a_2 + 5a_3$$

express the vector x as a linear combination of

    (a) e_1, a_2, a_3,   (b) e_1, e_2, a_3,   (c) e_1, e_2, e_3.

(Apply the exchange method in both the normal and the transposed forms.)

Solution. (a) $x = e_1 + 7a_3$
       (b) $x = e_1 + 7a_3$
       (c) $x = -\frac{11}{3}e_1 - \frac{7}{3}e_2 + \frac{7}{3}e_3$

2. Let $\{a_1, a_2, a_3\}$ be a basis of a 3-dimensional vector space and let

$$e_1 = 2a_1 + 7a_2 - 3a_3$$

$$e_2 = a_1 - 2a_2 + 5a_3$$

$$e_3 = 4a_1 + 3a_2 + 7a_3.$$

Show that it is not possible to express a_1, a_2, a_3 as linear combinations of e_1, e_2, e_3. What does this mean for e_1, e_2, e_3?

Solution. After two exchange steps, the only coefficient which might be used as a pivot is equal to zero. The vectors e_1, e_2, e_3 are linearly dependent and in fact $e_3 = e_1 + 2e_2$.

3.4 Convex Polyhedra

3.4.1 Convex Polyhedra

Definition 1. A 'convex polyhedron' is the convex hull of a finite set of vectors in a finite-dimensional real vector space.

As such the restriction to vector spaces of finite dimension is not actually needed in this definition. However it is natural to introduce it because, in any case, any convex polyhedron will always be contained in the finite-dimensional subspace spanned by the finite number of vectors of which it is the convex hull.

Example 1. In the vector space G_2 (Example 2.1;1) segments, triangles, rectangles are examples of convex polyhedra (cf. Example 2.6;1). In G_3, segments, triangles and rectangles are still convex polyhedra as are also, for example, tetrahedra, cubes and regular octahedra, icosahedra and dodecahedra.

Here we note once again that, in Linear Algebra, by these structures, we mean the corresponding sets of vectors (drawn from a fixed origin 0).

Example 2. Let $\{e_1, \ldots, e_n\}$ be a basis of the real vector space E. The set P of all vectors $x = \sum_{k=1}^{n} \alpha_k e_k$ in which $-1 \leqslant \alpha_k \leqslant +1$ $(k = 1, 2, \ldots, n)$ is a convex polyhedron. In fact P is the convex hull of the 2^n vectors $\sum_{k=1}^{n} \beta_k e_k$ where $\beta_k = \pm 1$ $(k = 1, 2, \ldots, n)$.

61

Firstly, if $n=1$, then $\alpha_1 e_1 = \frac{1}{2}(1+\alpha_1)e_1 + \frac{1}{2}(1-\alpha_1)(-e_1)$. The sum of the coefficients on the right-hand side is 1 and they are both positive or zero when $-1 \leqslant \alpha_1 \leqslant +1$. (cf. 2.6).

If $n=2$ and $x = \alpha_1 e_1 + \alpha_2 e_2$ where $-1 \leqslant \alpha_1, \alpha_2 \leqslant +1$ then, from the case $n=1$, there are representations

$$\alpha_1 e_1 = \lambda_1 e_1 + \mu_1(-e_1) \quad \text{where} \quad \lambda_1, \mu_1 \geqslant 0 \quad \text{and} \quad \lambda_1 + \mu_1 = 1$$

$$\alpha_2 e_2 = \lambda_2 e_2 + \mu_2(-e_2) \quad \text{where} \quad \lambda_2, \mu_2 \geqslant 0 \quad \text{and} \quad \lambda_2 + \mu_2 = 1.$$

It follows that

$$x = \alpha_1 e_1 + \alpha_2 e_2$$

$$= \lambda_1 \lambda_2(e_1+e_2) + \lambda_1 \mu_2(e_1-e_2) + \lambda_2 \mu_1(-e_1+e_2) + \mu_1 \mu_2(-e_1-e_2)$$

where $\lambda_1\lambda_2,\ \lambda_1\mu_2,\ \lambda_2\mu_1,\ \mu_1\mu_2 \geqslant 0$ and $\lambda_1\lambda_2 + \lambda_1\mu_2 + \lambda_2\mu_1 + \mu_1\mu_2 = 1$. For example, in Fig. 10,

$$x = \tfrac{2}{3}e_1 - \tfrac{1}{3}e_2 = \tfrac{5}{18}(e_1+e_2) + \tfrac{10}{18}(e_1-e_2) + \tfrac{1}{18}(-e_1+e_2) + \tfrac{2}{18}(-e_1-e_2).$$

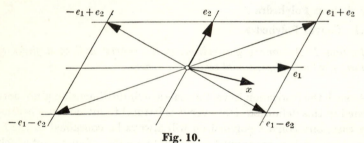

Fig. 10.

We can now see that it is possible to extend this result in a similar way to the cases $n=3,4,\ldots$. This shows that P is a subset of the convex hull. The converse, that every vector in the convex hull is also in P, is obvious. We will refer to this type of convex polyhedron as a *symmetric parallelepiped*.

The following diagram (Fig. 11) shows that some of the vectors in the finite set which spans a convex polyhedron may be superfluous.

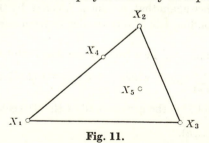

Fig. 11.

Only the end points of the vectors $x_1, \ldots, x_5$ (drawn from a common origin) are shown and they are supposed to lie in a plane. Clearly $K(x_1, \ldots, x_5) = K(x_1, x_2, x_3)$, so that x_4 and x_5 are superfluous. The remaining vectors x_1, x_2, x_3 correspond to the vertices of the triangle.

In the following, we will say what we mean by the vertex-vectors of a general convex polyhedron and then show that every convex polyhedron is the convex hull of its vertex-vectors.

Definition 2. A 'vertex-vector' of a convex polyhedron P is a vector of P which is not a convex linear combination of the other vectors in P.

In Fig. 11, x_4 and x_5 are not vertex-vectors because x_4 is a convex linear combination of x_1 and x_2, and x_5 of x_1, x_2 and x_3.

Theorem 1. If a convex polyhedron $P = K(x_1, \ldots, x_r)$, then all the vertex-vectors of P are included in the set $\{x_1, \ldots, x_r\}$.

Proof. Any vector $x_0 \in P \setminus \{x_1, \ldots, x_r\}$ is a convex linear combination of $x_1, \ldots, x_r \in P$ and therefore cannot be a vertex-vector.

It follows immediately from Theorem 1, that a convex polyhedron has only a finite number of vertex-vectors.

Theorem 2. With the notation of Theorem 1, a vector in the set $\{x_1, \ldots, x_r\}$ is a vertex-vector of P if and only if it is not a convex linear combination of the other vectors in the set. (It is assumed that $x_1, \ldots, x_r$ are distinct vectors.)

Proof. 1. It is clear that the condition is necessary.

2. Suppose for example that x_1 is not a vertex-vector of P. Then there is a representation $x_1 = \sum_{i=1}^{n} \lambda_i y_i$ where $\sum_{i=1}^{n} \lambda_i = 1$; $\lambda_i > 0$; $y_i \in P$ and $y_i \neq x_1$ for $i = 1, \ldots, n$.
Hence

$$x_1 = \sum_{i=1}^{n} \lambda_i \sum_{k=1}^{r} \mu_{ik} x_k, \tag{1}$$

where $\sum_{k=1}^{r} \mu_{ik} = 1$ and $\mu_{ik} \geqslant 0$ for $i = 1, \ldots, n$ and $k = 1, \ldots, r$. The coefficient of x_1 on the right-hand side of (1) is $\sum_{i=1}^{n} \lambda_i \mu_{i1}$ which is strictly less than 1. Since $\sum_{i=1}^{n} \lambda_i = 1$ and $\lambda_i > 0$, it could only be equal to 1 if

$$\mu_{11} = \mu_{21} = \ldots = \mu_{n1} = 1.$$

63

But this would mean that $y_i = x_1$ for all i, which is a contradiction of the assumptions.

Hence (1) can be solved for x_1, to obtain

$$x_1 = \left(1 - \sum_{i=1}^{n} \lambda_i \mu_{i1}\right)^{-1} \sum_{k=2}^{r} \sum_{i=1}^{n} \lambda_i \mu_{ik} x_k.$$

The coefficients on the right-hand side are all positive or zero. Their sum is

$$\left(1 - \sum_{i=1}^{n} \lambda_i \mu_{i1}\right)^{-1} \sum_{i=1}^{n} \lambda_i \sum_{k=2}^{r} \mu_{ik} = \left(1 - \sum_{i=1}^{n} \lambda_i \mu_{i1}\right)^{-1} \sum_{i=1}^{n} \lambda_i(1 - \mu_{i1}) = 1.$$

Therefore x_1 is a convex linear combination of $x_2, \ldots, x_r$.

Theorem 3. A convex polyhedron is the convex hull of its vertex-vectors.

Proof. Let $P = K(x_1, \ldots, x_r)$ and suppose that x_1 is not a vertex-vector. By Theorem 2, $\{x_1, \ldots, x_r\} \subseteq K(x_2, \ldots, x_r)$. Using Theorem 2.6;6, it follows that $P = K(x_2, \ldots, x_r)$. Now, if there is another of the vectors $x_2, \ldots, x_r$ which is not a vertex-vector, then this can also be left out in the same way, etc.

3.4.2 Simplexes

Suppose $P = K(x_0, \ldots, x_r)$ is a convex polyhedron. By Theorems 2.5;10 and 2.6;7, $P \subseteq N(P) = N(x_0, \ldots, x_r)$. This coset has dimension at most r, where the dimension of a coset is defined to be the dimension of the corresponding subspace. The subspace is in fact $L(x_1 - x_0, \ldots, x_r - x_0)$, (cf. the proof of Theorem 2.5;10.)

Definition 3. A convex polyhedron $P = K(x_0, \ldots, x_r)$ is said to be an 'r-dimensional simplex', if the dimension of the coset $N(x_0, \ldots, x_r)$ is equal to r.

If P is an r-dimensional simplex, it is easy to see that all of the vectors $x_0, \ldots, x_r$ are vertex-vectors. For instant, if x_r were not a vertex-vector then there would be a representation $x_r = \sum_{k=0}^{r-1} \lambda_k x_k$, where $\sum_{k=0}^{r-1} \lambda_k = 1$ (Theorem 2). But then $x_r \in N(x_0, \ldots, x_{r-1})$ (Theorem 2.5;10) and the dimension of the coset $N(x_0, \ldots, x_r)$ would not be greater than $r - 1$.

Example 3. In the vector space G_2 (Example 2.1;1) every simplex is one of the following types.

    (a) consisting of only one vector (dim. 0)

    (b) the convex hull of two distinct vectors (segment, dim. 1)

    (c) the convex hull of three vectors whose end points are not collinear (when they are drawn from a common origin) (triangle, dim. 2)

In the vector space G_3, the same cases appear as in G_2 and also

(d) the convex hull of four vectors whose end points are not coplanar (tetrahedron, dim. 3).

Example 4. In a real vector space E, suppose that $e_1, \ldots, e_r$ are linearly independent. Then $S = K(0, e_1, \ldots, e_r)$ is an r-dimensional simplex.

Theorem 4. If $S = K(x_0, \ldots, x_r)$ is an r-dimensional simplex, then every vector $x \in S$ can be written in exactly one way as a convex linear combination of $x_0, \ldots, x_r$.

Proof. Suppose that $x = \sum\limits_{k=0}^{r} \lambda_k x_k$ and $x = \sum\limits_{k=0}^{r} \mu_k x_k$, where

$$\sum_{k=0}^{r} \lambda_k = \sum_{k=0}^{r} \mu_k = 1.$$

Then

$$\sum_{k=1}^{r} (\lambda_k - \mu_k)(x_k - x_0) = 0$$

and, since the vectors $x_1 - x_0, \ldots, x_r - x_0$ are linearly independent, it follows that $\lambda_k = \mu_k$ for $k = 1, \ldots, r$ and hence that $\lambda_0 = \mu_0$.

If $r = n = \dim E$, then every vector $x \in E$ can be represented in exactly one way in the form $x = \sum\limits_{k=0}^{r} \lambda_k x_k$ where $\sum\limits_{k=0}^{r} \lambda_k = 1$. The numbers $\lambda_0, \ldots, \lambda_r$, which are uniquely determined by x and the condition $\sum\limits_{k=0}^{r} \lambda_k = 1$, are called *barycentric co-ordinates* in the vector space E with respect to the simplex S.

If $x_0, \ldots, x_r$ are the vertex-vectors of an r-dimensional simplex S_r, then, for $0 \leqslant s \leqslant r$, any $s+1$ of these vectors will be the vertex-vectors of an s-dimensional simplex S_s contained in S_r. We refer to S_s as an s-dimensional *face* of S_r. The number of s-dimensional faces of an r-dimensional simplex is $\binom{r+1}{s+1}$. In particular, there are $r+1$ $(r-1)$-dimensional faces and $r+1$ 0-dimensional faces (each consisting of just one vertex). S_r has just one r-dimensional face, viz. itself.

Theorem 5. The intersection of two faces of a simplex is either a face or it is empty.

Proof. 1. From Theorem 4, it follows that the intersection is empty when the two faces have no vertex-vectors in common.

2. If the two faces have vertex-vectors in common, their intersection is the simplex spanned by these vertex-vectors.

Theorem 6. *The intersection P of a coset N and a simplex S is either a convex polyhedron or it is empty.*

Proof. 1. If $P \neq \varnothing$, then, for each $y \in P$, we can construct a unique face S_y of S which contains y and has the least possible dimension. (By Theorem 5, S_y is uniquely determined as the intersection of all the faces which contain y.) If $P \cap S_y = \{y\}$, we will say that y is a distinguished vector of P. Since each face of S can be the S_y for at most one distinguished vector y and S has only a finite number of faces, it follows that there can only be a finite number of distinguished vectors.

2. We will now show that P is the convex hull of its distinguished vectors and this will prove the theorem.

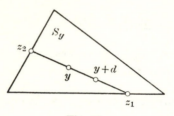

Fig. 12.

3. Suppose that $y \in P$ is not distinguished. Then there is a vector $d \in E$, $d \neq 0$ such that $y + d \in P \cap S_y$. For arbitrary real numbers μ_1, μ_2 it then follows that

$$z_1 = y + \mu_1 d \in N \quad \text{and} \quad z_2 = y - \mu_2 d \in N.$$

We will show in 4. that we may take μ_1, μ_2 to be positive and so choose them that z_1 and z_2 lie in proper faces of S_y (Fig. 12). Then

$$y = \frac{1}{\mu_1 + \mu_2} (\mu_2 z_1 + \mu_1 z_2)$$

i.e., y is a convex linear combination of z_1, $z_2 \in P$. If z_1, z_2 are not distinguished, we can repeat the process just described for each of them. After a finite number of steps (at the latest when we reach 0-dimensional faces of S_y), we obtain a representation of y as a convex linear combination of distinguished vectors of P.

4. It only remains to prove the assertion about the choice of μ_1 and μ_2. If $S_y = K(x_0, \ldots, x_s)$, then there are representations

$$y = \sum_{k=0}^{s} \lambda_k x_k \quad \text{and} \quad y+d = \sum_{k=0}^{s} (\lambda_k + \delta_k) x_k$$

where $\lambda_k > 0$ $(k=0,\ldots,s)$; $\sum_{k=0}^{s} \lambda_k = 1$ and $\sum_{k=0}^{s} \delta_k = 0$. Since $d \neq 0$, not all δ_k are zero and therefore there must be one which is strictly positive and one which is strictly negative. We now choose $\mu_1 > 0$ so that $\lambda_k + \mu_1 \delta_k \geqslant 0$, for $k=0,\ldots,s$, and at least one of these is zero—which is possible because $\lambda_k > 0$ for all k and $\delta_k < 0$ for some k. Hence

$$z_1 = y + \mu_1 d = \sum_{k=0}^{s} (\lambda_k + \mu_1 \delta_k) x_k \in S_y.$$

Since one of the coefficients is zero, z_1 actually belongs to a proper face of S_y. The choice of μ_2 is made in a similar way.

3.4.3 Convex Pyramids

Let E be a finite-dimensional real vector space.

Definition 4. A 'convex pyramid' $P \subseteq E$ is the positive hull of a finite number of vectors $x_1, \ldots, x_r \in E$ (cf. 2.6.2).

A convex pyramid is a convex cone and is therefore convex.

Example 5. In Example 2.6;5, the convex cones (1), (2) and (4) are convex pyramids in the vector space G_2. On the other hand (3) is not a convex pyramid.

In G_3, pyramids in the usual sense are convex pyramids providing they are convex and their origins correspond to the zero-vector.

Example 6. The set of all vectors $x \in E$, for which all the components (with respect to some basis) are positive, is a convex pyramid. It is the positive hull of the basis.

Every subspace $L \subseteq E$ (and hence E itself) is a convex pyramid. If $\{e_1, \ldots, e_r\}$ is a basis of L, then $L = P(e_1, \ldots, e_r, -e_1, \ldots, -e_r)$.

The theory of convex pyramids naturally develops along similar lines to that of convex polyhedra. We will therefore first look for the analogues of the vertex-vectors. These are what are known as the *edge-vectors*.

In order to simplify the definition of these, we will say that two vectors $x, y \in E$ are similar if each is a strictly positive multiple of the other, i.e., $y = \alpha x$ with $\alpha > 0$. (The zero-vector is therefore only similar to itself.)

*Definition 5. A vector x of a convex pyramid P is an 'edge-vector' of P if,
from $x = \sum\limits_{k=1}^{r} \lambda_k x_k$, $\lambda_k > 0$, $x_k \in P$, $x_k \neq 0$ $(k=1,\ldots,r)$, it follows that $x_1, \ldots, x_r$
are similar to x.*

If x is an edge-vector of P, then so is every vector which is similar to x.

Remembering Theorem 3, we could conjecture here that every convex pyramid is the positive hull of its edge-vectors. However, we can see that this is not always true, by considering the example of the convex pyramid E which has no edge-vectors at all.

Theorem 7. Every acute convex pyramid is the positive hull of its edge-vectors, where it is sufficient to choose just one representative from each class of similar edge-vectors. (See Definition 2.6 ;6.)

The proof of this theorem is directly similar to that of Theorem 3 and so we will not write it out again.

Problems

1. Prove that, if P, Q are convex polyhedra, then so are $K(P \cup Q)$ and $\lambda P + \mu Q$ (for all real λ, μ).

2. Let $\{e_1, \ldots, e_n\}$ be a basis of a real vector space. Show that $e_1, \ldots, e_n$ are the vertex-vectors of an $(n-1)$-dimensional simplex S.

3. In Problem 2, let $E = R_n$ and $e_i = (\delta_{i1}, \ldots, \delta_{in})$ $(i=1,\ldots,n)$. Find a condition on $\xi_1, \ldots, \xi_n$ in order that $x = (\xi_1, \ldots, \xi_n) \in S$.

4. Prove that, in a finite-dimensional vector space, the set of all those vectors all of whose components (with respect to a given basis) are positive or zero is a convex pyramid.

5. Prove that, if K is a convex polyhedron, then $P(K)$ is a convex pyramid.

CHAPTER 4

DETERMINANTS

4.1 Permutations

A *permutation* of a finite set of n elements is a mapping of the set onto itself. Since it is a mapping of a finite set *onto* itself, a permutation is automatically 1-1. If we denote the elements of the set A by $1, 2, \ldots, n$ and a permutation of A by ϕ, then $\phi(k)$ will take each of the values from 1 to n exactly once as k runs through the elements of A. (The use of small Greek letters to denote permutations will be the only departure from our usual convention of using small Roman letters for mappings.)

A convenient method of presenting a permutation is to write down the elements of A in a row and then, underneath each of these, to write down its image under the permutation. For example, if the permutation is written in the form

$$\phi_1 = \begin{pmatrix} 1 & 2 & 3 & 4 & 5 \\ 2 & 5 & 3 & 1 & 4 \end{pmatrix}$$

then this means that ϕ_1 is the mapping such that $\phi_1(1)=2$, $\phi_1(2)=5$, $\phi_1(3)=3$, etc.

The total number of permutations of a set of n elements is $n! = n(n-1) \ldots 3.2.1$, because there are n possibilities for the choice of $\phi_1(1)$, and then $(n-1)$ possibilities for $\phi_1(2)$ $(\phi_1(2) \neq \phi_1(1))$, etc. Finally $\phi_1(n)$ is completely determined by $\phi_1(1), \ldots, \phi_1(n-1)$.

Since permutations are mappings, they can be multiplied together (see 1.3.1). The product $\phi_2 \phi_1$ of the permutations ϕ_1 and ϕ_2 is given by $k \rightarrow \phi_2 \phi_1(k) = \phi_2(\phi_1(k))$. This multiplication of permutations is not commutative in general. For example, if ϕ_2 is the permutation

$$\phi_2 = \begin{pmatrix} 1 & 2 & 3 & 4 & 5 \\ 5 & 4 & 1 & 3 & 2 \end{pmatrix}$$

then

$$\phi_2 \phi_1 = \begin{pmatrix} 1 & 2 & 3 & 4 & 5 \\ 4 & 2 & 1 & 5 & 3 \end{pmatrix}$$

and

$$\phi_1 \phi_2 = \begin{pmatrix} 1 & 2 & 3 & 4 & 5 \\ 4 & 1 & 2 & 3 & 5 \end{pmatrix} \neq \phi_2 \phi_1.$$

On the other hand, the multiplication of permutations is associative (Theorem 1.3;1).

The identity permutation ϵ is given by $\epsilon(k)=k$ $(k=1,\ldots,n)$ and we see that $\phi\epsilon=\epsilon\phi=\phi$ for all permutations ϕ.

Since a permutation ϕ is a 1-1 mapping, it has an inverse ϕ^{-1} for which $\phi^{-1}\phi=\phi\phi^{-1}=\epsilon$ (see 1.3.2). This can be found by interchanging the rows in the above representation of ϕ. For example,

$$\phi_1^{-1} = \begin{pmatrix} 2 & 5 & 3 & 1 & 4 \\ 1 & 2 & 3 & 4 & 5 \end{pmatrix} = \begin{pmatrix} 1 & 2 & 3 & 4 & 5 \\ 4 & 1 & 3 & 5 & 2 \end{pmatrix}$$

These rules show that the permutations of a set of n elements together with their multiplication form a group, which is known as the symmetric group S_n. (See [25] pp. 61–64.)

If ψ is any permutation of A, then, as ϕ runs through all the permutations of A, so do $\phi\psi$ and $\psi\phi$ because, for any permutation θ, $\theta=(\theta\psi^{-1})\psi=\psi(\psi^{-1}\theta)$. Also ϕ^{-1} runs through all the permutations as ϕ does, because $\theta=(\theta^{-1})^{-1}$ for all permutations θ.

A transposition is a permutation which interchanges two of the elements and leaves all the rest fixed. Thus the permutation ϕ is a transposition when there are two elements k_1, $k_2 \in A$ such that $\phi(k_1)=k_2$, $\phi(k_2)=k_1$ and $\phi(k)=k$ for all $k \in A$, $k_1 \neq k \neq k_2$.

We will denote this transposition briefly by $(k_1 k_2)$. Obviously, every permutation can be written as a product of transpositions and in fact in many different ways. For example,

$$\phi_1 = (1\ 4)(1\ 5)(1\ 2) = (2\ 5)(1\ 5)(4\ 5)$$

Theorem 1. *It is not possible to write a permutation both as a product of an odd number of transpositions and as a product of an even number of transpositions.*

Proof. The nature of the elements to be permuted is not relevant to the statement of the theorem. We will assume that they are real variables and denote them by $\xi_1,\ldots,\xi_n$ (instead of $1,\ldots,n$).

The function

$$\Delta(\xi_1,\ldots,\xi_n) = (\xi_1-\xi_2)(\xi_1-\xi_3)\ldots(\xi_1-\xi_n)$$
$$\cdot(\xi_2-\xi_3)\ldots(\xi_2-\xi_n)$$
$$\vdots$$
$$\cdot(\xi_{n-1}-\xi_n)$$

has the special property that, if $\xi_1,\ldots,\xi_n$ are permuted, then either it is unchanged or it is simply multiplied by -1. Any transposition (for example of ξ_1 and ξ_2) has the latter effect. Hence a permutation which is a product of an odd number of transpositions will have the effect of multiplying Δ by -1 and a product of an even number of transpositions will leave Δ unchanged. Consequently no permutation can be both a product of an odd number of transpositions and a product of an even number of transpositions.

In view of Theorem 1, we can make the following Definition.

Definition 1. A permutation is said to be 'even' (odd) if it is the product of an even (odd) number of transpositions. The 'characteristic' ch(ϕ) of a permutation ϕ is equal to $+1$ if ϕ is even and is equal to -1 if ϕ is odd.

Theorem 2. 1. $\text{ch}(\epsilon) = 1$
          2. $\text{ch}(\phi_2 \phi_1) = \text{ch}(\phi_2)\,\text{ch}(\phi_1)$
          3. $\text{ch}(\phi^{-1}) = \text{ch}(\phi)$

Proof. The identity permutation is the product of no transpositions and is therefore even. A representation of $\phi_2 \phi_1$ as a product of transpositions can be found by multiplying such representations of ϕ_1 and ϕ_2. The second rule follows from this. Finally $\phi \cdot \phi^{-1} = \epsilon$, and by the first and second rules, $\text{ch}(\phi) \cdot \text{ch}(\phi^{-1}) = 1$, from which the third rule follows.

Exercises

1. Decide whether the following two permutations are odd or even.

$(a)\ \phi = \begin{pmatrix} 1 & 2 & 3 & 4 & 5 \\ 5 & 2 & 4 & 1 & 3 \end{pmatrix}$ $(b)\ \psi = \begin{pmatrix} 1 & 2 & 3 & 4 & 5 \\ 4 & 3 & 1 & 5 & 2 \end{pmatrix}$

Solution. (a) ϕ is odd (b) ψ is even.

2. Calculate the product $\phi\psi$ of the permutations in Exercise 1.

Solution. $\phi\psi = \begin{pmatrix} 1 & 2 & 3 & 4 & 5 \\ 1 & 4 & 5 & 3 & 2 \end{pmatrix}$.

Problems

1. Show that, if ϕ is a permutation, then there is a power ϕ^r of ϕ which is equal to the identity permutation. (Hint: consider the countably infinite set of powers $\phi, \phi^2, \phi^3, \ldots$ and show that there are two of these with different exponents but which represent the same permutation.)

2. Prove that the set of all even permutations of n symbols is a group with multiplication. This is known as the alternating group A_n (A_n is a subgroup of the symmetric group S_n of all permutations). Is this also true for the set of all odd permutations?

4.2 Determinants

4.2.1 The Concept of a Determinant

Let E_n be an n-dimensional vector space and let $\{e_1, \ldots, e_n\}$ be a basis of E_n which will be kept fixed in the following. We consider functions $D(x_1, \ldots, x_n)$, whose variables $x_1, \ldots, x_n$ are vectors in E_n and whose values are scalars. Thus, corresponding to each ordered set $\{x_1, \ldots, x_n\} \subseteq E_n$, there is a scalar value of the function $D(x_1, \ldots, x_n)$.

Definition 1. $D(x_1, \ldots, x_n)$ is said to be a 'determinant' in E_n (with respect to the basis $\{e_1, \ldots, e_n\}$) if the following conditions are satisfied.

D1. If a permutation ϕ is applied to the variables, then $D(x_1, \ldots, x_n)$ is multiplied by the characteristic of ϕ. That is to say, $D(\phi(x_1), \ldots, \phi(x_n)) = \mathrm{ch}(\phi) \cdot D(x_1, \ldots, x_n)$. (In particular, $D(x_1, \ldots, x_n)$ is multiplied by -1 if two vectors are interchanged.)

D2. $D(x_1, \ldots, x_n)$ is linear in each variable x_k. This means, for example, when $k = 1$,

$$D(\alpha x_1 + \beta y_1, x_2, \ldots, x_n) = \alpha D(x_1, x_2, \ldots, x_n) + \beta D(y_1, x_2, \ldots, x_n)$$

for arbitrary scalars α, β (cf. Definition 6.1;1).

D3. $D(x_1, \ldots, x_n)$ is normalized, i.e., $D(e_1, \ldots, e_n) = 1$.

The *dimension* of a determinant is the dimension n of the vector space E_n or equivalently the number of variables in $D(x_1, \ldots, x_n)$.

Of course it is not claimed initially that there exists a determinant in E_n with respect to the basis $\{e_1, \ldots, e_n\}$. However, in the following, we will prove that there is in fact exactly one determinant.

Theorem 1. Let ϕ be a mapping of the basis $\{e_1, \ldots, e_n\}$ into itself (ϕ is not necessarily a permutation). Then, if there is a determinant D,

$$D(\phi(e_1), \ldots, \phi(e_n)) \begin{cases} = \mathrm{ch}(\phi), \text{ if } \phi \text{ is a permutation} \\ = 0, \text{ if two of the vectors } \phi(e_1), \ldots, \phi(e_n) \text{ are equal.} \end{cases}$$

Proof. If ϕ is a permutation, the assertion follows from D3 by substituting the basis vectors $e_1, \ldots, e_n$ for $x_1, \ldots, x_n$ in D1. On the other hand, if ϕ is not a permutation, then there are two distinct basis vectors e_{k_1}, e_{k_2} such that $\phi(e_{k_1}) = \phi(e_{k_2})$. Thus interchanging the $\phi(e_{k_1})$ and $\phi(e_{k_2})$ in $D(\phi(e_1), \ldots, \phi(e_n))$ will clearly not alter this expression. However, by D1, it must also be multiplied by -1. Hence it must be equal to zero.

Now suppose that the vectors $x_1, \ldots, x_n$ are represented in terms of the basis vectors $e_1, \ldots, e_n$

$$x_i = \sum_{k=1}^{n} \xi_{ik} e_k \qquad (i = 1, \ldots, n)$$

Then, for a determinant D, it follows from D2 that

$$D(x_1, \ldots, x_n) = \sum_{k=1}^{n} \xi_{1k} D(e_k, x_2, \ldots, x_n) = \ldots.$$

$$= \sum_{k_1=1}^{n} \cdots \sum_{k_n=1}^{n} \xi_{1k_1} \xi_{2k_2} \cdots \xi_{nk_n} D(e_{k_1}, e_{k_2}, \ldots, e_{k_n}).$$

(Since the n summation indices vary independently from 1 to n, it is necessary to denote them by different symbols which we have taken to be $k_1, \ldots, k_n$.) By Theorem 1, all those terms, for which the mapping $i \to k_i$ ($i = 1, \ldots, n$) is not a permutation, drop out, and we have

$$D(x_1, \ldots, x_n) = \sum_{\phi} \xi_{1, \phi(1)} \xi_{2, \phi(2)} \cdots \xi_{n, \phi(n)} D(\phi(e_1), \ldots, \phi(e_n))$$

or, alternatively, $D(x_1, \ldots, x_n) = \sum_{\phi} \operatorname{ch}(\phi) \xi_{1, \phi(1)} \cdots \xi_{n, \phi(n)},$ \hfill (1)

where the summations are taken over all permutations ϕ of the indices $1, \ldots, n$, and $\phi(e_k) = e_{\phi(k)}$. Thus we have expressed $D(x_1, \ldots, x_n)$ in terms of the components of $x_1, \ldots, x_n$, which means that there can be at most one determinant. In order to show that a determinant actually exists, we must verify that the function (1) satisfies the conditions D1, D2 and D3.

D1. For a permutation ψ, it follows from (1) that

$$D(\psi(x_1), \ldots, \psi(x_n)) \overset{1}{=} \sum_{\phi} \operatorname{ch}(\phi) \xi_{\psi(1), \phi(1)} \cdots \xi_{\psi(n), \phi(n)}$$

$$\overset{2}{=} \sum_{\phi} \operatorname{ch}(\phi) \xi_{1, \phi\psi^{-1}(1)} \cdots \xi_{n, \phi\psi^{-1}(n)}$$

$$\overset{3}{=} \operatorname{ch}(\psi) \sum_{\theta} \operatorname{ch}(\theta) \xi_{1, \theta(1)} \cdots \xi_{n, \theta(n)}$$

$$\overset{4}{=} \operatorname{ch}(\psi) D(x_1, \ldots, x_n).$$

In this the equalities indicated are true for the following reasons.

1. Equation (1) and $\psi(x_k) = x_{\psi(k)}$.

2. If $\psi(k) = i$, then $k = \psi^{-1}(i)$ and hence $\xi_{\psi(k), \phi(k)} = \xi_{i, \phi\psi^{-1}(i)}$. Also $\psi(k)$ runs through all the values $1, \ldots, n$ as k does.

3. $\theta = \phi\psi^{-1}$ runs through all the permutations as ϕ does and $\operatorname{ch}(\phi) = \operatorname{ch}(\psi) \cdot \operatorname{ch}(\theta)$.

4. Equation (1).

D2. If the components of y_1 are denoted by η_{1k} then

$$D(\alpha x_1 + \beta y_1, x_2, \ldots, x_n) = \sum_{\phi} \operatorname{ch}(\phi) (\alpha \xi_{1, \phi(1)} + \beta \eta_{1, \phi(1)}) \xi_{2, \phi(2)} \cdots \xi_{n, \phi(n)}$$

$$= \alpha D(x_1, x_2, \ldots, x_n) + \beta D(y_1, x_2, \ldots, x_n)$$

73

D3. $D(e_1, \ldots, e_n) = \sum\limits_{\phi} \text{ch}(\phi)\, \delta_{1,\phi(1)} \cdots \delta_{n,\phi(n)} = 1$ since the product of the Kronecker deltas $\delta_{k,\phi(k)}$ (see Example 3.1;2) is only non-zero when ϕ is the identity permutation.

Hence we have now proved the following theorem.

Theorem 2. For a given basis $\{e_1, \ldots, e_n\}$, there exists exactly one determinant in E_n. It is represented by (1) by using the components of the vectors $x_1, \ldots, x_n$.

In (1) the determinant $D(x_1, \ldots, x_n)$ appears as a sum of $n!$ terms corresponding to the $n!$ permutations ϕ. Each of these terms is the product of the characteristic $\text{ch}(\phi)$ with n components of the vectors $x_1, \ldots, x_n$. In fact exactly one component from each of the vectors appears as a factor and each of the values $1, \ldots, n$ appears exactly once as a second index $(\phi(1), \ldots, \phi(n)$ are all distinct).

Example 1. If $n=2$, then

$$D(x_1, x_2) = \xi_{11}\xi_{22} - \xi_{12}\xi_{21}$$

If $n=3$, then

$$D(x_1, x_2, x_3) = \xi_{11}\xi_{22}\xi_{33} - \xi_{11}\xi_{23}\xi_{32} + \xi_{12}\xi_{23}\xi_{31} - \xi_{13}\xi_{22}\xi_{31}$$
$$+ \xi_{13}\xi_{21}\xi_{32} - \xi_{12}\xi_{21}\xi_{33}.$$

4.2.2 Properties of the Determinant

Theorem 3. If one of the vectors $x_1, \ldots, x_n$ is the zero-vector, then

$$D(x_1, \ldots, x_n) = 0.$$

Proof. If for example $x_k = 0$, then $x_k = 0x_k$ and, by D2,

$$D(x_1, \ldots, x_k, \ldots, x_n) = D(x_1, \ldots, 0x_k, \ldots, x_n) = 0D(x_1, \ldots, x_n) = 0.$$

Theorem 4. If two of the vectors $x_1, \ldots, x_n$ are equal, then $D(x_1, \ldots, x_n) = 0$.

Proof. Transposing the two equal vectors does not change D. On the other hand, by D1, it also changes the sign of D because a transposition is an odd permutation.

Theorem 5. If one of the vectors $x_1, \ldots, x_n$ is a linear combination of the others, then $D(x_1, \ldots, x_n) = 0$.

Proof. Suppose for example that $x_1 = \sum\limits_{k=2}^{n} \alpha_k x_k$.

Then, by D2,

$$D(x_1, \ldots, x_n) = \sum_{k=2}^{n} \alpha_k D(x_k, x_2, \ldots, x_n).$$

The determinants in the summation all vanish because of Theorem 4.

Theorem 6. The value of the determinant is unaltered when one of the vectors $x_1, \ldots, x_n$ is modified by adding to it a linear combination of the others.

Proof. For example, adding the linear combination y of $x_2, \ldots, x_n$ onto x_1, we obtain

$$D(x_1+y, x_2, \ldots, x_n) = D(x_1, x_2, \ldots, x_n) + D(y, x_2, \ldots, x_n)$$
$$= D(x_1, x_2, \ldots, x_n) \text{ by D2 and Theorem 5.}$$

Theorem 7. $D(x_1, \ldots, x_n) = 0$ *if and only if the vectors $x_1, \ldots, x_n$ are linearly dependent.*

Proof. One half of the assertion has already been proved in Theorem 5. Suppose now that $x_1, \ldots, x_n$ are linearly independent, so that by Theorem 3.2;3 they form a basis of E_n and $e_1, \ldots, e_n$ can be written in the form

$$e_i = \sum_{k=1}^{n} \alpha_{ik} x_k \qquad (i = 1, \ldots, n). \tag{2}$$

Hence

$$1 \overset{1}{=} D(e_1, \ldots, e_n) \overset{2}{=} \sum_{k_1=1}^{n} \cdots \sum_{k_1=1}^{n} \alpha_{1k_1} \cdots \alpha_{nk_n} D(x_{k_1}, \ldots x_{k_n})$$

$$\overset{3}{=} \sum_{\phi} \alpha_{1,\phi(1)} \cdots \alpha_{n,\phi(n)} D(\phi(x_1), \ldots, \phi(x_n))$$

$$\overset{4}{=} \sum_{\phi} \operatorname{ch}(\phi) \, \alpha_{1,\phi(1)} \cdots \alpha_{n,\phi(n)} D(x_1, \ldots, x_n)$$

and therefore $D(x_1, \ldots, x_n) \neq 0$.

The indicated equalities are true for the following reasons.

1. D3
2. (2) and D2
3. Restriction to the terms for which $i \to k_i (i = 1, \ldots, n)$ is a permutation (Theorem 4)
4. D1

Since the vectors $x_1, \ldots, x_n$ are determined by their components ξ_{ik} (with respect to the particular basis $\{e_1, \ldots, e_n\}$), the determinant can also be written as a function of ξ_{ik}. We do this as follows

$$D(x_1, \ldots, x_n) = \begin{vmatrix} \xi_{11} & \xi_{12} & \cdots & \xi_{1n} \\ \cdots & \cdots & \cdots & \cdots \\ \cdots & \cdots & \cdots & \cdots \\ \cdots & \cdots & \cdots & \cdots \\ \xi_{n1} & \xi_{n2} & \cdots & \xi_{nn} \end{vmatrix} = \det(\xi_{ik}) = \det X$$

where X is a shortened notation for the square array of components ξ_{ik}. Such an array of scalars will later be referred to as a *matrix* (cf. 5.2.2) and the quantities ξ_{ik} as the elements of the matrix. Det X will also be called the 'determinant of the matrix X'. The explicit expression for det X in terms of the elements ξ_{ik} is given by (1).

Theorems 3–7 can now be reformulated, in such a way that each of the vectors $x_1, \ldots, x_n$ corresponds to a row of the matrix X. Thus, for example, Theorem 7 reads: det $X = 0$ if and only if the rows of the matrix X are linearly dependent. (The rows of X are to be thought of as elements of the space R_n. By Theorem 3.1;4, the rows of X are linearly dependent if and only if the vectors $x_1, \ldots, x_n \in E$ are linearly dependent.)

The array of numbers which is constructed from X by writing the rows as columns and vice versa is again a matrix which is called the *transpose* of X and is denoted by X' (cf. Definition 5.2;3). Naturally, we can also form the determinant of X'

$$\det X' = \begin{vmatrix} \xi_{11} & \xi_{21} & \cdots & \xi_{n1} \\ \cdots & \cdots & \cdots & \cdots \\ \cdots & \cdots & \cdots & \cdots \\ \cdots & \cdots & \cdots & \cdots \\ \xi_{1n} & \xi_{2n} & \cdots & \xi_{nn} \end{vmatrix}$$

In connection with this determinant, we now have the following Duality Theorem.

Theorem 8. det $X' = $ det X. *That is, a determinant is unaltered by writing the rows as columns and vice versa.*

Proof. By (1)

$$\det X' = \sum_\phi \operatorname{ch}(\phi)\, \xi_{\phi(1),1} \cdots \xi_{\phi(n),n}$$
$$= \sum_\phi \operatorname{ch}(\phi)\, \xi_{1,\phi^{-1}(1)} \cdots \xi_{n,\phi^{-1}(n)}$$
$$= \sum_\psi \operatorname{ch}(\psi)\, \xi_{1,\psi(1)} \cdots \xi_{n,\psi(n)} = \det X,$$

because $\psi = \phi^{-1}$ runs through all the permutations as ϕ does and $\text{ch}(\psi) = \text{ch}(\phi)$.

Thus Theorems 3–7 could also be formulated in terms of the columns of the matrix X. For example, Theorem 6 reads: The value of the determinant is unaltered when one of the columns (thought of as an element of R_n) is modified by adding a linear combination of the others.

There is a further important property which we are not yet ready to prove but which can be found in Theorem 5.2;14.

Problems

1. What sign does the term $\xi_{1n}\xi_{2(n-1)} \cdots \xi_{n1}$ have in the representation 4.2;(1) of the determinant?

2. Prove that, if the rows of a determinant are linearly dependent (independent), then so are the columns.

3. Let the square matrix X be partitioned as follows:

$$X = \begin{pmatrix} X_1 & Y \\ 0 & X_2 \end{pmatrix}$$

where X_1, X_2 are square matrices and the 0 in the bottom left-hand corner consists of all zeros. Show that $\det X = \det X_1 . \det X_2$.

4. Show that the 'Vandermonde Determinant'

$$\begin{vmatrix} 1 & 1 & \dots & 1 \\ \xi_1 & \xi_2 & \dots & \xi_n \\ \xi_1^2 & \xi_2^2 & \dots & \xi_n^2 \\ \dots & \dots & \dots & \dots \\ \xi_1^{n-1} & \xi_2^{n-1} & \dots & \xi_n^{n-1} \end{vmatrix}$$

is equal to

$$\prod_{i,\, k=1}^{n} (\xi_i - \xi_k). \qquad (i > k).$$

(Hint: Subtract the first column from each of the others, expand about the first row and subtract, from each row, ξ_1 times the preceding row.)

5. Evaluate the determinants

$$D_1 = \begin{vmatrix} -\lambda & 1 \\ -\alpha_0 & -(\alpha_1 + \lambda) \end{vmatrix} \quad \text{and} \quad D_2 = \begin{vmatrix} -\lambda & 1 & 0 \\ 0 & -\lambda & 1 \\ -\alpha_0 & -\alpha_1 & -(\alpha_2 + \lambda) \end{vmatrix}$$

and, using these, make a conjecture for the value of

$$D_n = \begin{vmatrix} -\lambda & 1 & 0 & \cdots & 0 \\ 0 & -\lambda & 1 & \cdots & 0 \\ \cdots & \cdots & \cdots & \cdots & \cdots \\ 0 & 0 & 0 & \cdots & 1 \\ -\alpha_0 & -\alpha_1 & -\alpha_2 & \cdots & -(\alpha_{n-1}+\lambda) \end{vmatrix}.$$

Prove this conjecture!

6. Let α_{ik} and β_{ik} $(i,k=1,\ldots,n)$ be arbitrary scalars. Prove that there exist in general at most n scalars λ such that $\det(\alpha_{ik}+\lambda\beta_{ik})=0$.

4.3 Numerical Evaluation of Determinants

For given elements ξ_{ik}, the value of the determinant $\det X = D(x_1,\ldots,x_n)$ can be computed directly from the fundamental formula 4.2;(1). However this requires $n!(n-1)$ multiplications and, for higher dimensions n, the work involved becomes almost too much to cope with. Nevertheless it will be realized that there are many partial products which appear in several of the terms of (1) and any practical method of calculation will try to ensure that these are not recalculated anew each time. We will now show several ways of doing this.

4.3.1 Expansion of a Determinant about a Row or a Column

We investigate the sum of those terms in the representation 4.2;(1) of the determinant which contain the factor ξ_{11}. This sum has the value

$$\xi_{11} \sum_{\phi}{}^* \mathrm{ch}\,(\phi)\, \xi_{2,\phi(2)} \cdots \xi_{n,\phi(n)}$$

where $\sum\limits_{\phi}{}^*$ does not mean the sum over all $n!$ permutations but only over those for which $\phi(1)=1$. If we restrict ϕ to the elements $2,\ldots,n$ we obtain a permutation ϕ^* of these $n-1$ elements and obviously $\mathrm{ch}\,(\phi)=\mathrm{ch}\,(\phi^*)$. That is, ξ_{11} is multiplied by the $(n-1)$-dimensional determinant D_{11} which is obtained from the original by erasing the row and column which contain ξ_{11}. We call D_{11} the *cofactor* of ξ_{11}.

We now investigate more generally the sum of all those terms which have the element ξ_{ik} as a factor. If we shift the ith row of the determinant $(i-1)$ times one place upwards and the kth column $(k-1)$ times one place to the left then ξ_{ik} will be in the top left-hand corner of the determinant. At the same time, we will have made $i+k-2$ transpositions of rows and columns so that the determinant will be multiplied by $(-1)^{i+k}$ (D1 and Theorem 4.2;8). As before we refer to the $(n-1)$-dimensional determinant

D_{ik}, which is obtained from the original by erasing the row and column which contain ξ_{ik}, as the cofactor of ξ_{ik}. It follows that altogether ξ_{ik} is multiplied by $(-1)^{i+k} D_{ik}$.

Noting that, in 4.2;(1) each term contains exactly one factor from each row and each column, we can now obtain the following representations of the determinant.

Theorem 1

$$\det X = \sum_{k=1}^{n} (-1)^{i+k} \xi_{ik} D_{ik} \qquad (i = 1, \ldots, n)$$

(*Expansion about the ith row*)

$$\det X = \sum_{i=1}^{n} (-1)^{i+k} \xi_{ik} D_{ik} \qquad (k = 1, \ldots, n).$$

(*Expansion about the kth column*)

Example 1. The expansion of the 3-dimensional determinant

$$D = \begin{vmatrix} 2 & 1 & 3 \\ 1 & 2 & 4 \\ 0 & 1 & 2 \end{vmatrix}$$

about the second row is

$$D = -1 \begin{vmatrix} 1 & 3 \\ 1 & 2 \end{vmatrix} + 2 \begin{vmatrix} 2 & 3 \\ 0 & 2 \end{vmatrix} - 4 \begin{vmatrix} 2 & 1 \\ 0 & 1 \end{vmatrix}$$

4.3.2 Numerical Computation of Determinants

Suppose $\xi_{11} \neq 0$. If we subtract ξ_{i1}/ξ_{11} times the first row from the ith row for each $i = 2, 3, \ldots, n$, then by Theorem 4.2;6, the value of the determinant is not altered. The new form of the determinant is

$$\det X = \begin{vmatrix} \xi_{11} & \xi_{12} & \cdots & \xi_{1n} \\ 0 & \xi_{22}^{*} & \cdots & \xi_{nn}^{*} \\ \cdots & \cdots & \cdots & \cdots \\ \cdots & \cdots & \cdots & \cdots \\ \cdots & \cdots & \cdots & \cdots \\ 0 & \xi_{n2}^{*} & \cdots & \xi_{nn}^{*} \end{vmatrix}$$

where $\xi_{ik}^{*} = \xi_{ik} - \dfrac{\xi_{i1} \xi_{1k}}{\xi_{11}}$.

If we expand about the first column (Theorem 1), we obtain $\det X = \xi_{11} D^{*}$, where D^{*} is the $(n-1)$-dimensional determinant with elements ξ_{ik}^{*}.

We now observe, that we would obtain the elements ξ_{ik}^*, if we applied the exchange method to the given determinant (considered as a tableau) with ξ_{11} as the pivot but leaving out the corresponding pivotal row and column.

In place of ξ_{11}, any non-zero element ξ_{ik} can be used as the pivot but then it is necessary to multiply by $(-1)^{i+k}$.

The determinant D^* can be dealt with in a similar manner and so on, so that we obtain the following rules for the computation of a determinant.

1. We choose an element $\xi_{ik} \neq 0$ to be the pivot and carry out the exchange method ignoring the new pivotal row and column.

2. The resulting $(n-1)$-dimensional determinant is dealt with in the same way, and so on.

3. If we reach a determinant in which a row or a column consists of all zeros, then $D = 0$ (Theorem 4.2;3).

4. Otherwise, we form the product of all the pivots used (including the element of the last 1-dimensional determinant). The determinant is equal to this product multiplied by $+1$ or -1 according to whether the sum of the row and column indices of all the pivots is even or odd.

Example 2

$$D = \begin{array}{|cccc|}
1 & 2 & 3 & 1* \\
2 & 1 & 2 & 2 \\
1 & 1 & 4 & 3 \\
1 & 1 & 2 & 1 \\
\hline
-1 & -2 & -3 & * \\
\end{array} \qquad (i+k = 5)$$

1st step

$$\begin{array}{|ccc|}
0 & -3 & -4 \\
-2 & -5 & -5 \\
0 & -1 & -1* \\
\hline
0 & -1 & * \\
\end{array} \qquad (i+k = 6)$$

2nd step

$$\begin{array}{|cc|}
0 & 1 \\
-2* & 0 \\
\hline
* & 0 \\
\end{array} \qquad (i+k = 3)$$

3rd step $\boxed{1}$ $(i+k = 2)$

The product of all the pivots is $1.(-1).(-2).1 = +2$ and therefore $D = +2$ since the sum of all the row and column indices is even.

Exercises

1. Evaluate the determinant

$$\varDelta = \begin{vmatrix} 1 & 2 & 3 \\ -1 & -3 & 1 \\ 5 & 3 & 4 \end{vmatrix}$$

Solution. $\varDelta = 39$.

2. Evaluate the determinant

$$\varDelta = \begin{vmatrix} 2 & 5 & 1 & 3 \\ -1 & 2 & -2 & 1 \\ 11 & 3 & 5 & 5 \\ -1 & -3 & 1 & -1 \end{vmatrix}$$

Solution. $\varDelta = -86$.

3. Decide if the following three vectors from R_3 are linearly independent:

$$x_1 = (2, -1, 5), \qquad x_2 = (3, 7, 1), \qquad x_3 = (2, 1, 1).$$

Solution. They are linearly independent because the determinant formed from their components is equal to $-42 \neq 0$.

4. Decide if the following four vectors from R_4 are linearly dependent:

$$x_1 = (1, -1, 3, 1), \qquad x_2 = (2, 1, 3, 5),$$

$$x_3 = (1, 0, 2, 2), \qquad x_4 = (1, -1, -1, 2).$$

Solution. They are linearly dependent, because the corresponding determinant is equal to zero.

CHAPTER 5

LINEAR MAPPINGS OF
VECTOR SPACES, MATRICES

5.1 Linear Mappings

5.1.1 Linear Mappings

Definition 1. Let E and F be vector spaces which are both real or both complex. A mapping f of E into F is said to be 'linear' if

L1. $f(x_1 + x_2) = f(x_1) + f(x_2)$

and L2. $f(\alpha x) = \alpha f(x)$,

for all x_1, x_2, $x \in E$ and all scalars α.

The conditions L1 and L2 can be combined in

L. $f(\alpha_1 x_1 + \alpha_2 x_2) = \alpha_1 f(x_1) + \alpha_2 f(x_2)$,

for all x_1, $x_2 \in E$ and all scalars α_1, α_2.

Theorem 1. If f is a linear mapping, then $f(0) = 0$ and the image of a linear combination of vectors is the corresponding linear combination of their images.

Proof. From L1, it follows that

$$f(0) + f(0) = f(0+0) = f(0) \quad \text{and therefore} \quad f(0) = 0.$$

The second assertion follows by repeated application of L.

Thus a linear mapping of E into F can also be defined as a mapping in which every linear relationship between vectors in E corresponds to the same linear relationship between their images in F.

In the general theory of algebraic systems, mappings which preserve the algebraic structure in this way are known as *homomorphisms*. Thus the linear mappings of a vector space could also be described as the homomorphisms of the vector space.

Theorem 2. If f is a linear mapping of E into F, then the image f(E) is a subspace of F.

Proof. If y_1, $y_2 \in f(E)$, then $y_1 = f(x_1)$ and $y_2 = f(x_2)$, for some x_1, $x_2 \in E$.

For arbitrary scalars α_1, α_2, it follows from L that

$$\alpha_1 y_1 + \alpha_2 y_2 = f(\alpha_1 x_1 + \alpha_2 x_2) \in f(E).$$

Hence by Theorem 2.4;1, it follows that $f(E)$ is a subspace of F.

Theorem 3. *There exists exactly one linear mapping of E into F which maps each vector of a given basis of E onto a given image in F.*

Proof. Let $B \subseteq E$ be a basis of E and let g be mapping of B into F. If f is a linear mapping of E into F, such that $f(e)=g(e)$ for all $e \in B$, then, by Theorem 1,

$$x = \sum_{e \in B} \xi_e e \text{ is mapped onto } f(x) = \sum_{e \in B} \xi_e g(e). \qquad (*)$$

Therefore there exists at most one linear mapping f which satisfies the conditions. Conversely, the mapping f defined by (*) is linear for any given vectors $g(e) \in F$ ($e \in B$) and also, for each $e \in B$, $f(e)=g(e)$ (cf. Problem 1). Finally we note that $f(E) \subseteq F$ is the linear hull of the image of the basis $g(B) \subseteq F$.

Theorem 4. *A linear mapping f of E into F is 1-1 if and only if there is a basis of E which is mapped 1-1 onto a linearly independent set in F. Then this is also true for every basis of E.*

Proof. 1. Suppose B is a basis of E and that $f(B) \subseteq F$ is linearly dependent. Then there is a relation $\sum_{e \in B} \xi_e f(e)=0$ in which not all the coefficients ξ_e are zero. Then $x = \sum_{e \in B} \xi_e e \neq 0$ and $f(x)=0=f(0)$ and f is not 1-1.

2. Suppose B is a basis of E, that f is 1-1 on B, and that $f(B) \subseteq F$ is linearly independent. Further suppose that $f(x_1)=f(x_2)$ for two vectors $x_1, x_2 \in E$.
If $x_i = \sum_{e \in B} \xi_{ie} e$ ($i=1,2$), then

$$f(x_1)-f(x_2) = \sum_{e \in B} (\xi_{1e} - \xi_{2e}) f(e) = 0$$

and therefore $\xi_{1e}=\xi_{2e}$ for all $e \in B$. Hence $x_1=x_2$ and f is 1-1.

It is clear that if B is a basis of E and f is 1-1, then $f(B)$ is a basis of $f(E)$.

A 1-1 linear mapping of a vector space E *onto* a vector space F is known as an *isomorphism*. By Theorem 6, the inverse mapping of an isomorphism is again an isomorphism. Two vector spaces are said to be *isomorphic*, if each can be mapped isomorphically onto the other.

Theorem 5. *A linear mapping f of a vector space E into a vector space F is an isomorphism of E onto F if and only if there is a basis B of E which is mapped 1-1 onto a basis of F. This is then also the case for all bases of E. An isomorphism maps any linearly independent set onto a linearly independent set.*

Proof. 1. Let f be a linear mapping of E into F, let B be a basis of E, and suppose that f maps B 1-1 onto a basis of F. Then $f(E) = L(f(B)) = F$ and, by Theorem 4, f is 1-1. Hence f is an isomorphism.

2. Let f be an isomorphism of E onto F and let B be a basis of E. Then $f(B)$ spans the vector space F and, by Theorem 4, it is linearly independent. Hence $f(B)$ is a basis of F.

3. Let f be an isomorphism of E onto F and let A be a linearly independent subset of E. $L(A)$ is mapped linearly and 1–1 into F by f and, by Theorem 4, A goes into a linearly independent set.

Theorem 6. *The product of two linear mappings and the inverse of a 1-1 linear mapping are again linear.*

Proof. 1. Suppose that E, F and G are three vector spaces, f is a linear mapping of E into F and g is a linear mapping of F into G. Then

$$gf(\alpha_1 x_1 + \alpha_2 x_2) = g(\alpha_1 f(x_1) + \alpha_2 f(x_2))$$
$$= \alpha_1 gf(x_1) + \alpha_2 gf(x_2).$$

Hence gf is a linear mapping of E into G.

2. Let f be linear and 1-1. Then

$$f^{-1}(\alpha_1 f(x_1) + \alpha_2 f(x_2)) = f^{-1}(f(\alpha_1 x_1 + \alpha_2 x_2))$$
$$= \alpha_1 x_1 + \alpha_2 x_2$$
$$= \alpha_1 f^{-1}(f(x_1)) + \alpha_2 f^{-1}(f(x_2))$$

and f^{-1} is therefore linear.

Example 1. Let L be a subspace of the vector space E, and suppose that to each vector $x \in E$ there is assigned the coset $L + x \in E/L$. Then this defines a linear mapping of E onto the quotient space E/L (the proof of this is the substance of Problem 2). This mapping is referred to as the *canonical mapping* of E onto E/L. The image of a vector $x \in E$ under this mapping is called the *canonical image* of x and is often denoted simply by $\bar{x}$.

5.1.2 The Kernel of a Linear Mapping

Theorem 7. *If f is a linear mapping of E into F, then the inverse image $K = f^{-1}(0)$ of the zero-vector of F is a subspace of E. For $x_1, x_2 \in E, f(x_1) = f(x_2)$ if and only if $x_1 - x_2 \in K$.*

Proof. 1. If x, $y \in f^{-1}(0)$, then $f(x)=f(y)=0$ and hence, by Definition 1, $f(\alpha x+\beta y)=0$, for all scalars α, β. That is, $\alpha x+\beta y \in f^{-1}(0)$. Further, since $0 \in f^{-1}(0)$, $f^{-1}(0)\neq\varnothing$ and hence, by Theorem 2.4;1, K is a subspace of E.

    2. If $f(x_1)=f(x_2)$, then $f(x_1-x_2)=0$ and $x_1-x_2 \in f^{-1}(0)$.

    3. If $x_1-x_2 \in f^{-1}(0)$, then $f(x_1-x_2)=0$ and $f(x_1)=f(x_2)$.

Definition 2. The subspace $K=f^{-1}(0)$ of E is known as the 'kernel' of the linear mapping f and will be denoted by ker f.

The second part of Theorem 7 can now be reformulated as follows.
$f(x_1)=f(x_2)$ *if and only if x_1 and x_2 belong to the same coset of the kernel of f.* (*Cf. Theorem 2.5;2.*)

Theorem 8. A linear mapping f is 1-1 if and only if $f^{-1}(0)=\{0\}$, i.e., if its kernel consists of only the zero-vector.

Proof. 1. If f is 1-1, then, from $f(x)=0=f(0)$, it follows that $x=0$. Hence $f^{-1}(0)=\{0\}$.

    2. Suppose $f^{-1}(0)=\{0\}$ and $f(x_1)=f(x_2)$. Then $f(x_1-x_2)=0$, i.e., $x_1-x_2 \in f^{-1}(0)$ and $x_1=x_2$. Hence f is 1-1.

Since all the vectors in a coset of the kernel K of the mapping f have the same image under f, a mapping $\bar{f}$ of the quotient space E/K into the vector space F can be defined by setting $\bar{f}(K+x)=f(x)$. In order to find the image $\bar{f}(N)$ of an arbitrary coset N of K, it is necessary to choose any element $x \in N$ and to put $\bar{f}(N)=f(x)$. It follows from the above that $\bar{f}(N)$ is independent of the choice of x.

Theorem 9. The mapping $\bar{f}$ is a 1-1 linear mapping of the quotient space E/K onto the image $f(E) \subseteq F$. E/K and $f(E)$ are therefore isomorphic.

Proof. 1. Let $N_1=K+x_1$ and $N_2=K+x_2$ be two cosets of K. Then, for arbitrary scalars α_1, α_2 (cf. 2.5;(1)/(2))

$$\bar{f}(\alpha_1 N_1+\alpha_2 N_2) = \bar{f}(K+\alpha_1 x_1+\alpha_2 x_2)$$
$$= f(\alpha_1 x_1+\alpha_2 x_2)$$
$$= \alpha_1 f(x_1)+\alpha_2 f(x_2) = \alpha_1\bar{f}(N_1)+\alpha_2\bar{f}(N_2).$$

Hence $\bar{f}$ is linear.

    2. Suppose $\bar{f}(K+x_1)=\bar{f}(K+x_2)$. Then $f(x_1)=f(x_2)$, and therefore, by Theorem 7, $x_1-x_2 \in K$ and $K+x_1=K+x_2$ (Theorem 2.5;2). Hence $\bar{f}$ is 1-1.

3. It is easy to see that $\bar{f}$ maps E/K onto $f(E) \subseteq F$.

5.1.3 The Vector Space of the Linear Mappings of E into F

If f_1 and f_2 are linear mappings of a vector space E into a vector space F, then a new mapping g of E into F can be defined by setting

$$g(x) = f_1(x) + f_2(x) \text{ for all } x \in E. \tag{1}$$

It is easy to verify that this mapping is linear. We call g the sum of the mappings f_1 and f_2 and correspondingly write

$$g = f_1 + f_2. \tag{2}$$

If α is a scalar, a further mapping h of E into F can be defined by setting

$$h(x) = \alpha f_1(x) \text{ for all } x \in E. \tag{3}$$

This is also linear and we write

$$h = \alpha f_1. \tag{4}$$

Thus (2) and (4) define an addition and a multiplication by scalars in the set $\mathscr{L}(E,F)$ of all linear mappings of E into F. It is now easy to verify that the axioms for a vector space (Definition 2.1;1) are satisfied. For example, the zero element is the linear mapping $0 \in \mathscr{L}(E,F)$ which maps every vector $x \in E$ onto the zero-vector of F. That is, $0(x) = 0 \in F$ for all $x \in E$. Because, for any mapping $f \in \mathscr{L}(E,F)$, $(f+0)(x) = f(x) + 0(x) = f(x)$ for all $x \in E$ and hence $f + 0 = f$. The additive inverse element $(-f)$ of a mapping $f \in \mathscr{L}(E,F)$ in the sense of Axiom A3 (p. 10), is defined by $(-f)(x) = -[f(x)]$.

Theorem 10. The set $\mathscr{L}(E,F)$ of linear mappings of the vector space E into the vector space F is a vector space with the rules of composition defined by (1), ..., (4). The simple proof of this is the substance of Problem 3.

Problems

1. Fill in the details of the proof of Theorem 3.
2. Show that the canonical mapping (Example 1) of a vector space E onto the quotient space E/L is linear.
3. Prove Theorem 10.
4. Show that if f is a linear mapping of a finite-dimensional vector space E, then $f(E)$ is also finite-dimensional and $\dim f(E) \leqslant \dim E$.
5. Let f be a linear mapping of a vector space E into itself. Is the mapping $x \to g(x)$ where $g(x) = x + f(x)$ also linear?

6. To each function $x(\tau) \in C$ (Example 2.1;7), there is assigned the function $y(\tau) = x(\tau) + x(-\tau) \in C$. Does this define a linear mapping of C into itself? What is its kernel? Answer the same questions for $z(\tau) = x(\tau) - x(-\tau)$.

7. The direct product of two vector spaces E_1 and E_2 (Problem 2.1;4) is mapped into itself by $f((x_1, x_2)) = (x_1, 0)$. Is f linear? What is its kernel?

8. Let K be a convex subset (cone, polyhedron, or pyramid) of a real vector space E and let f be a linear mapping of E. Show that $f(K) \subseteq f(E)$ is also a convex set (cone, polyhedron, or pyramid). Is it possible for $f(K)$ to be acute, if K is not acute? (see Theorem 2.6;8).

9. Prove that, if L_1 and L_2 are subspaces of a vector space, then the quotient spaces $(L_1 + L_2)/L_1$ and $L_2/(L_1 \cap L_2)$ are isomorphic. Derive a new proof for Theorem 3.2;7.

10. Let L be a subspace of the vector space E. Show that the set of all linear mappings of E into a vector space F for which $f(L) = \{0\}$ is a subspace of $\mathscr{L}(E, F)$. More generally, show that the same is true for those mappings for which $f(L) \subseteq M$, where M is any subspace of F.

11. Prove that the inverse image of a subspace under a linear mapping of a vector space E is a subspace of E.

5.2 Linear Mappings of Finite-Dimensional Vector Spaces, Matrices

5.2.1 The Rank of a Linear Mapping

Theorem 1. Suppose E is a finite-dimensional vector space. A vector space F (with the same scalars) is isomorphic to E if and only if F is also finite-dimensional and $\dim F = \dim E$.

Proof. Theorem 5.1;5.

Definition 1. Let E be a finite-dimensional vector space and let f be a linear mapping of E into a vector space F. Then the 'rank' of f is the dimension of the image $f(E) \subseteq F$.
(If E is finite-dimensional, then so is $f(E)$.)

Theorem 2. If K is the kernel of the linear mapping f, then the rank of f is equal to $\dim (E/K) = \dim E - \dim K$.

Proof. By Theorem 5.1;9, $f(E)$ and E/K are isomorphic. Hence the assertion follows directly from Theorems 1 and 3.2;5.

Theorem 3. A linear mapping f of the finite-dimensional vector space E is 1-1 if and only if its rank is equal to $\dim E$.

Proof. 1. If f is 1-1, then, by Theorem 5.1;8, $K = \{0\}$. Hence the rank of f is equal to $\dim E$ (Theorem 2).

2. If the rank of f is equal to $\dim E$, then, by Theorem 2, $K = \{0\}$ and hence f is 1-1 (Theorem 5.1;8).

Theorem 4. Suppose $\dim E = \dim F$. *A linear mapping f of E into F is 1-1 if and only if it is onto F.*

Proof. 1. If f is 1-1, then, by Theorem 3, $\dim E = \dim f(E) = \dim F$ and hence $f(E) = F$ by Theorem 3.2;4.

2. If $f(E) = F$, then the rank of f is equal to $\dim E$ and, by Theorem 3, f is 1-1.

5.2.2 Linear Mappings and Matrices

We will now investigate how the components of a vector can be used to express the components of its image under a linear mapping. Suppose therefore that $\{e_1, \ldots, e_n\}$ is a basis of E and that $\{f_1, \ldots, f_m\}$ is a basis of F. Suppose further that f is a linear mapping of E into F which, in accordance with Theorem 5.1;3, is given by

$$f(e_k) = \sum_{i=1}^{m} \alpha_{ik} f_i \qquad (k = 1, 2, \ldots, n). \tag{1}$$

Then the vector $x \in E$, $x = \sum_{k=1}^{n} \xi_k e_k$, is mapped onto

$$f(x) = \sum_{k=1}^{n} \xi_k f(e_k) = \sum_{i=1}^{m} \left[\sum_{k=1}^{n} \alpha_{ik} \xi_k \right] f_i.$$

The components η_i of $f(x)$ are therefore

$$\eta_i = \sum_{k=1}^{n} \alpha_{ik} \xi_k \qquad (i = 1, \ldots, m). \tag{2}$$

Thus we have the following result.

Theorem 5. Under a linear mapping, the components of the image of a vector x are linear functionals in the components of x (cf. 6.1;(1)).
We note that the coefficients $\alpha_{1k}, \ldots, \alpha_{mk}$ are the components of $f(e_k)$ $(k = 1, \ldots, n)$.

For given bases, the linear mapping f is uniquely determined by the coefficients a_{ik} which, as in an exchange tableau (cf. 3.3), we usually arrange in the form of a rectangular array.

$$A = \begin{pmatrix} \alpha_{11} & \alpha_{12} & \cdots & \alpha_{1n} \\ \alpha_{21} & \alpha_{22} & \cdots & \alpha_{2n} \\ \cdots & \cdots & \cdots & \cdots \\ \cdots & \cdots & \cdots & \cdots \\ \alpha_{m1} & \alpha_{m2} & \cdots & \alpha_{mn} \end{pmatrix} = \alpha_{ik}.$$

We refer to a rectangular array of numbers like this as a *matrix*. The matrix A has m rows and n columns. The mn scalars α_{ik} are known as the *elements* of the matrix. The matrix A represents the linear mapping f with respect to the two given bases of E and F, and we will say that A is the matrix of f with respect to these bases. Referring to a matrix which has m rows and n columns as an $m \times n$ matrix, we can easily verify the following theorem.

Theorem 6. Let E and F be vector spaces of dimensions n and m in which fixed bases are given. Then each linear mapping f of E into F corresponds to an $m \times n$ matrix $A = (\alpha_{ik})$ such that f is given by (2). Conversely (2) represents a linear mapping of E into F for any $m \times n$ matrix $A = (\alpha_{ik})$. The correspondence between the linear mappings of E into F and the $m \times n$ matrices is 1-1.

Now, suppose that F is mapped by a further linear mapping g into a finite-dimensional vector space G. Let the relation corresponding to (2) for g be

$$\zeta_h = \sum_{i=1}^{m} \beta_{hi} \eta_i \qquad (h = 1, \ldots, p) \tag{3}$$

where p is the dimension of G and $\zeta_1, \ldots, \zeta_p$ are components with respect to a given basis of G. Thus the mapping g corresponds to the matrix

$$B = \begin{pmatrix} \beta_{11} & \cdots & \cdots & \beta_{1m} \\ \cdots & \cdots & \cdots & \cdots \\ \cdots & \cdots & \cdots & \cdots \\ \cdots & \cdots & \cdots & \cdots \\ \beta_{p1} & \cdots & \cdots & \beta_{pm} \end{pmatrix} = (\beta_{hi}).$$

From (2) and (3), it follows by substitution that

$$\zeta_h = \sum_{k=1}^{n} \left[\sum_{i=1}^{m} \beta_{hi} \alpha_{ik} \right] \xi_k$$

$$= \sum_{k=1}^{n} \gamma_{hk} \xi_k \qquad (h = 1, \ldots, p), \tag{4}$$

89

where $$\gamma_{hk} = \sum_{i=1}^{m} \beta_{hi}\alpha_{ik} \qquad (h = 1,\ldots,p; \ k = 1,\ldots,n). \tag{5}$$

Thus the coefficients γ_{hk} of the product mapping gf can be calculated from the coefficients α_{ik} of f and β_{hi} of g. The mapping gf is represented by the matrix

$$C = \begin{pmatrix} \gamma_{11} & \cdots & \cdots & \gamma_{1n} \\ \cdots & \cdots & \cdots & \cdots \\ \cdots & \cdots & \cdots & \cdots \\ \cdots & \cdots & \cdots & \cdots \\ \gamma_{p1} & \cdots & \cdots & \gamma_{pn} \end{pmatrix} = (\gamma_{hk}).$$

Definition 2. The matrix $C=(\gamma_{hk})$ *where* γ_{hk} *is given by* (5), *is called the 'product'* $C=BA$ *of the matrices* $A=(\alpha_{ik})$ *and* $B=(\beta_{hi})$.

Thus the element γ_{hk} of the product matrix $C=BA$ is obtained when the hth row of B is 'multiplied' into the kth column of A. We 'multiply' a row into a column containing the same number of elements by taking the sum of the products of the elements of the row with the corresponding elements of the column.

Example 1. On multiplying the row $(1,2,3)$ into the column

$$\begin{pmatrix} -1 \\ 2 \\ 1 \end{pmatrix},$$

we obtain the number $1.(-1)+2.2+3.1=6$.

Hence the matrix product $C=BA$ only has a meaning when the row length (the number of elements in a row) of B is equal to the column length of A. The column length (or the number of rows) of the product matrix is then equal to that of B and its row length is equal to that of A.

Example 2

Let $B = \begin{pmatrix} 1 & 2 & 1 \\ 2 & 1 & 2 \end{pmatrix}$ and $A = \begin{pmatrix} 1 & 2 & -1 & 2 \\ 0 & 1 & 2 & 1 \\ 3 & 1 & 1 & 0 \end{pmatrix}$

Then $BA = \begin{pmatrix} 4 & 5 & 4 & 4 \\ 8 & 7 & 2* & 5 \end{pmatrix}$, where, for example, the element marked with an asterisk is obtained by multiplying the second row of B into the third row of A:

$$2.(-1)+1.2+2.1 = 2.$$

Multiplication of matrices is not commutative, i.e., the equation $AB=BA$ is not true in general, even when both products have a meaning.

Example 3

If
$$A = \begin{pmatrix} 1 & 2 \\ 0 & 1 \end{pmatrix}; \quad B = \begin{pmatrix} 2 & 1 \\ -1 & 0 \end{pmatrix},$$

then
$$AB = \begin{pmatrix} 0 & 1 \\ -1 & 0 \end{pmatrix} \quad \text{and} \quad BA = \begin{pmatrix} 2 & 5 \\ -1 & -2 \end{pmatrix},$$

hence $AB \neq BA$.

In view of Definition 2, the statements of formulae (4) and (5) can be combined into the following theorem.

Theorem 7. The matrix of the product of two linear mappings is the product of their matrices (in the same order).

Naturally it is assumed here that the same basis of the intermediate space F is used for both mappings.

Theorem 8. Multiplication of matrices is associative.

Proof. Let $A=(\alpha_{hi})$, $B=(\beta_{ik})$, $C=(\gamma_{kl})$ where the row length of A is equal to the column length of B and similarly for B and C. Then

$$[(AB)C]_{hl} = \sum_k \left(\sum_i \alpha_{hi}\beta_{ik} \right) \gamma_{kl} = \sum_{i,k} \alpha_{hi}\beta_{ik}\gamma_{kl}$$

and the same is true for $[A(BC)]_{hl}$. (For simplicity, we have denoted the element of $(AB)C$ with indices h and l by $[(AB)C]_{hl}$.) The matrices $(AB)C$ and $A(BC)$ therefore have the same elements and hence they are equal.

The formulae (2), (3) and (4) can also be expressed more clearly in terms of matrices. If we denote the matrix which consists of the one column

$$\begin{pmatrix} \xi_1 \\ \vdots \\ \xi_n \end{pmatrix}$$

by ξ and similarly the matrix

$$\begin{pmatrix} \eta_1 \\ \vdots \\ \eta_m \end{pmatrix}$$

by η then (2) can be written in the form

$$\eta = A\xi. \tag{6}$$

Similarly we can write (3) in the form

$$\zeta = B\eta. \tag{7}$$

Now, using the associative law, we have

$$\zeta = B(A\xi) = (BA)\xi, \tag{8}$$

which is just the same as (4).

Definition 3. The 'transpose' of a matrix $A = (\alpha_{ik})$ is the matrix $A' = (\alpha'_{ik})$ where $\alpha'_{ik} = \alpha_{ki}$. Thus A' is found from A by making the rows into columns and vice versa. A matrix A is said to be 'symmetric' if $A = A'$.

For every matrix A, $(A')' = A$. A symmetric matrix obviously has the same number of rows as columns and is therefore an example of what is generally known as a *square* matrix.

Referring to the line of those elements α_{ii} of A which have equal indices as the *main diagonal* of A (and the elements α_{ii} as the *diagonal elements*), we can think of a symmetric matrix as one which is unaltered when it is reflected in the main diagonal.

A square matrix is said to be *diagonal* if $\alpha_{ik} = 0$ for all elements with $i \neq k$.

Example 4. If $A = \begin{pmatrix} 2 & 1 & 3 \\ 1 & 2 & 0 \end{pmatrix}$, then $A' = \begin{pmatrix} 2 & 1 \\ 1 & 2 \\ 3 & 0 \end{pmatrix}$

The matrix $B = \begin{pmatrix} 1 & 2 & 3 \\ 2 & 0 & 4 \\ 3 & 4 & 2 \end{pmatrix}$ is symmetric.

Theorem 9. The transpose of a product of matrices is the product of the transposes of these matrices in the reversed order, i.e., $(BA)' = A'B'$.

Proof. $[(BA)']_{ki} = (BA)_{ik} = \sum_j \beta_{ij} \alpha_{jk}$

$$(A'B')_{ki} = \sum_j (A')_{kj} (B')_{ji} = \sum_j \beta_{ij} \alpha_{jk}.$$

For the matrices ξ, η, ζ which appear in (6) and (7), $\xi' = (\xi_1, \ldots, \xi_n)$, etc. Hence, in view of Theorem 9, (6) and (7) can also be written as follows:

$$\eta' = \xi' A' \tag{9}$$

$$\zeta' = \eta' B' \tag{10}$$

From these it follows by substitution that

$$\zeta' = (\xi' A') B' = \xi'(BA)' \tag{11}$$

which is (8) in transposed form.

5.2.3 The Vector Space of $m \times n$ Matrices

Suppose that the linear mappings f_1 and f_2 of E into F correspond to the matrices A_1 and A_2. In connection with 5.1.3, we are interested to find the matrix B which corresponds to the sum $g = f_1 + f_2$ of the mappings, but first we must explain how matrices are to be added and how they are to be multiplied by scalars.

Definition 4. Let $A = (\alpha_{ik})$ and $B = (\beta_{ik})$ be $m \times n$ matrices. The sum $C = A + B$ of these matrices is defined to be the $m \times n$ matrix $C = (\alpha_{ik} + \beta_{ik})$. If μ is a scalar, then μA is defined to be the $m \times n$ matrix $\mu A = (\mu \alpha_{ik})$.

Example 5. Let $A = \begin{pmatrix} 1 & 2 & 3 \\ 2 & 1 & 0 \end{pmatrix}$ and $B = \begin{pmatrix} -1 & 2 & 1 \\ 1 & -1 & 2 \end{pmatrix}$

Then $A + B = \begin{pmatrix} 0 & 4 & 4 \\ 2 & 0 & 2 \end{pmatrix}$ and $3A = \begin{pmatrix} 3 & 6 & 9 \\ 6 & 3 & 0 \end{pmatrix}$

Clearly the distributive law holds, $(A + B)C = AC + BC$ and $C(A + B) = CA + CB$.

Further it is easy to see that, in view of Definition 4, the following theorem holds.

Theorem 10. The set $\mathscr{M}_{m,n}$ of all $m \times n$ matrices together with the operations introduced by Definition 4 is a vector space of dimension mn.

The dimension mn comes from the fact that this vector space is obviously isomorphic to the mn-dimensional space of mn-tuples R_{mn}.

Now suppose that for $x \in E$,

$$y_1 = f_1(x),\, y_2 = f_2(x) \quad \text{and} \quad y = y_1 + y_2 = (f_1 + f_2)(x) = g(x). \tag{12}$$

Writing these in the same way as formula (6), we have

$$\eta_1 = A_1 \xi,\, \eta_2 = A_2 \xi,\, \eta = \eta_1 + \eta_2 = (A_1 + A_2) \xi \tag{13}$$

where η_1, η_2, η denote the matrices formed from the components of y_1, y_2, y.

Therefore $B = A_1 + A_2$ is the matrix of the mapping $g = f_1 + f_2$. Similarly we can show that the matrix of the mapping αf_1 is αA for any scalar α. If we now also take into consideration the fact that the correspondence between linear mappings and matrices (for given bases of E and F) is 1-1 (Theorem 6), we have the following theorem.

Theorem 11. Let E and F be vector spaces of dimensions n and m. Then the vector space $\mathscr{L}(E,F)$ of the linear mappings of E into F is isomorphic to the vector space $\mathscr{M}_{m,\,n}$ and therefore has the dimension mn.

The last part of the theorem follows from Theorems 1 and 10.

5.2.4 Column and Row Ranks of a Matrix

The rank of a linear mapping f (cf. Definition 1) corresponds to a property of the corresponding matrix A as follows. From (1) the kth column of the matrix $A = (\alpha_{ik})$ consists of the components of the vector $f(e_k)$. The rank of the mapping f by Definition 1 is equal to the maximum number of linearly independent vectors amongst the vectors $f(e_k)$. By Theorem 3.1;4, this is now equal to the maximum number of linearly independent columns of the matrix A. (Columns of length m are considered to be linearly dependent or independent as elements of the m-dimensional vector space R_m.)

The maximum number of independent columns in a matrix is called the *column rank* of the matrix.

Theorem 12. If the linear mapping f corresponds to the matrix A, then the rank of f is equal to the column rank of A. The mapping f is 1-1 if and only if the column rank of A is equal to the dimension of E (i.e., equal to the number of columns of A).

The proof of the second part follows from Theorem 3.

If E and F have the same dimension, then A has the same number of rows as columns and is therefore square. Hence, from Theorem 4.2;7, we have

Theorem 13. If E and F have the same dimension, then f is 1-1 if and only if $\det A \neq 0$, i.e., if A is non-singular. (cf. Def. 5.3;1).

Like the column rank, we can also introduce the *row rank* of a matrix as the maximum number of linearly independent rows. We will later prove (Theorem 6.2;10) that the row and column ranks are equal, so that we will be able to refer to *the rank* of a matrix.

5.2.5 The Determinant of a Matrix Product

If A and B are square $n \times n$ matrices, then their product AB exists and is again an $n \times n$ matrix. For the determinants of these matrices, we have the following result.

Theorem 14. $\det AB = \det A . \det B$

Proof. If $A = (\alpha_{ik})$, $B = (\beta_{kl})$ and $C = AB = (\gamma_{il})$ then, by 4.2;(1) and the definition of the matrix product

$$\det C = \sum_{\phi} \operatorname{ch}(\phi)\, \gamma_{1\phi(1)} \cdots \gamma_{n\phi(n)}$$

$$= \sum_{k_1=1}^{n} \cdots \sum_{k_n=1}^{n} \alpha_{1,\, k_1} \cdots \alpha_{n,\, k_n} \sum_{\phi} \operatorname{ch}(\phi)\, \beta_{k_1,\, \phi(1)} \cdots \beta_{k_n,\, \phi(n)}$$

where the summation $\sum\limits_{\phi}$ is the determinant whose ith row $(i=1,\ldots,n)$ is equal to the k_ith row of B. Hence, by Theorem 4.2;4, all the terms in which the mapping $i \to k_i$ is not a permutation are zero. In the case of a permutation, if we put $k_i = \psi(i)$, the last expression becomes

$$\det C = \sum_{\psi} \alpha_{1,\, \psi(1)} \cdots \alpha_{n,\, \psi(n)} \sum_{\phi} \operatorname{ch}(\phi)\, \beta_{1,\, \phi\psi^{-1}(1)} \cdots \beta_{n,\, \phi\psi^{-1}(n)}$$

or, putting $\theta = \phi\psi^{-1}$,

$$\det C = \sum_{\psi} \operatorname{ch}(\psi)\, \alpha_{1,\, \psi(1)} \cdots \alpha_{n,\, \psi(n)} \sum_{\theta} \operatorname{ch}(\theta)\, \beta_{1,\, \theta(1)} \cdots \beta_{n,\, \theta(n)}$$

$$= \det A . \det B.$$

For the formal details of this proof, compare the verification of D1 in 4.2.1 on the basis of 4.2;(1).

Exercises

1. Multiply the following two matrices

$$A = \begin{pmatrix} 1 & 2 & 1 & -2 \\ 3 & 1 & -1 & 2 \\ 4 & 3 & -2 & 1 \end{pmatrix} \qquad B = \begin{pmatrix} 2 & 4 \\ 1 & -3 \\ 2 & 1 \\ -1 & 1 \end{pmatrix}$$

Solution.

$$AB = \begin{pmatrix} 8 & -3 \\ 3 & 10 \\ 6 & 6 \end{pmatrix}$$

2. Let E, F, G be vector spaces. Suppose E is mapped into F by the matrix A and F is mapped into G by the matrix B (as in 5.2;(2)), where

$$A = \begin{pmatrix} 1 & 2 & -1 & 3 \\ -1 & 2 & 3 & -2 \\ 1 & 5 & -2 & 1 \end{pmatrix}, \qquad B = \begin{pmatrix} 1 & -2 & 3 \\ 1 & 2 & -1 \end{pmatrix}.$$

What can be said about the dimensions of E, F and G? If the vector $x \in E$ has components $(2, 1, 1, -1)$, what are the components of the image of x under the product mapping?

Solution. The image of x is $(2, 6)$. The dimensions of E, F and G are 4, 3 and 2 respectively.

Problems

1. What is the dimension of the subspace of $\mathscr{L}(E, F)$ described in Problem 5.1;10 when E and F are finite-dimensional? (Hint: Consider the quotient spaces E/L and F/M.)

2. Let E be a finite-dimensional vector space and let f be a linear mapping of E into itself. Show that, if the kernel of f is equal to the image of E under f, then $\dim E$ is even. Can such mappings exist? Can such a mapping be 1-1?

3. Show that, if f and g are linear mappings of a finite-dimensional vector space, then

$$|\operatorname{rank} f - \operatorname{rank} g| \leqslant \operatorname{rank}(f+g) \leqslant \operatorname{rank} f + \operatorname{rank} g.$$

4. Show that $A'A$ and AA' are symmetric for all matrices A.

5. Suppose that the matrices A and B are partitioned as follows:

$$A = \begin{pmatrix} A_{11} & A_{12} \\ A_{21} & A_{22} \end{pmatrix} \qquad B = \begin{pmatrix} B_{11} & B_{12} \\ B_{21} & B_{22} \end{pmatrix}.$$

Show that, provided the matrix products are defined,

$$AB = \begin{pmatrix} A_{11}B_{11} + A_{12}B_{21}, & A_{11}B_{12} + A_{12}B_{22} \\ A_{21}B_{11} + A_{22}B_{21}, & A_{21}B_{12} + A_{22}B_{22} \end{pmatrix}.$$

How can this result be generalized?

5.3 Linear Mappings of a Vector Space into Itself (Endomorphisms)

The domain E and the range F of a linear mapping can in special cases be the same space. The mapping f is then a linear mapping (homomorphism) of a vector space E into itself. In the language of general algebra, such a mapping is known as an *endomorphism* of E. The endomorphisms of E are also known as its *linear transformations*.

5.3.1 The Algebra of Endomorphisms of a Vector Space

From Theorem 5.1;10, it follows that the set of all endomorphisms of a vector space is itself a vector space, if addition of endomorphisms and multiplication by scalars are defined as in 5.1.3. For any two endomorphisms

f, g of E, their products fg and gf are always defined and, by Theorem 5.1;6, they are also endomorphisms of E. Thus we have the following binary operations defined on the set of all endomorphisms of a vector space.

1. Addition of endomorphisms.
2. Multiplication by scalars.
3. Multiplication of endomorphisms.

The set is a vector space with 1 and 2, a ring with 1 and 3, and an *algebra* with 1, 2 and 3. (See [24] pp. 239 and 370.)

5.3.2 The Group of Automorphisms of a Vector Space

We will now restrict our attention to those endomorphisms of a vector space E which are 1-1 and onto E (if E is finite-dimensional, then by Theorem 5.2;4 just one of these assumptions is sufficient), i.e., the isomorphisms of E onto itself. In the language of general algebra, these are known as the *automorphisms* of E.

We denote the set of all automorphisms of the vector space E by $A(E)$. Clearly

1. If $f \in A(E)$ and $g \in A(E)$ then $fg \in A(E)$ (i.e., multiplication is a binary operation in $A(E)$).
2. The identity mapping e of E is in $A(E)$.
3. If $f \in A(E)$, then $f^{-1} \in A(E)$.

Together with the associativity of the multiplication of mappings, properties 1, 2 and 3 show that we have the following Theorem. (See [25] pp. 1–3.)

Theorem 1. The set of all automorphisms of a vector space E (i.e., the 1-1 linear mappings of E onto itself) is a group with the usual multiplication of mappings.

(Note, however, that the automorphisms of E do not form a vector space because the sum of two automorphisms need not be an automorphism.)

5.3.3 Linear Groups

Now suppose that E is finite-dimensional. For a given basis of E, each automorphism $f \in A(E)$ is represented by an $n \times n$ matrix (or, as we will also say, an n-dimensional square matrix) whose determinant is not zero (Theorem 5.2;13). Conversely every matrix of this kind represents an automorphism of E and the correspondence between the automorphisms and the matrices is 1-1.

The matrix I which corresponds to the identity mapping $e \in A(E)$ has the property that

$$IA = AI = A \text{ for all } n \times n \text{ matrices } A, \tag{1}$$

because e satisfies the relationship $ef = fe = f$, for every endomorphism f of E. The matrix I is uniquely determined by (1). If for example I^* were another matrix with the same property, then $I = II^* = I^*$. Now it is easy to verify that (1) is satisfied by the matrix $I = (\delta_{ik})$ where δ_{ik} is the Kronecker delta (see Example 3.1;2). Because, for any matrix $A = (\alpha_{kl})$,

$$[IA]_{il} = \sum_k \delta_{ik} \alpha_{kl} = \alpha_{il},$$

i.e., $IA = A$ and similarly $AI = A$.
This matrix

$$I = (\delta_{ik}) = \begin{pmatrix} 1 & 0 & 0 & \ldots & 0 \\ 0 & 1 & 0 & \ldots & 0 \\ \ldots & \ldots & \ldots & \ldots & \ldots \\ \ldots & \ldots & \ldots & \ldots & \ldots \\ 0 & 0 & 0 & \ldots & 1 \end{pmatrix}$$

is referred to as the n-dimensional *identity matrix*. Its determinant is given by

$$\det I = 1. \tag{2}$$

This can be shown either by direct calculation or by using Theorem 5.2;14 to get $\det I = \det I^2 = [\det I]^2$ and hence (2) because, by Theorem 5.2;13, $\det I \neq 0$.

Finally $IA = A$ for all matrices A which have the same number of rows as I has columns, and similarly $AI = A$ whenever this product has a meaning.

Further suppose that the matrix of the automorphism $f \in A(E)$ is A and that the matrix of the inverse automorphism $f^{-1} \in A(E)$ is $\overline{A}$. Since $f^{-1}f = ff^{-1} = e$,

$$A\overline{A} = \overline{A}A = I. \tag{3}$$

The matrix $\overline{A}$ is uniquely determined by (3) when A is given. Because, if A^* is another matrix which also satisfies (3), then

$$\overline{A} = I\overline{A} = A^* A\overline{A} = A^*I = A^*$$

The matrix $\overline{A}$ which is uniquely determined by (3) for the given matrix A, is referred to as the *inverse matrix* of A and is usually denoted by A^{-1}.

Not every square matrix has an inverse and in fact we have the following theorem.

Theorem 2. A square matrix A has an inverse if and only if $\det A \neq 0$.

Proof. 1. If $\det A \neq 0$, then A represents an automorphism of a vector space. The matrix of the inverse automorphism is then the inverse of A.

2. Suppose A has an inverse, then $AA^{-1} = I$ and, by Theorem 5.2;14, it follows that

$$\det A . \det A^{-1} = \det I = 1$$

and hence $\det A \neq 0$.

As a secondary result, we also see that

$$\det A^{-1} = [\det A]^{-1}. \tag{4}$$

By Theorems 2 and 4.2;8, if A has an inverse then so has A' and

$$(A')^{-1} = (A^{-1})'$$

because, by Theorem 5.2;9, $(A^{-1})' A' = (A A^{-1})' = I' = I$.

Definition 1. A square matrix A which has an inverse (i.e., $\det A \neq 0$) is said to be 'non-singular'. All other square matrices are said to be 'singular'.

Theorem 3. The set of all non-singular $n \times n$ matrices is a group with the usual matrix multiplication and is known as the 'general linear group' of degree n and denoted by GL_n. It is isomorphic to the group $A(E)$ of automorphisms of any vector space E of dimension n.

(Two groups G_1 and G_2 are said to be isomorphic if there is a 1-1 mapping t of G_1 onto G_2 such that $t(AB) = t(A) t(B)$ for all $A, B \in G_1$. Cf. the definition of isomorphic vector spaces in 5.1.1 and also [25] p. 19.)

Proof. 1. The group properties:

1.1. By Theorem 5.2;14, if $A \in GL_n$ and $B \in GL_n$, then $AB \in GL_n$ and hence matrix multiplication is a binary operation in the set GL_n.

1.2. Matrix multiplication is associative (Theorem 5.2;8).

1.3. The identity matrix $I \in GL_n$ is the neutral element and, by (4), for each matrix $A \in GL_n$, the inverse A^{-1} is also in GL_n.

2. If we choose a basis of the vector space E of dimension n, then every automorphism $f \in A(E)$ corresponds to a matrix $A \in GL_n$. In this way, a 1-1 mapping is defined from $A(E)$ onto GL_n (Theorems 5.2; 6/13). The fact that this is an isomorphism is a consequence of Theorem 5.2;7.

Problems

1. Show that, if A and B are non-singular $n \times n$ matrices, then AB is also non-singular and $(AB)^{-1} = B^{-1} A^{-1}$.

2. Show that, if A is a non-singular $n \times n$ matrix, then $(A')^{-1} = (A^{-1})'$.

3. Let f be an endomorphism of a finite-dimensional vector space E, let

L be a subspace of E and let $f(L) \subseteq L$. Suppose that a basis of E is constructed which contains a basis of L. What can be said about the matrix A of f in the representation 5.2;(6)?

4. Prove that, if f is an endomorphism of E, then $f^2 = 0$ if and only if $f(E) \subseteq \ker f$.

5. Prove that, if f is an endomorphism of E, then

$\ker f \subseteq \ker f^2 \subseteq \ker f^3 \subseteq \ldots$ and $f(E) \supseteq f^2(E) \supseteq f^3(E) \supseteq \ldots$

6. An endomorphism f of a vector space E is said to be a projection if $f^2 = f$. Show that, if f is a projection, then

$$E = \ker f \oplus f(E) \quad \text{and} \quad f(x) = x \text{ for all } x \in f(E).$$

7. Show that the set $A(E)$ of all automorphisms of a finite-dimensional vector space E spans the vector space $\mathscr{L}(E, E)$ of endomorphisms of E.

8. Show that the mapping f given by $f(x) = dx/d\tau$ is an endomorphism of the vector space P_n (Example 2.1;5). What is the matrix of f with respect to the basis $e_k(\tau) = \tau^k$ $(k = 0, \ldots, n)$? What is the kernel of f? Is f an automorphism?

9. Prove that, if A, B are $n \times n$ matrices and λ is a scalar, then $AB - \lambda I$ is non-singular if and only if $BA - \lambda I$ is non-singular.
(Hint: $\lambda(AB - \lambda I)^{-1} = A(BA - \lambda I)^{-1}B - I$.)

5.4 Change of Basis

5.4.1 The Transformation of Vector Components by a Change of Basis

Let E be an n-dimensional vector space and suppose that $\{e_1, \ldots, e_n\}$ and $\{f_1, \ldots, f_n\}$ are two bases of E. Denote the components of the vector $x \in E$ with respect to the first basis by $\xi_1, \ldots, \xi_n$ and with respect to the second by $\eta_1, \ldots, \eta_n$. Thus

$$x = \sum_{k=1}^{n} \xi_k e_k = \sum_{k=1}^{n} \eta_k f_k \tag{1}$$

or, in matrix form, (cf. 5.2;(6)–(10))

$$x = \boldsymbol{\xi}' \mathbf{e} = \boldsymbol{\eta}' \mathbf{f}. \tag{2}$$

For convenience of writing we have extended the concept of a matrix here to include matrices such as

$$\mathbf{e} = \begin{pmatrix} e_1 \\ \vdots \\ e_n \end{pmatrix},$$

whose elements are vectors. It is not difficult to verify that our applications of this extended concept do not lead to any mistakes.

Naturally the two bases are connected by relationships of the form

$$f_i = \sum_{k=1}^{n} \sigma_{ik} e_k \qquad (i = 1, \ldots, n), \tag{3}$$

i.e.,

$$\mathbf{f} = S\mathbf{e} \quad \text{where} \quad S = (\sigma_{ik}). \tag{4}$$

By Theorem 3.1;2, the coefficients σ_{ik} and hence the matrix S are uniquely determined by the two bases. Since each row of S consists of the components of a vector f_i (with respect to the basis $\{e_1, \ldots, e_n\}$), these rows are linearly independent (Theorem 3.1;4) and hence $\det S \neq 0$, (Theorem 4.2;7). Conversely, if $\det S \neq 0$, if $\{e_1, \ldots, e_n\}$ is a basis and if $f_1, \ldots, f_n$ are defined by (3) or (4), then $\{f_1, \ldots, f_n\}$ is also a basis. We express this in the next theorem.

Theorem 1. An $n \times n$ matrix S corresponds to a change of basis if and only if it is non-singular.

Now from (2) and (4) it follows that

$$x = \boldsymbol{\xi}' \mathbf{e} = (\boldsymbol{\eta}' S) \mathbf{e},$$

so that x is represented as a linear combination of the basis vectors e_k in two ways. The coefficients of these two representations must be identical, i.e., $\boldsymbol{\xi}' = \boldsymbol{\eta}' S$. If we change from this relationship between matrices to their transposes, it follows by Theorem 5.2;9 that

$$\boldsymbol{\xi} = S' \boldsymbol{\eta},$$

i.e., that

$$\boldsymbol{\eta} = S^* \boldsymbol{\xi} \quad \text{where} \quad S^* = (S')^{-1} = (S^{-1})'. \tag{5}$$

Theorem 2. The transformation equations (4) and (5) for the basis vectors and the vector components correspond to the same change of basis if and only if $S^* = (S^{-1})' = (S')^{-1}$.

Proof. One part of the theorem has already been proved. If $S^* = (S')^{-1}$, then (5) can be written in the form $\boldsymbol{\xi}' = \boldsymbol{\eta}' S$. Multiplying on the right by $\mathbf{e}$, we have $\boldsymbol{\xi}' \mathbf{e} = \boldsymbol{\eta}' S\mathbf{e} = \boldsymbol{\eta}' \mathbf{f}$. Hence ξ_k and η_k are the components of the same vector with respect to the two bases.

5.4.2 Change of Basis and Linear Mappings

Let f be a linear mapping of a vector space E of dimension n into a vector space F of dimension m, which, as in 5.2;(6), is represented by

$$\boldsymbol{\eta} = A\boldsymbol{\xi} \tag{6}$$

with respect to given bases of E and F.

Now suppose that new bases of E and F are introduced. Suppose that the new vector components, which we will denote by $\boldsymbol{\xi}^*$ in E and $\boldsymbol{\eta}^*$ in F, are connected with the original ones as in (5) by the relationships

$$\boldsymbol{\xi} = S\boldsymbol{\xi}^*; \qquad \boldsymbol{\eta} = T\boldsymbol{\eta}^*. \tag{7}$$

(The matrix S^* in (5) is here denoted by S.)

Then, by substituting in (6), we have $T\boldsymbol{\eta}^* = AS\boldsymbol{\xi}^*$ and, multiplying on the left by T^{-1},

$$\boldsymbol{\eta}^* = (T^{-1}AS)\,\boldsymbol{\xi}^*. \tag{8}$$

The mapping f is therefore represented by the matrix

$$A^* = T^{-1}AS \tag{9}$$

with respect to the new bases.

In particular, if $E = F$, then $S = T$ (all vectors and images under f will be referred to the same basis of E) and

$$A^* = S^{-1}AS. \tag{10}$$

Square matrices A and A^* are said to be *similar* if there is a matrix S such that (10) is satisfied. Since from (10) it follows that $A = SA^*S^{-1}$, the relationship of similarity is symmetric.

5.5 Numerical Inversion of Matrices

Let $A = (\alpha_{ik})$ be a non-singular $n \times n$ matrix which therefore has an inverse A^{-1}. Choose any n-dimensional vector space E and any basis $\{e_1, \ldots, e_n\}$ of E. The equation

$$f_i = \sum_{k=1}^{n} \alpha_{ik} e_k \qquad (i = 1, \ldots, n), \qquad \text{i.e., } \mathbf{f} = A\mathbf{e} \tag{1}$$

defines n new vectors, which by Theorem 5.4;1 again form a basis of E. Multiplying (1) on the left by A^{-1}, we have

$$\mathbf{e} = A^{-1}\mathbf{f}. \tag{2}$$

Now the relations (1) are represented in the normal interpretation by the exchange tableau (cf. 3.3; (1),(2))

$$
\begin{array}{c|ccc}
 & e_1 \ \cdots \ \cdots \ e_n \\
\hline
\begin{array}{c} f_1 = \\ \vdots \\ \vdots \\ f_n = \end{array} & A
\end{array}
\tag{3}
$$

where, for the sake of brevity, we have written A in place of the individual coefficients α_{ik} in the tableau. Since $\{f_1, \ldots, f_n\}$ is a basis of E, $e_1, \ldots, e_n$ can be exchanged with $f_1, \ldots, f_n$ (Theorem 3.2;1). If we carry out this exchange by using the exchange method, we will express the vectors e_i as linear combinations of the vectors f_k and hence by (2) we will find the matrix A^{-1}.

Example 1

$$A = \begin{pmatrix} 1 & 2 & 1 \\ 0 & 1 & 1 \\ 1 & -1 & 1 \end{pmatrix}$$

1st Tableau

	e_1	e_2	e_3
$f_1 =$	1	2	1*
$f_2 =$	0	1	1
$f_3 =$	1	-1	1
	-1	-2	*

2nd Tableau

	e_1	e_2	f_1
$e_3 =$	-1	-2	1
$f_2 =$	-1*	-1	1
$f_3 =$	0	-3	1
	*	-1	1

3rd Tableau

	f_2	e_2	f_1
$e_3 =$	1	-1	0
$e_1 =$	-1	-1	1
$f_3 =$	0	-3*	1
	0	*	$\frac{1}{3}$

CH. 5: LINEAR MAPPINGS OF VECTOR SPACES, MATRICES

4th Tableau

	f_2	f_3	f_1
$e_3 =$	1	$\frac{1}{3}$	$-\frac{1}{3}$
$e_1 =$	-1	$\frac{1}{3}$	$\frac{2}{3}$
$e_2 =$	0	$-\frac{1}{3}$	$\frac{1}{3}$

In order to find A^{-1} from the last tableau, we must re-order the vectors e_i and f_k into the natural order of their indices, i.e., we must interchange the corresponding rows and columns of the tableau. We find

$$A^{-1} = \tfrac{1}{3} \begin{pmatrix} 2 & -3 & 1 \\ 1 & 0 & -1 \\ -1 & 3 & 1 \end{pmatrix}$$

which we can easily verify by direct calculation of AA^{-1} or $A^{-1}A = I$.

The final re-ordering of the rows and columns of the tableau is not necessary if we always use coefficients in the main diagonal as the pivots (providing this is possible), i.e., if we exchange each e_i with the f_i having the same index.

Exercises

1. Calculate the inverses of the following matrices:

$$A = \begin{pmatrix} 1 & 2 & 3 \\ 2 & -1 & 4 \\ 1 & 0 & 2 \end{pmatrix} \qquad B = \begin{pmatrix} 3 & 1 & -4 & 2 \\ 0 & 5 & -3 & 1 \\ 4 & 2 & 1 & 2 \\ -3 & 3 & 0 & -1 \end{pmatrix}$$

Solution.

$$A^{-1} = \begin{pmatrix} -2 & -4 & 11 \\ 0 & -1 & 2 \\ 1 & 2 & -5 \end{pmatrix} \qquad B^{-1} = \begin{pmatrix} -32 & 39 & -11 & -47 \\ -9 & 11 & -3 & -13 \\ 8 & -10 & 3 & 12 \\ 69 & -84 & 24 & 101 \end{pmatrix}$$

2. Let $\{e_1, e_2, e_3\}$ and $\{f_1, f_2, f_3\}$ be bases of the vector space E. Let the corresponding change of basis be given by the equations $f_i = \sum_{k=1}^{3} \alpha_{ik} e_k \ (i = 1, 2, 3)$ where

$$A = (\alpha_{ik}) = \begin{pmatrix} 2 & 0 & 1 \\ 2 & 5 & -4 \\ 3 & -1 & 2 \end{pmatrix}$$

Express the components η_i of a vector $x \in E$ with respect to the second basis in terms of the components ξ_k of x with respect to the first basis.

Solution. $\qquad\qquad \eta_i = \sum_{k=1}^{n} \beta_{ik} \xi_k \qquad (i = 1, 2, 3),$

where $\qquad\qquad (\beta_{ik}) = A^* = \begin{pmatrix} 6 & -16 & -17 \\ -1 & 3 & 3 \\ -5 & 14 & 15 \end{pmatrix}.$

5.6 The Exchange Method and Matrix Calculation

The tableau

	e_1 $\cdots$ $\cdots$ e_n
$f_1 =$	
$\vdots$	A
$\vdots$	
$f_m =$	

$\qquad\qquad\qquad\qquad\qquad\qquad\qquad\qquad\qquad$ (1)

represents in horizontal interpretation the relations

$$f_i = \sum_{k=1}^{n} \alpha_{ik} e_k \qquad (i = 1, \ldots, m), \text{ i.e., } \mathbf{f} = A\mathbf{e}. \qquad (2)$$

For later use and as an exercise in calculation with matrices, we will now investigate how an exchange of a whole group of vectors f_i with the corresponding group of vectors e_k is carried out. For this we will assume that the first r of the f_i are to be exchanged with the first r of the e_k. If this is not the case to start with, we can clearly re-order the f_i's and the e_k's and correspondingly the rows and columns of the tableau until it is. We now write the tableau (1) in a slightly different form

	e^1	e^2
$f^1 =$	A_1	B
$f^2 =$	C	D

$\qquad\qquad\qquad\qquad\qquad\qquad\qquad\qquad\qquad$ (3)

where e^1 means $e_1, \ldots, e_r$; e^2 means $e_{r+1}, \ldots, e_n$ and similarly for f^1 and f^2. Also A_1, B, C, D are parts of the matrix A where A_1 is an $r \times r$ matrix, B is an $r \times (n-r)$ matrix, C is an $(m-r) \times r$ matrix and D is an $(m-r) \times (n-r)$ matrix.

The relations (2) which are represented by the tableau (3) can now be written in the form

$$\mathbf{f^1} = A_1 \mathbf{e^1} + B\mathbf{e^2} \tag{4.1}$$

$$\mathbf{f^2} = C\mathbf{e^1} + D\mathbf{e^2} \tag{4.2}$$

where

$$\mathbf{e^1} = \begin{pmatrix} e_1 \\ \vdots \\ e_r \end{pmatrix}, \text{ etc.}$$

We now assume that A_1^{-1} exists (otherwise e^1 could not be exchanged with f^1) and multiply on the left by A_1^{-1} to obtain

$$\mathbf{e^1} = A_1^{-1} \mathbf{f^1} - A_1^{-1} B\mathbf{e^2} \tag{5.1}$$

Substituting in (4.2), we further have

$$\mathbf{f^2} = CA_1^{-1} \mathbf{f^1} + (D - CA_1^{-1} B)\, \mathbf{e^2} \tag{5.2}$$

These relations (5.1) and (5.2) are now represented in the normal interpretation by the tableau

	f^1	e^2	
$e^1 =$	A_1^{-1}	$-A_1^{-1}B$	(6)
$f^2 =$	CA_1^{-1}	$D - CA_1^{-1}B$	

This is identical with the tableau which we obtain by exchanging $e_1, \ldots, e_r$ with $f_1, \ldots, f_r$ in individual exchange steps. Thus the passage from (3) to (6) represents the net result of r exchange steps.

Problems

1. Show that the method described in this section is identical with the exchange method when $A_1 = (\alpha_{11})$.
2. Exchange e^2 and f^2 in (6), and hence represent A^{-1} as a partitioned matrix.
3. Carry out the arguments of 5.6 for tableaux in vertical interpretation.

CHAPTER 6

LINEAR FUNCTIONALS

6.1 Linear Functionals and Cosets

6.1.1 Linear Functionals

The fundamental scalars of a vector space (i.e., the real or the complex numbers) themselves form a vector space (S_R or S_K) of dimension 1 (see Examples 2.1;11, 17), which we will now collectively denote by S. Consequently we can consider the linear mappings $x \rightarrow f(x)$ of the vector space E into the vector space S. These mappings are just those real- or complex-valued functions f on E for which

$$\text{L. } f(\alpha_1 x_1 + \alpha_2 x_2) = \alpha_1 f(x_1) + \alpha_2 f(x_2)$$

for all $x_1, x_2 \in E$ and all scalars, α_1, α_2.

Definition 1. A 'linear functional' on a vector space E is a linear mapping of E into the 1-dimensional vector space of its scalars.

The linear functional f, which is given by $f(x) = 0$ for all $x \in E$, will be denoted by $f = 0$ and hence the statement $f \neq 0$ will mean that there is a vector $x \in E$ such that $f(x) \neq 0$. If $f \neq 0$, then f maps E onto S.

If E is finite-dimensional, then, in the case of a linear functional f, the representation 5.2;(2) of f becomes

$$f(x) = \sum_{k=1}^{m} \alpha_k \xi_k. \tag{1}$$

It will be seen that these are the functions which were introduced in Example 2.1;9.

Theorem 1. Let E be a vector space and let x_0 be a non-zero vector in E. Then there is a linear functional f on E such that $f(x_0) \neq 0$.

Proof. Since $x_0 \neq 0$, there is a basis B of E which contains x_0. Hence the theorem follows from Theorem 5.1;3.

Theorem 2. Let L be a subspace of E and $x_0 \in E \setminus L$. Then there is a linear functional f on E such that $f(x) = 0$ for all $x \in L$ and $f(x_0) \neq 0$.

Proof. The canonical image $\bar{x}_0$ of x_0 in the quotient space E/L is not the zero-element (cf. Example 5.1;1). Hence, by Theorem 1, there is a linear functional $\bar{f}$ on E/L for which $\bar{f}(\bar{x}_0) \neq 0$. We define a mapping f of E into S by $f(x) = \bar{f}(\bar{x})$ for each $x \in E$, where $\bar{x}$ is the canonical image of x. Since $x \rightarrow \bar{x}$ is a linear mapping of E onto E/L (Problem 5.1;2), it follows that f is a linear functional on E (Theorem 5.1;6). It is easy to verify that f satisfies the requirements of the theorem.

6.1.2 Hyperplanes

Definition 2. Let H be a coset in a vector space E and let L be the subspace corresponding to H. Then H is said to be a 'hyperplane' of E, if the quotient space E/L has dimension 1.

Thus a coset is a hyperplane if and only if the corresponding subspace is a hyperplane. If $\dim E = n$, then a subspace L of E is a hyperplane if and only if $\dim L = n - 1$. (Theorem 3.2;5).

Theorem 3. A coset $H \subseteq E$ is a hyperplane if and only if E is the only coset which contains H as a proper subset.

Proof. 1. We will first deal with the case when H is a subspace.

1.1. Suppose the subspace L is a hyperplane. Then $\dim E/L = 1$ and hence the subspace $\{0\} = L/L$ of E/L is properly contained in only one subspace, viz. E/L. Hence, by Theorem 2.5;7, E is the only subspace which properly contains L.

1.2. Suppose that the subspace L of E is not properly contained in any subspace except E itself. Then the subspace $\{0\} = L/L$ of E/L is not properly contained in any subspace except E/L. Hence $\dim E/L = 1$ and L is a hyperplane of E.

2. We now prove the theorem for arbitrary hyperplanes.

2.1. Suppose H is a hyperplane of E. Then so is the subspace L corresponding to H so that E is the only subspace which properly contains L. Hence E is the only coset which contains H as a proper subset (cf. Problem 2.5;5).

2.2. Suppose H is a coset of E and E is the only coset which contains H as a proper subset. Then E is also the only subspace which properly contains the subspace L corresponding to H. Hence L is a hyperplane and so also is H.

Theorem 4. Let $f \neq 0$ be a linear functional on the vector space E. Then $L = f^{-1}(0) = \{x; x \in E, f(x) = 0\}$ is a subspace and a hyperplane of E.
(By Definition 5.1;2, L is the kernel of the linear functional f.)

Proof. 1. By Theorem 5.1;7, L is a subspace.

2. By Theorem 5.1;9, the vector spaces E/L and $f(E) = S$ are isomorphic. By Theorem 5.2;1, $\dim E/L = \dim S = 1$ and therefore L is a hyperplane.

Theorem 5. *Let $f \neq 0$ be a linear functional on the vector space E and let α be a scalar. Then $H = f^{-1}(\alpha) = \{x; \, x \in E, f(x) = \alpha\}$ is a hyperplane.*

Proof. Since $f \neq 0$, H is not empty. Suppose therefore that $x_0 \in H$. Then $L = H - x_0 = f^{-1}(0)$ is a subspace and H is a coset of L. By Theorem 4, L is a hyperplane and therefore H is also a hyperplane (cf. Problem 3).

Theorem 6. *Given any hyperplane H of E, then there exists a linear functional $f \neq 0$ on E and a scalar α such that $H = f^{-1}(\alpha)$. f is uniquely determined by H except for a non-zero scalar factor.*

Proof. 1. Let L be the subspace corresponding to H. By Theorems 2 and 3, there is a linear functional $f \neq 0$ on E such that $f(x) = 0$ for all $x \in L$. By Theorem 4, $L^* = f^{-1}(0)$ is a hyperplane and $L^* \supseteq L$. Since L is a hyperplane itself, it follows that $L^* = L$. Now, $H = f^{-1}(\alpha)$, where $\alpha = f(x_0)$, and x_0 is any vector in H.

2. If $H = f^{-1}(\alpha) = g^{-1}(\beta)$, then $L = f^{-1}(0) = g^{-1}(0)$ is the subspace corresponding to H. If $z_0 \in E \setminus L$, then every $x \in E$ can be written uniquely in the form $x = y + \lambda z_0$ where $y \in L$ (cf. Problem 4). Hence $f(x) = \lambda f(z_0)$ and $g(x) = \lambda g(z_0)$, i.e., $f = \dfrac{f(z_0)}{g(z_0)} g$.

In the sense of analytic geometry, Theorem 6 states that every hyperplane has an equation of the form $f(x) = \alpha$ where $f \neq 0$. Theorem 5 states that every equation of this form is the equation of a hyperplane.

6.1.3 Systems of Linear Equations and Cosets

Theorem 7. *Every coset M of E is the intersection of the family $\mathscr{M}$ of all those hyperplanes H which contain M.*

Proof. 1. Clearly $M \subseteq \bigcap(\mathscr{M})$ (Theorem 1.2;3).

2. Suppose $y_0 \notin M$ (if $M = E$, the theorem is trivially true) and $x_0 \in M$. If L is the subspace corresponding to M, then $y_0 - x_0 \notin L$. By Theorem 2, there is a linear functional f on E such that $f(x) = 0$ for all $x \in L$ and $f(y_0 - x_0) \neq 0$. We put $f(x_0) = \alpha$ and consider the hyperplane $H = f^{-1}(\alpha)$. For each $x \in M$, $f(x) = f(x_0) = \alpha$ and hence $H \supseteq M$, i.e., $H \in \mathscr{M}$. Since $f(y_0) \neq f(x_0)$, $y_0 \notin H$ and therefore also $y_0 \notin \bigcap(\mathscr{M})$. Hence $\bigcap(\mathscr{M}) \subseteq M$.

If f is a linear functional on the vector space E and α is a scalar, then the equation $f(x) = \alpha$ is known as a *linear equation*. A vector $x_0 \in E$ is a solution of this equation, if $f(x_0) = \alpha$. By Theorem 6, every hyperplane is the set of solutions of a linear equation. Accordingly, Theorem 7 can also be stated as follows.

Theorem 8. Every coset is the set of solutions of a system of linear equations. (viz. the system consisting of the equations of the hyperplanes in the family $\mathscr{M}$. If $M = E$, then $\mathscr{M} = \varnothing$. The theorem is still true in this case because E is the set of solutions of the equation $0(x) = 0$.)

By Theorems 5 and 2.5;9, the following converse of Theorem 8 is also true.

Theorem 9. The set of solutions of a system of linear equations is either empty or it is a coset.

It is in fact possible for the set of solutions to be empty, for example when there is an equation $f(x) = \alpha$ in which f is the zero functional and $\alpha \neq 0$.

Example 1. In the vector space G_3 (Example 2.1;1), the set of solutions of the linear equation (with respect to some basis)

$$f(x) = \alpha_1 \xi_1 + \alpha_2 \xi_2 + \alpha_3 \xi_3 = \beta \qquad (f \neq 0)$$

is a plane, i.e., when the solution vectors are drawn from a fixed point, the set of their end points is a plane. A system of two linear equations has as its solution set either a straight line or a plane or it is empty. For three equations, the solution set can either be empty or a single vector or a straight line or a plane. If we allow $f = 0$, the solution set can also be the whole space, viz. when the system has only the one equation $0(x) = 0$.

Problems

1. Let a linear functional f be given in the form (1). What happens to the coefficients α_k under the change of basis $\boldsymbol{\xi} = T\boldsymbol{\eta}$?

2. Show that, if $k \in C$ (Example 2.1;7), then $f(x) = \int_{-1}^{+1} k(\tau) x(\tau) \, d\tau$ is a linear functional on C.

3. Show that, if f is a linear functional on a vector space, α is a scalar and $f^{-1}(\alpha) = H$, then $L = f^{-1}(0)$ is the subspace corresponding to the coset H.

4. Suppose f is a linear functional on the vector space E, $L = f^{-1}(0)$ and $z_0 \in E \setminus L$. Show that every vector $x \in E$ can be written in exactly one way in the form $x = y + \lambda z_0$ where $y \in L$.

5. Suppose E_1 and E_2 are vector spaces and L_1 is a hyperplane of E_1. Prove that the set of all pairs (x_1, x_2), where $x_1 \in L_1$ and $x_2 \in E_2$, is a hyperplane in the direct product of E_1 and E_2 (Problem 2.1;4).

6. Prove that a subspace L of the n-dimensional vector space E is a hyperplane if and only if $\dim L = n - 1$.

7. Let $k(\sigma, \tau)$ be a continuous scalar valued function of the real variables σ, τ in the interval $-1 \leqslant \sigma, \tau \leqslant +1$. Show that, if $g \in C$, then the set of all $x \in C$ which satisfy the first order linear integral equation

$$\int\limits_{-1}^{+1} k(\sigma, \tau) x(\tau) \, \mathrm{d}\tau = g(\sigma)$$

for all σ in the interval $-1 \leqslant \sigma \leqslant +1$, is either empty or it is a coset in C. What is the equation which characterizes the corresponding subspace?

8. Let $f_i(x) = \sum\limits_{k=1}^{n} \alpha_{ik} \xi_k \ (i = 1, \ldots, m)$ be linear functionals on the n-dimensional vector space E. Prove that $f_1, \ldots, f_m$ are linearly dependent if and only if the rows of the matrix $A = (\alpha_{ik})$ are linearly dependent.

6.2 Duality in Finite-Dimensional Vector Spaces

In this section we will restrict the discussion to finite-dimensional vector spaces.

6.2.1 The Dual Space of a Vector Space

By Theorem 5.1;10 the linear functionals on a vector space E themselves form a vector space $\mathscr{L}(E, S)$, which from now on we will denote by E^*. Thus, the elements of E^* are the linear functionals on E. If f_1, f_2 are two of these, then the linear combination

$$f = \alpha_1 f_1 + \alpha_2 f_2 \tag{1}$$

is defined by

$$f(x) = \alpha_1 f_1(x) + \alpha_2 f_2(x) \text{ for all } x \in E.$$

(cf. 5.1.3).

Definition 1. The vector space E^* is known as the 'dual space' of E.

The next theorem follows immediately from Theorem 5.2;11.

Theorem 1. If E is finite-dimensional, then so is E^* and the two dimensions are equal.

111

If L is a subspace of E, we consider the set $L^0 \subseteq E^*$ which is given by

$$L^0 = \{f; f \in E^*, f(x) = 0 \text{ for all } x \in L\}.$$

Thus L^0 is the set of all linear functionals on E which take the value 0 on all the vectors of L.

Theorem 2. L^0 is a subspace of E^* and

$$\dim L + \dim L^0 = \dim E.$$

Proof. 1. It is easy to see that L^0 is a subspace.

2. Corresponding to each linear functional $\bar{f}$ on E/L (i.e., $\bar{f} \in (E/L)^*$), there is a linear functional f on E given by $f(x) = \bar{f}(L + x)$ and $f \in L^0$. It is easy to verify that this defines an isomorphism of $(E/L)^*$ onto L^0. Hence by Theorems 1, 5.2;1 and 3.2;5,

$$\dim L^0 = \dim (E/L)^* = \dim E/L = \dim E - \dim L.$$

6.2.2 Bilinear Forms

Let E and F be finite-dimensional vector spaces having the same scalars, and suppose that f is a mapping which assigns a scalar $f(x,y)$ to each pair of vectors x, y where $x \in E$ and $y \in F$.

Definition 2. The mapping f is a 'bilinear form' on the pair of spaces (E,F) if
   1. *for each $y_0 \in F$, the mapping $x \to f(x,y_0)$ is a linear functional on E, and*
   2. *for each $x_0 \in E$, the mapping $y \to f(x_0,y)$ is a linear functional on F.*

Choosing bases of E and F and denoting the corresponding vector components by ξ_i and η_k respectively, we can easily verify that

$$f(x,y) = \sum_{i,k} \alpha_{ik} \xi_i \eta_k \tag{2}$$

is a bilinear form on (E,F) for all α_{ik}. On the other hand, every bilinear form f on (E,F) can be represented in the form (2), because, from Definition 2 and $x = \sum_i \xi_i e_i$; $y = \sum_k \eta_k f_k$, it follows that (2) is satisfied by $\alpha_{ik} = f(e_i, f_k)$.

6.2.3 Dual Pairs of Spaces

Definition 3. A pair (E,F) of finite-dimensional vector spaces with a bilinear form $f(x,y)$ is said to be a 'dual pair of spaces' if
   1. *from $f(x,y_0) = 0$ for all $x \in E$, it follows that $y_0 = 0$ and*

2. *from* $f(x_0, y) = 0$ *for all* $y \in F$, *it follows that* $x_0 = 0$.

The form $f(x, y)$ is known as the *scalar product* of the dual pair of spaces and in the following it will be denoted by $\langle x, y \rangle$.

Example 1. Suppose that E and F are 2-dimensional and $f(x, y) = \xi_1 \eta_2$. Then this is not a dual pair of spaces, because, for the vector $x_0 \in E$ with $\xi_1 = 0$, $\xi_2 = 1$, $f(x_0, y) = 0$ for all $y \in F$. But $x_0 \neq 0$.

However if we put $f(x, y) = \xi_1 \eta_1 + \xi_2 \eta_2$, then (E, F) becomes a dual pair of spaces, because, if $f(x, y) = 0$ for all $y \in F$, then, with $\eta_1 = 1$, $\eta_2 = 0$, we have $\xi_1 = 0$ and, with $\eta_1 = 0$, $\eta_2 = 1$, we have $\xi_2 = 0$, i.e., $x = 0$. The same argument is still valid when the roles of x and y are interchanged.

Example 2. Let E and F be two vector spaces of the same dimension n and let ξ_i and η_k be the vector components with respect to given bases of E and F. Then (E, F) is a dual pair of spaces with the scalar product

$$\langle x, y \rangle = \sum_{k=1}^{n} \xi_k \eta_k.$$

Example 3. A vector space E and its dual space E^* is a dual pair of spaces when, for $x \in E$ and $f \in E^*$, the scalar product is defined by $\langle x, f \rangle = f(x)$. Because $f(x) = 0$ for all $x \in E$ is just the definition of $f = 0$ and, if $f(x) = 0$ for all $f \in E^*$, then, from Theorem 6.1;1, it follows that $x = 0$. Finally, it is easy to verify that $\langle x, f \rangle$ is a bilinear form.

Now suppose that (E, F) is a dual pair of spaces. Corresponding to each $x \in E$ we have the linear functional f_x on F which is given by $f_x(y) = \langle x, y \rangle$. In this way we define a mapping g of E into the dual space F^* of F and we assert that g is both linear and 1-1. It is linear because

$$f_{\alpha_1 x_1 + \alpha_2 x_2}(y) = \langle \alpha_1 x_1 + \alpha_2 x_2, y \rangle$$
$$= \alpha_1 \langle x_1, y \rangle + \alpha_2 \langle x_2, y \rangle$$
$$= \alpha_1 f_{x_1}(y) + \alpha_2 f_{x_2}(y)$$

for all $y \in F$, and it is 1-1 because $f_{x_1} = f_{x_2}$ means $\langle x_1, y \rangle = \langle x_2, y \rangle$ for all $y \in F$, and therefore $\langle x_1 - x_2, y \rangle = 0$ for all $y \in F$, and hence, by Definition 3, $x_1 = x_2$.

Thus g is a 1-1 linear mapping of E into F^*. Hence $\dim E \leqslant \dim F^*$ (Theorem 5.2;3). Similarly, $\dim F \leqslant \dim E^*$. Since further, by Theorem 1, $\dim E = \dim E^*$ and $\dim F = \dim F^*$, it follows that the vector spaces E, E^*, F, F^* all have the same dimension and that g is an isomorphism of E onto F^* (Theorem 5.2;4). Hence we have the following *Duality Theorem*.

Theorem 3. *If (E,F) is a dual pair of spaces, then, corresponding to each linear functional f on F there exists exactly one $x \in E$ such that $f(y) = \langle x,y \rangle$ for all $y \in F$. Similarly, corresponding to each linear functional f on E, there exists exactly one $y \in F$ such that $f(x) = \langle x,y \rangle$ for all $x \in E$. These correspondences are isomorphisms between E and F^* and F and E^*.*

As a corollary to this we have

Theorem 4. *The vector spaces in a dual pair of spaces have the same dimension.*

6.2.4 Orthogonality in Dual Pairs of Spaces

Definition 4. *Vectors $x \in E$ and $y \in F$ of a dual pair of spaces (E,F) are said to be 'orthogonal' if $\langle x,y \rangle = 0$.*

More generally, a vector $y \in F$ is said to be orthogonal to the subset X of E if y is orthogonal to every vector $x \in X$. We see immediately that y is then also orthogonal to $L(X)$.

We now start with a subset X of E and investigate the set $X^\dagger$ of all those vectors of F which are orthogonal to X, i.e.,

$$X^\dagger = \{y; y \in F, \quad y \text{ orthogonal to } X\}.$$

The set $X^\dagger$ is a subspace of F, because, if $\langle x,y_1 \rangle = 0$ and $\langle x,y_2 \rangle = 0$, then $\langle x, \alpha_1 y_1 + \alpha_2 y_2 \rangle = 0$ for all scalars α_1, α_2. Further $0 \in X^\dagger$ so that $X^\dagger \neq \varnothing$.

It is easy to see that, if $X_1 \subseteq X_2$, then $X_2^\dagger \subseteq X_1^\dagger$.

Since the roles of E and F can be interchanged, $(X^\dagger)^\dagger = X^{\dagger\dagger}$ is a subspace of E and we have the following theorem.

Theorem 5. $X^{\dagger\dagger} = L(X)$.

Proof. 1. Suppose $z \in L(X)$. Then $\langle z,y \rangle = 0$ for all $y \in X^\dagger$ and hence $z \in X^{\dagger\dagger}$.

2. Suppose $z \in E$ and $z \notin L(X)$. Then, by Theorem 6.1;2, there is a linear functional f on E such that $f(z) \neq 0$ and $f(x) = 0$ for all $x \in L(X)$. Hence by Theorem 3, there is a $y \in F$ such that $\langle z,y \rangle \neq 0$ and $\langle x,y \rangle = 0$ for all $x \in X$. Hence $y \in X^\dagger$ and $z \notin X^{\dagger\dagger}$.

If in particular $X = L$ is a subspace of E, then $L^{\dagger\dagger} = L$. The subset $L^\dagger$ of F consists of all those $y \in F$ which are orthogonal to L. Similarly $L = (L^\dagger)^\dagger$ consists of all those $x \in E$ which are orthogonal to $L^\dagger$. Thus the connection between L and $L^\dagger$ is completely symmetric. We will say that L and $L^\dagger$ are *dual subspaces.* ($L^\dagger$ is also known as the *total orthogonal subspace* or the *orthogonal complement* of L.)

Theorem 6. If (E,F) is a dual pair of spaces, then, corresponding to each subspace L of E, there exists a unique orthogonal complement $L^\dagger$ in F and conversely. $(L^\dagger)^\dagger = L$ and each of the subspaces L and $L^\dagger$ consists of all the vectors which are orthogonal to the other. The sum of the dimensions of L and $L^\dagger$ is equal to the dimension of E (and of F).

Proof. It only remains to prove the last part of the theorem. By Theorem 3, $L^\dagger$ is isomorphic to the subspace L^0 of E^*. Hence, by Theorem 2, $\dim L^\dagger = \dim L^0 = \dim E - \dim L$.

Theorem 7. Let $f_1, \ldots, f_r$ and g be linear functionals on the finite-dimensional vector space E such that, for all those $x \in E$, for which $f_1(x), \ldots, f_r(x)$ are simultaneously zero, $g(x)$ is also zero. Then g is a linear combination of $f_1, \ldots, f_r$.

Proof. Using the scalar product $\langle x, f \rangle = f(x)$ of the dual pair of spaces (E, E^*), we can write the hypothesis of the theorem as follows. If $\langle x, f_k \rangle = 0$ for $k = 1, \ldots, r$, then $\langle x, g \rangle = 0$. That is, if $x \in \{f_1, \ldots, f_r\}^\dagger$, then $x \in \{g\}^\dagger$, i.e., $\{f_1, \ldots, f_r\}^\dagger \subseteq \{g\}^\dagger$. If we go over to the dual subspace, we have by Theorem 5

$$L(g) = \{g\}^{\dagger\dagger} \subseteq \{f_1, \ldots, f_r\}^{\dagger\dagger} = L(f_1, \ldots, f_r).$$

Hence g is a linear combination of $f_1, \ldots, f_r$.

Example 4. Suppose that two different 2-dimensional subspaces L_1 and L_2 of the vector space G_3 are given by

$$f_i(x) = \alpha_{i1}\xi_1 + \alpha_{i2}\xi_2 + \alpha_{i3}\xi_3 = 0 \qquad (i = 1, 2).$$

Then $L = L_1 \cap L_2$ is a 1-dimensional subspace. Let a further 2-dimensional subspace M be given by $g(x) = \beta_1\xi_1 + \beta_2\xi_2 + \beta_3\xi_3 = 0$. Now if L is contained in M, then by Theorem 7, g is a linear combination of f_1 and f_2, i.e., $g = \lambda_1 f_1 + \lambda_2 f_2$ or $\beta_k = \lambda_1\alpha_{1k} + \lambda_2\alpha_{2k}$ $(k = 1, 2, 3)$. This is the well-known theory of a pencil of planes in Analytic Geometry.

6.2.5 Dual Linear Mappings

Let (E_1, F_1) and (E_2, F_2) be two dual pairs of spaces and let f be a linear mapping of E_1 into E_2. Then corresponding to each $y_2 \in F_2$, there is a linear functional h_{y_2} on E_1 given by $h_{y_2}(x_1) = \langle f(x_1), y_2 \rangle$. Hence, by Theorem 3, there exists a unique $g(y_2) \in F_1$, such that $h_{y_2}(x_1) = \langle x_1, g(y_2) \rangle$ for all $x_1 \in E_1$. The mapping $y_2 \to g(y_2)$ of F_2 into F_1 is linear, as can be seen from the following argument.

$$\langle x_1, g(\alpha_1 y_1 + \alpha_2 y_2)\rangle = \langle f(x_1), \alpha_1 y_1 + \alpha_2 y_2\rangle$$
$$= \alpha_1 \langle f(x_1), y_1\rangle + \alpha_2 \langle f(x_1), y_2\rangle$$
$$= \alpha_1 \langle x_1, g(y_1)\rangle + \alpha_2 \langle x_1, g(y_2)\rangle$$
$$= \langle x_1, \alpha_1 g(y_1) + \alpha_2 g(y_2)\rangle \text{ for all } x_1 \in E_1.$$

Hence $g(\alpha_1 y_1 + \alpha_2 y_2) = \alpha_1 g(y_1) + \alpha_2 g(y_2)$.

Definition 5. Let (E_1, F_1) and (E_2, F_2) be two dual pairs of spaces and let f te a linear mapping of E_1 into E_2. The unique linear mapping g of F_2 into F_1 which is given by

$$\langle f(x_1), y_2\rangle = \langle x_1, g(y_2)\rangle \text{ for all } x_1 \in E_1, y_2 \in F_2 \tag{3}$$

is called the 'dual mapping' of f. Because of the symmetry of (3), f is also the dual mapping of g and is therefore uniquely determined by g.

Theorem 8. Dual linear mappings have the same rank. (Cf. Definition 5.2;1.)

Proof. We consider the subspace $H = [f(E_1)]^\dagger$ of F_2. For $y_2 \in H$, $\langle x_1, g(y_2)\rangle = \langle f(x_1), y_2\rangle = 0$ for all $x_1 \in E_1$. Hence $g(y_2) = 0$. From this it follows that H is contained in the kernel of g. Hence

$$\text{rank of } g = \dim g(F_2) = \dim F_2 - \dim (\text{kernel of } g)$$
$$\leqslant \dim F_2 - \dim H = \dim f(E_1) = \text{rank of } f.$$

Similarly rank of $f \leqslant$ rank of g, and the assertion follows.

6.2.6 Dual Bases, The Rank of a Matrix

Let (E, F) be a dual pair of spaces. The bases $\{e_1, \ldots, e_n\}$ of E and $\{f_1, \ldots, f_n\}$ of F are said to be dual if

$$\langle e_i, f_k\rangle = \delta_{ik} \text{ for all } i, k = 1, \ldots, n \tag{4}$$

where δ_{ik} is the Kronecker delta (cf. Example 3.1;2).

Corresponding to each basis of E, there is exactly one dual basis of F (and conversely). If, for example, $e_1, \ldots, e_n$ are given, then we choose f_1 to be non-zero and orthogonal to $e_2, \ldots, e_n$. By Theorem 6, f_1 is uniquely determined except for a scalar factor. This factor is uniquely determined by the condition $\langle e_1, f_1\rangle = 1$. Because of the orthogonality we further have $\langle e_i, f_1\rangle = 0$, for $i = 2, \ldots, n$. We choose $f_2, \ldots, f_n$ analogously. If $\sum\limits_{k=1}^{n} \lambda_k f_k = 0$, then $\sum\limits_{k=1}^{n} \lambda_k \langle e_i, f_k\rangle = \sum\limits_{k=1}^{n} \lambda_k \delta_{ik} = \lambda_i = 0$ for $i = 1, \ldots, n$. Thus $f_1, \ldots, f_n$ are linearly independent and hence they are a basis of F.

If the scalar product of a dual pair of spaces is referred to dual bases, then, from (2) and (4), it follows that

$$\langle x, y \rangle = \sum_{k=1}^{n} \xi_k \eta_k = \boldsymbol{\xi}' \boldsymbol{\eta}. \tag{5}$$

Conversely, if the scalar product is given by (5), then the bases which are involved are dual.

Now suppose that, with the notation of 6.2.5, f and g are dual mappings. Suppose that with respect to given bases of E_1 and E_2, f is given by $\boldsymbol{\xi}_2 = A\boldsymbol{\xi}_1$. (Here for example $\boldsymbol{\xi}_1$ is the matrix which has just one column consisting of the components $\xi_{11}, \ldots, \xi_{1n}$ of $x_1 \in E_1$.) Suppose further that, with respect to the dual bases, g is given by $\boldsymbol{\eta}_1 = B\boldsymbol{\eta}_2$. In view of (5), the duality condition (3) now reads

$$\boldsymbol{\xi}_1' A' \boldsymbol{\eta}_2 = \boldsymbol{\xi}_1' B \boldsymbol{\eta}_2.$$

Since this is valid for all $x_1 \in E_1$ and all $y_2 \in F_2$, i.e., for all values of ξ_{1i} and η_{2k}, it follows that $B = A'$.

Theorem 9. Dual mappings are represented by transposed matrices with respect to dual bases. Conversely, transposed matrices represent dual mappings when they are referred to dual bases in dual pairs of spaces (which is always possible).

Proof. It only remains to prove the second part. If A is an $m \times n$ matrix, we start with two vector spaces E_1 and E_2 of dimensions n and m. Then, with respect to any bases of E_1 and E_2, A represents a linear mapping f of E_1 into E_2. Now suppose that F_1 and F_2 form dual pairs of spaces with E_1 and E_2 (e.g., we could put $F_i = E_i^*$).

If we now refer A' to the bases of F_1 and F_2 which are dual to the chosen bases of E_1 and E_2, then A' represents the dual mapping g of f. Because by (5), for $x_1 \in E_1$ and $y_2 \in F_2$,

$$\langle f(x_1), y \rangle = \boldsymbol{\xi}_1' A' \boldsymbol{\eta}_2 = \langle x_1, g(y_2) \rangle.$$

In view of Theorem 5.2;12, it now follows from Theorem 8 that transposed matrices have the same column rank. If we now define the row rank of a matrix in analogy with the column rank (cf. 5.2.4) to be the maximum number of linearly independent rows, then obviously the column rank of A' is equal to the row rank of A. The column rank of a matrix is therefore always equal to the row rank so that we usually refer simply to the *rank* of a matrix, i.e.,

Theorem 10. The row and column ranks of a matrix are equal, so that both can be referred to simply as the 'rank' of the matrix.

Theorem 11. If the matrix A has rank r, then it contains a determinant of dimension r whose value is not zero and every determinant of dimension greater than r which is contained in A has the value zero.

('Determinants which are contained in A' are obtained by choosing a number of the rows of A and the same number of columns and taking those elements of A which belong both to one of the chosen rows and to one of the chosen columns).

Proof. 1. If A has rank r, then it has r linearly independent rows. These form a submatrix of A which also has rank r, and therefore has r linearly independent columns. The determinant formed from these r partial columns of A has a value not equal to zero.

2. If a determinant which is contained in A has dimension greater than r, then its rows are linearly dependent, because the corresponding rows of A are linearly dependent. Hence the determinant has the value zero (Theorem 4.2;7).

6.2.7 Numerical Calculation of the Rank of a Matrix

Let $A = (\alpha_{ik})$ be an $m \times n$ matrix whose rank is to be calculated. We start with a vector space E of dimension n and choose a basis $\{e_1, \ldots, e_n\}$ of E. If we put

$$\mathbf{f} = A\mathbf{e}, \text{ i.e., } f_i = \sum_{k=1}^{n} \alpha_{ik} e_k \qquad (i = 1, \ldots, m) \qquad (6)$$

then, by Theorem 3.1;4, the row rank (and hence the rank) of A is equal to the maximum number of linearly independent vectors in the set $\{f_1, \ldots, f_m\}$ and hence to the dimension of $L(f_1, \ldots, f_m)$. We start now with the tableau which expresses the relations (6) in normal interpretation. This is the same tableau as 5.6;(1). Now we exchange vectors f_i with vectors e_k until it is no longer possible to do so, i.e., until there are no more suitable pivots available. Note that it does not matter which pair of vectors we exchange first and which we exchange second, etc., because the only restriction lies in the availability of pivots. With a possible re-ordering of the rows and of the columns, the last tableau will have the form (see 5.6;(3) for the notation)

	f^1	e^2
$e^1 =$	B	C
$f^2 =$	D	O

where the bottom right-hand corner contains only zeros (otherwise it would be possible to make a further exchange). The vectors in the group f^1 are linearly independent because otherwise it would not have been possible to exchange them. It also follows from the tableau that each of the vectors in the group f^2 is a linear combination of those in the group f^1. Thus the dimension of $L(f_1,\ldots,f_m)$ and hence the rank of A is equal to the number of vectors f_i which have been exchanged. In a practical calculation, it is clearly not necessary to carry each pivotal row and column into the rest of the subsequent calculation. In the next Example, we will show the full calculation on the left and the calculation reduced to its essentials on the right.

Example 5. To find the rank of the matrix

$$A = \begin{pmatrix} 1 & 2 & 1 & 2 \\ 2 & 1 & 2 & 1 \\ 1 & 1 & 1 & 1 \end{pmatrix}$$

1st Tableau

	e_1	e_2	e_3	e_4
$f_1 =$	1*	2	1	2
$f_2 =$	2	1	2	1
$f_3 =$	1	1	1	1
	*	-2	-1	-2

2nd Tableau

	f_1	e_1	e_3	e_4
$e_1 =$	1	-2	-1	-2
$f_2 =$	2	-3	0	-3
$f_3 =$	1	-1*	0	-1
	1	*	0	-1

-3	0	-3
-1*	0	-1
*	0	-1

3rd Tableau

	f_1	f_3	e_3	e_4
$e_1 =$	-1	2	-1	0
$f_2 =$	-1	3	0	0
$e_2 =$	1	-1	0	-1

$$\boxed{\begin{array}{cc} 0 & 0 \end{array}}$$

The vector f_2 cannot be exchanged either with e_3 or e_4. The number of possible exchange steps is 2 and therefore the rank of A is equal to 2.

Exercises

1. Find the ranks of the following two matrices,

$$A = \begin{pmatrix} 1 & 2 & 3 & 4 \\ -1 & -3 & 1 & 5 \\ -1 & -4 & 5 & 14 \end{pmatrix} \qquad B = \begin{pmatrix} 1 & -1 & 2 \\ -1 & 3 & 2 \\ -2 & 2 & -4 \\ -3 & 5 & -2 \\ -1 & -1 & -6 \end{pmatrix}$$

Solution. Rank A = rank B = 2.

2. Find the dimension of the subspace of R_5 which is spanned by the following four vectors.

$$x_1 = (2, -1, 3, 5, -2), \qquad x_2 = (3, -2, 5, -1, 3)$$
$$x_3 = (5, -3, 8, 4, 1), \qquad x_4 = (1, 0, 1, 11, -7).$$

Solution. The dimension is 2.

3. Let
$$\begin{aligned} f_1(x) &= \xi_1 + 2\xi_2 + 3\xi_3 + 4\xi_4 \\ f_2(x) &= -\xi_1 - 3\xi_2 + \xi_3 + 5\xi_4 \\ f_3(x) &= -\xi_1 - 4\xi_2 + 5\xi_3 + 14\xi_4 \\ g(x) &= \xi_1 + 3\xi_2 - \xi_3 - 5\xi_4 \end{aligned}$$

be linear functionals on the vector space R_4. Is there a vector $x \in R_4$ such that $g(x) \neq 0$ and $f_1(x) = f_2(x) = f_3(x) = 0$?

Solution. No, because g is a linear combination of f_1, f_2, f_3.

Problems

1. The definition of the dual space can be extended to arbitrary vector spaces in an obvious way. Show that the dual space F_0^* of the vector space F_0 (Example 2.1;3) is isomorphic to the space F.

2. Let L be a subspace of the vector space E. What special property characterizes the linear functionals $f \in L^0 \subseteq E^*$?

3. Show that, if L is a subspace of the vector space E, then L^* is isomorphic to $E^*/L^\dagger$.

4. Suppose that L is the subspace of R_5 which is spanned by the vectors $z_1 = (2,1,1,2,0)$ and $z_2 = (1,2,3,1,1)$. Find a basis for the dual subspace $L^0 \subseteq R_5^*$.

5. Calculate the basis of R_n^* which is dual to the basis $e_i = (\delta_{i1}, \ldots, \delta_{in})$ $(i = 1, \ldots, n)$ of R_n.

6. Let t be an endomorphism of the finite-dimensional vector space E. What is the image of a linear functional $f \in E^*$ under the mapping which is dual to t?

7. Let (E,F) be a dual pair of spaces. Calculate the dual subspaces $\{0\}^\dagger$ and $E^\dagger$.

8. Let (E,F) be a dual pair of spaces and suppose that two bases of E are connected by $\hat{\mathbf{e}} = S\mathbf{e}$ (see 5.4;(4)). What is the connection between the dual bases?

9. Show that $(L_1 + L_2)^\dagger = L_1^\dagger \cap L_2^\dagger$ and $(L_1 \cap L_2)^\dagger = L_1^\dagger + L_2^\dagger$.

10. Let (E,F) be a dual pair of spaces and suppose $E = L_1 \oplus L_2$. Show that $F = L_1^\dagger \oplus L_2^\dagger$.

11. Show that the dual mapping of a product of two linear mappings is the product of the dual mappings in the reverse order.

12. Let $f(x,y) = \sum_{i,k} \alpha_{ik} \xi_i \eta_k$ be a bilinear form on two vector spaces of the same dimension. Find a necessary and sufficient condition on the α_{ik} so that (E,F) is a dual pair of spaces with the scalar product $\langle x,y \rangle = f(x,y)$.

13. Show that the bilinear form $f(x,y)$ in Problem 12 is a product of a linear functional $g(x)$ on E and a linear functional $h(y)$ on F if and only if the rank of $A = (\alpha_{ik})$ is equal to 0 or 1.

14. What is the dual mapping of a projection? (see Problem 5.3;6).

6.3 Linear Functionals which are Positive on a Convex Set

6.3.1 The Separation Theorem

In 2- or 3-dimensional space, we have the intuitive idea that any convex set K which does not contain the origin lies completely to one side of some line or plane through the origin (i.e., some hyperplane). If $f(x) = 0$ is the equation of this hyperplane, then $f(x) \geqslant 0$ for all $x \in K$ (multiplying f by -1 if necessary).

We will now show that this state of affairs also obtains in all *finite-dimensional* real vector spaces.

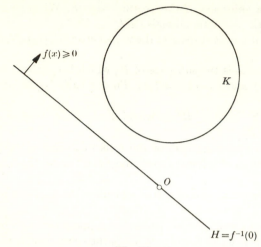

Fig. 13.

Theorem 1. Let E be a real vector space of finite dimension n and let K be a convex subset of E such that $0 \notin K$. Then there is a linear functional $f \neq 0$ on E such that $f(x) \geqslant 0$ for all $x \in K$. We say that f is positive on K (although strictly speaking we should say 'non-negative').

Proof. 1. We consider the family $\mathscr{L}$ of all subspaces L of E which have the following property.

A: There is a linear functional $f \neq 0$ on L such that

$$f(x) \geqslant 0 \text{ for all } x \in L \cap K.$$

We need to show that $E \in \mathscr{L}$.

2. If $K = \varnothing$, the theorem is trivially true. Suppose that $x_0 \in K$. If $\lambda x_0 \in K$, then $\lambda > 0$ because $0 \notin K$. Hence a linear functional $f \neq 0$ is defined on the 1-dimensional subspace $L(x_0)$ by $f(\lambda x_0) = \lambda$ so that $L(x_0)$ has property A. Hence $L(x_0) \in \mathscr{L}$ and $\mathscr{L}$ is not empty.

3. In 4. we will show that, if $L \in \mathscr{L}$ and $L \neq E$, then there is an $\hat{L} \in \mathscr{L}$ such that $\dim \hat{L} = \dim L + 1$. Since $\mathscr{L}$ is not empty, it follows that $E \in \mathscr{L}$ and the theorem will be proved.

4. Suppose $L \in \mathscr{L}$, $L \neq E$ and f is a linear functional on L as in property A. If t is a vector in $E \setminus L$, we form the subspace $\hat{L} = L(L, t)$. Then $\dim \hat{L} = \dim L + 1$ and every vector $\hat{x} \in \hat{L}$ can be written uniquely in the form $\hat{x} = x + \lambda t$, where $x \in L$. We distinguish three cases.

1st case: For all $\hat{x} = x + \lambda t \in \hat{L} \cap K, \lambda \geqslant 0$. Then $\hat{f}(\hat{x}) = \lambda$ is a linear functional on $\hat{L}, \hat{f} \neq 0$ and $\hat{f}(\hat{x}) \geqslant 0$ for all $\hat{x} \in \hat{L} \cap K$. Hence $\hat{L} \in \mathscr{L}$.

2nd case: For all $\hat{x} = x + \lambda t \in \hat{L} \cap K$, $\lambda \leqslant 0$. Then we put $\hat{f}(\hat{x}) = -\lambda$.

3rd case: If neither of the first two cases holds, then there exist

$$\hat{x}_1 = x_1 + \lambda_1 t \in \hat{L} \cap K \quad \text{where} \quad x_1 \in L \quad \text{and} \quad \lambda_1 > 0$$

and $\qquad \hat{x}_2 = x_2 - \lambda_2 t \in \hat{L} \cap K$ where $x_2 \in L$ and $\lambda_2 > 0$.

We consider the two sets of real numbers

$$M_1 = \left\{ \mu_1; \text{ there exists } x_1 \in L, \lambda_1 > 0 \quad \text{such that} \right.$$

$$\left. x_1 + \lambda_1 t \in \hat{L} \cap K \quad \text{and} \quad \mu_1 = -\frac{f(x_1)}{\lambda_1} \right\}$$

$$M_2 = \left\{ \mu_2; \text{ there exists } x_2 \in L, \lambda_2 > 0 \text{ such that} \right.$$

$$\left. x_2 - \lambda_2 t \in \hat{L} \cap K \quad \text{and} \quad \mu_2 = +\frac{f(x_2)}{\lambda_2} \right\}.$$

Both M_1 and M_2 are not empty. Further $\mu_1 \leqslant \mu_2$ for all $\mu_1 \in M_1$ and all $\mu_2 \in M_2$. Because, for the corresponding $x_1, \lambda_1, x_2, \lambda_2$,

$$\frac{1}{\lambda_1 + \lambda_2} (\lambda_1 x_2 + \lambda_2 x_1) \in L \cap K$$

therefore $\lambda_1 f(x_2) + \lambda_2 f(x_1) \geqslant 0$ and hence

$$\mu_1 = -\frac{f(x_1)}{\lambda_1} \leqslant \frac{f(x_2)}{\lambda_2} = \mu_2.$$

It follows that there is a real number θ such that $\mu_1 \leqslant \theta \leqslant \mu_2$ for all $\mu_1 \in M_1$ and all $\mu_2 \in M_2$.

We now consider the linear functional $\hat{f}$ on $\hat{L}$ given by

$$\hat{f}(x + \lambda t) = f(x) + \lambda \theta.$$

Clearly $\hat{f} \neq 0$ and it is easy to see that $\hat{f}(\hat{x}) \geqslant 0$ for all $\hat{x} \in \hat{L} \cap K$. Hence in this case also $\hat{L} \in \mathscr{L}$ and the proof is complete.

Theorem 2. Let P be a convex pyramid in a real vector space E of dimension n and let $x_0 \in E \setminus P$. Then there exists a symmetric parallelepiped W in E such that $(x_0 + W) \cap P = \varnothing$. (Cf. Example 3.4;2.)

Proof. Suppose $P = P(a_1, \ldots, a_r)$ (see Definition 3.4;4). A linear mapping g of the r-dimensional space R_r onto the linear hull $L(P)$ of P is given by

$g(\lambda_1, \ldots, \lambda_r) = \sum\limits_{k=1}^{r} \lambda_k a_k$. We choose a basis $\{e_1, \ldots, e_n\}$ of E and denote the components of $x = \sum\limits_{k=1}^{r} \lambda_k a_k$ by $\xi_1, \ldots, \xi_n$. Since each component is a linear functional on E, it follows that for each $x \in L(P)$, $\xi_i = h_i(\lambda_1, \ldots, \lambda_r)$ $(i = 1, \ldots, r)$, where h_i is a linear functional on R_r (Theorem 5.1;6). We denote the components of x_0 by ξ_{0i} and construct the quantity

$$\theta = \min_{x \in P} \max_{1 \leqslant i \leqslant n} |\xi_i - \xi_{0i}|$$

$$= \min_{\lambda_k \geqslant 0} \max_{1 \leqslant i \leqslant n} |h_i(\lambda_1, \ldots, \lambda_r) - \xi_{0i}| \tag{1}$$

where the maxima are taken over all the values $i = 1, \ldots, n$, and the minima are over all vectors $x \in P$ and all non-negative values of λ_k. The existence of the minimum is not obvious and will be proved later in 9.2.1. If we assume this for the time being, then, since $x_0 \notin P$, the maximum in (1) is strictly positive for all $x \in P$ and hence $\theta > 0$. The significance of θ lies in the fact that for $x \in P$ the difference $x - x_0$ always has a component whose absolute value is not less than θ. Thus the theorem is satisfied by the symmetric parallelepiped

$$W = \{x; x = \sum_{i=1}^{n} \xi_i e_i, |\xi_i| \leqslant \theta/2, (i = 1, \ldots, n)\}.$$

Theorem 3. Let P be a convex pyramid in a finite-dimensional real vector space E and let $x_0 \in E \setminus P$. Then there exists a linear functional f on E such that $f(x_0) < 0$ and $f(x) \geqslant 0$ for all $x \in P$.

Proof. Suppose W is a symmetric parallelepiped in E for which $(x_0 + W) \cap P = \varnothing$ (Theorem 2). Then, by Theorem 2.6;1, $K = P - W - x_0$ is convex and $0 \notin K$. By Theorem 1, there is a linear functional $f \neq 0$ on E such that $f(x) \geqslant 0$ for all $x \in K$. Therefore for $p \in P$ and $w \in W$

$$f(p) \geqslant f(x_0 + w). \tag{2}$$

From this it follows that $f(p) \geqslant 0$ for all $p \in P$. Otherwise f would take every negative value on P in contradiction of (2). Further, since $0 \in P$, it also follows from (2) that $f(x_0 + w) \leqslant 0$ for all $w \in W$. Since W certainly contains a vector w_0 for which $f(w_0) > 0$, it follows that $f(x_0) > 0$.

Theorem 3 is often referred to as the 'Separation Theorem' because the hyperplane $H = f^{-1}(0)$ 'separates' the pyramid P and the vector x_0.

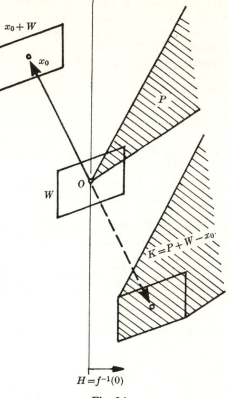

Fig. 14.

6.3.2 Dual Convex Pyramids

Let E and F be two finite-dimensional real vector spaces which form a dual pair of spaces with the scalar product $\langle x, y \rangle$. Let P be a convex pyramid in E. The subset P^0 of F given by

$$P^0 = \{y \, ; \, y \in F, \, \langle x, y \rangle \geqslant 0 \text{ for all } x \in P\}$$

is called the *dual pyramid* of P.

If $F = E^*$ is the dual space of E, then P^0 is the set of all linear functionals on E which take no negative values on P.

The set P^0 is itself a convex pyramid, because, if P is generated by the vectors $a_1, \ldots, a_r$, then for each vector $y \in F$ the following two statements are equivalent

$$\langle x, y \rangle \geqslant 0 \text{ for all } x \in P \tag{3}$$

$$\langle a_k, y \rangle \geqslant 0 \text{ for all } k = 1, \ldots, r \tag{4}$$

Now, however, (4) is a system of finitely many linear homogeneous inequalities whose solutions form a convex pyramid. (We will show this in Theorem 7.4;1.) Now, corresponding to P^0, we can construct the dual pyramid $(P^0)^0$ or briefly P^{00} in E. We have $P^{00} = \{x; x \in E, \langle x,y \rangle \geqslant 0$ for all $y \in P^0\}$ and the following duality theorem holds.

Theorem 4. $P^{00} = P$. *Duality is a symmetric relationship.*

Proof. 1. If $x_0 \in P$, then $\langle x_0, y \rangle \geqslant 0$ for all $y \in P^0$ and hence $x_0 \in P^{00}$. Thus $P \subseteq P^{00}$.

     2. If $x_0 \notin P$, then by Theorem 6.3;3 there exists a linear functional g on E such that $g(x_0) < 0$ and $g(x) \geqslant 0$ for all $x \in P$. Therefore, by Theorem 6.2;3, there exists a vector $y_0 \in F$ such that $\langle x_0, y_0 \rangle < 0$ and $\langle x, y_0 \rangle \geqslant 0$ for all $x \in P$. Hence $y_0 \in P^0$ and $x_0 \notin P^{00}$. Therefore $P^{00} \subseteq P$.

Example 1. If $P = \{0\}$, then $P^0 = F$ and, if $P = E$, then $P^0 = \{0\}$. If P is a subspace of E then P^0 is the dual subspace of P (cf. Theorem 6.2;6).

Example 2. If $F = E^*$ and if $P^0 \subseteq E^*$ consists of the multiples λf ($\lambda \geqslant 0$) of a linear functional $f \neq 0$ on E, then $P \subseteq E$ is the 'halfspace' $P = \{x; x \in E, f(x) \geqslant 0\}$.

Theorem 5. If P, Q are convex pyramids in E, then $(P \cap Q)^0 = P^0 + Q^0$ and $(P+Q)^0 = P^0 \cap Q^0$.

Proof. 1. If $y \in (P+Q)^0$, then $\langle x,y \rangle \geqslant 0$ for all $x \in P+Q$ and, in particular therefore, for all $x \in P$. Hence $y \in P^0$, and similarly $y \in Q^0$, i.e., $y \in P^0 \cap Q^0$.

     2. If $y \in P^0 \cap Q^0$, then $y \in P^0$ and therefore $\langle x,y \rangle \geqslant 0$ for all $x \in P$. Similarly $\langle x,y \rangle \geqslant 0$ for all $x \in Q$. Hence $\langle x,y \rangle \geqslant 0$ for all $x \in P+Q$, and therefore $y \in (P+Q)^0$. This proves the second part of the theorem. The first part now follows with the help of Theorem 4.

$$(P \cap Q)^0 = (P^{00} \cap Q^{00})^0 = (P^0 + Q^0)^{00} = P^0 + Q^0$$

Example 3. If $F = E^*$ and $P^0 \subseteq E^*$ is a convex pyramid $P^0 = P(f_1, \ldots, f_r)$, then $P \subseteq E$ is the set of all solutions of the system of linear inequalities $f_k(x) \geqslant 0$ ($k = 1, \ldots, r$), i.e., the intersection of the halfspaces corresponding to each of the individual inequalities of the system. (Example 2 and Theorem 5.)

Theorem 6. If $P \subseteq Q$, then $P^0 \supseteq Q^0$.

Proof. If $y \in Q^0$, then $\langle x,y \rangle \geqslant 0$ for all $x \in Q$ and therefore also for all $x \in P$. Hence $y \in P^0$.

Theorem 7. Let $f_1, \ldots, f_r$ and f_0 be linear functionals on the finite-dimensional real vector space E such that, for all those $x \in E$ for which $f_1(x), \ldots, f_r(x)$ are simultaneously non-negative, $f_0(x)$ is also non-negative. Then there exist non-negative coefficients $\lambda_1, \ldots, \lambda_r$ such that $f_0 = \sum_{k=1}^{r} \lambda_k f_k$.

Proof. Let E^* be the dual space of E and let $P^0 \subseteq E^*$ be the convex pyramid generated by $\{f_1, \ldots, f_r\}$ and Q^0 that generated by f_0. Then $P = \{x; x \in E, f_k(x) \geqslant 0 \ (k=1, \ldots, r)\}$ is the dual pyramid of P^0 and $Q = \{x; x \in E, f_0(x) \geqslant 0\}$ is the dual pyramid of Q^0. The hypothesis of the theorem means that $P \subseteq Q$. Therefore, by Theorem 6, $Q^0 \subseteq P^0$ and hence $f_0 \in P^0$ and the assertion follows.

If we choose a basis of E (components ξ_l), then we have representations as follows.

$$f_k(x) = \sum_l \alpha_{kl} \xi_l \quad \text{or} \quad \mathbf{f}(x) = A\boldsymbol{\xi}$$

and

$$f_0(x) = \sum_l \gamma_l \xi_l \quad \text{or} \quad \mathbf{f}_0(x) = \boldsymbol{\gamma}'\boldsymbol{\xi}.$$

Now, writing $C \geqslant 0$ to mean that all the elements of the matrix C are non-negative, we may restate Theorem 7 in the following form.

Theorem 8. Let A and $\boldsymbol{\gamma}'$ be real matrices with the same number of columns and suppose that whenever $A\boldsymbol{\xi} \geqslant 0$, it follows that $\boldsymbol{\gamma}'\boldsymbol{\xi} \geqslant 0$. Then there exists a matrix $\boldsymbol{\eta}' \geqslant 0$ such that $\boldsymbol{\gamma}' = \boldsymbol{\eta}'A$. That is, $\boldsymbol{\gamma}'$ is a linear combination of the rows of A with non-negative coefficients.

6.3.3 The Minimax Theorem

Theorem 9. Let E and F be finite-dimensional real vector spaces, let $f(x,y)$ $(x \in E, y \in F)$ be a bilinear form and let $X \subseteq E$ and $Y \subseteq F$ be non-empty convex polyhedra. Then

$$\min_{y \in Y} \max_{x \in X} f(x,y) = \max_{x \in X} \min_{y \in Y} f(x,y). \tag{5}$$

This theorem is known as the *Minimax Theorem* and is due to J. von Neumann. The right-hand side of the equation (5) is calculated by first finding, for each $x \in X$, a $y \in Y$ such that $f(x,y)$ is minimal. This minimal value is dependent on x. We then find an $x \in X$ such that the corresponding minimum is as large as possible. The left-hand side is calculated in an analogous manner.

The fact that the minimum on the left-hand side and the maximum on the right-hand side both exist is not obvious and will not in fact be proved until a later section (9.2.2). Nevertheless we will continue with the present theorem, assuming for the time being that the existence of both sides is known.

Proof of the Minimax Theorem. 1. Suppose that the value of $f(x,y)$ on the left-hand side of (5) is taken at $x=x_1$ and $y=y_1$ and that the value on the right-hand side is taken at $x=x_2$, $y=y_2$. Then

$$\min_{y\in Y}\max_{x\in X} f(x,y) = f(x_1,y_1) \geqslant f(x_2,y_1) \geqslant f(x_2,y_2) = \max_{x\in X}\min_{y\in Y} f(x,y).$$

Hence it only remains to prove that the left-hand side of (5) cannot be strictly greater than the right-hand side. In other words, that there is no real number θ such that

$$\min_{y\in Y}\max_{x\in X} f(x,y) > \theta > \max_{x\in X}\min_{y\in Y} f(x,y). \qquad (6)$$

2. We now prove that (6) is impossible for the special case in which the following extra conditions are satisfied.

(a) E and F form a dual pair of spaces with $\langle x,y\rangle = f(x,y)$

(b) Each of the polyhedra X and Y is a subset of a hyperplane in E and F respectively, which are not subspaces.

(c) $\theta = 0$

From (c) and (6), it follows that

$$\max_{x\in X} f(x,y) > 0 \text{ for all } y \in Y,$$

and

$$\min_{y\in Y} f(x,y) < 0 \text{ for all } x \in X,$$

or, alternatively,

for each $y \in Y$, there is an $x \in X$ such that $f(x,y) > 0$

and

for each $x \in X$, there is a $y \in Y$ such that $f(x,y) < 0$.

Now, in terms of the convex pyramids $P=P(X) \subseteq E$ and $Q=P(Y) \subseteq F$ generated by X and Y (see Problem 3.4;5), this means that

for each $y \in Q$, $y \neq 0$, there is an $x \in P$ such that $f(x,y) > 0$ $\qquad (7)$

and

for each $x \in P$, $x \neq 0$, there is a $y \in Q$ such that $f(x,y) < 0$. $\qquad (8)$

That is,

$$(-P^0) \cap Q = \{0\} \qquad (9)$$

and

$$P \cap Q^0 = \{0\}. \qquad (10)$$

From (10), it follows by Theorem 5 (and Example 1) that

$$P^0 + Q = F. \tag{11}$$

From (9) and (11), it follows by Theorem 2.6;9 that $Q = -Q$ which is a contradiction of (b). Thus we have proved that (6) is impossible with the extra conditions (a), (b) and (c).

3. We show that the general case can be reduced to the case in which condition (a) is satisfied. Let H and K be the following subspaces of E and F.

$$H = \{x; f(x, y) = 0 \text{ for all } y \in F\}$$

$$K = \{y; f(x, y) = 0 \text{ for all } x \in E\}.$$

We construct the quotient spaces $\bar{E} = E/H$ and $\bar{F} = F/K$ and, on these the bilinear form $\bar{f}(\bar{x}, \bar{y}) = f(x, y)$, where $\bar{x}$, $\bar{y}$ are the canonical images of x, y (see Example 5.1;1). Now $\bar{E}$ and $\bar{F}$ become a dual pair of spaces with the scalar product $\langle \bar{x}, \bar{y} \rangle = \bar{f}(\bar{x}, \bar{y})$. If, for example, $\bar{f}(\bar{x}_0, \bar{y}) = 0$, for all $\bar{y} \in \bar{F}$, then $f(x_0, y) = 0$ for all $y \in F$, i.e., $x_0 \in H$ or $\bar{x}_0 = 0$.

Further let $\bar{X} \subseteq \bar{E}$ and $\bar{Y} \subseteq \bar{F}$ be the canonical images of X and Y (i.e., the images of X and Y under the canonical mappings, for example $\bar{X}$ consists of all cosets N of H for which $N \cap X \neq \varnothing$). They are again convex polyhedra and the range of the values of $\bar{f}$ on $\bar{x} \in \bar{X}$, $\bar{y} \in \bar{Y}$ is the same as that of f on $x \in X$, $y \in Y$. Hence the impossibility of (6) will be proved for the bilinear form f and the polyhedra X and Y, as soon as it is proved for $\bar{f}$, $\bar{X}$ and $\bar{Y}$.

4. It remains to show that the case in which the extra condition (a) is satisfied can be reduced to the case in which (a), (b) and (c) are all satisfied. We choose bases of E and F and denote the components of a vector $x \in E$ and a vector $y \in F$ with respect to these bases by $\xi_1, \ldots, \xi_n$ and $\eta_1, \ldots, \eta_m$. The mappings h and k which are given by

$$h(x) = (\xi_1, \ldots, \xi_n, 1) \in R_{n+1}$$

and

$$k(y) = (\eta_1, \ldots, \eta_m, 1) \in R_{m+1}$$

map E onto the hyperplane $H \subseteq R_{n+1}$ with the equation $\xi_{n+1} = 1$ and F onto the hyperplane $K \subseteq R_{m+1}$ with the equation $\eta_{m+1} = 1$.

We define the bilinear form

$$f^*(\xi_1, \ldots, \xi_{n+1}; \eta_1, \ldots, \eta_{m+1}) = f(x, y) - \theta \xi_{n+1} \eta_{m+1}$$

on R_{n+1} and R_{m+1} where x and y have the components $\xi_1, \ldots, \xi_n$ and $\eta_1, \ldots, \eta_m$. It is easy to verify that R_{n+1} and R_{m+1} then form a dual pair of spaces. The sets

$$X^* = h(X) \subseteq H \subseteq R_{n+1} \quad \text{and} \quad Y^* = k(Y) \subseteq K \subseteq R_{m+1}$$

are again convex polyhedra and from (6), we have the same relationship between f^*, X^*, Y^* but with 0 in the place of θ. Hence (a), (b) and (c) are now satisfied. Thus (6) is impossible in all cases and this completes the proof of Theorem 9.

Problems

1. Let $\{e_1, \ldots, e_n\}$ be a basis of the real vector space E. What is the dual convex pyramid in E^* of the convex pyramid $P(e_1, \ldots, e_n)$ in E? What is the dual basis of $\{e_1, \ldots, e_n\}$?

2. Show that every convex pyramid is the intersection of the half spaces which contain it. (A half space is a set $H = \{x; f(x) \geqslant 0\}$, where f is a non-zero linear functional.)

3. Let K_1, K_2 be non-empty disjoint convex sets in a real finite-dimensional vector space E. Prove that there is a hyperplane H in E which separates K_1 and K_2, i.e., there is a linear functional $f \neq 0$ on E and a real number α such that $f(x) \geqslant \alpha$ for all $x \in K_1$ and $f(x) \leqslant \alpha$ for all $x \in K_2$. (Hint: Consider the set $K = K_1 - K_2$.)

4. Let (E_1, F_1) and (E_2, F_2) be dual pairs of spaces, let f be a linear mapping of E_1 into E_2 and let g be the dual linear mapping. Show that, if P is a convex pyramid in E_2, then $[f^{-1}(P)]^0 = g(P^0)$.

CHAPTER 7

SYSTEMS OF LINEAR
EQUATIONS AND INEQUALITIES

7.1 The Solutions of a System of Linear Equations

In 6.1.3, we briefly mentioned systems of linear equations in arbitrary vector spaces. We will now study more closely the case of finite-dimensional vector spaces and finite systems of equations. Such a system has the form

$$\sum_{k=1}^{n} \alpha_{ik}\xi_k + \beta_i = 0 \quad (i = 1, \ldots, m), \quad \text{i.e., } A\boldsymbol{\xi} + \boldsymbol{\beta} = 0 \tag{1}$$

(m equations with n unknowns $\xi_1, \ldots, \xi_n$).

7.1.1 Homogeneous Systems

The system (1) is said to be *homogeneous*, if $\beta_i = 0$ for all $i = 1, \ldots, m$. A homogeneous system always has the solution $\xi_1 = \ldots = \xi_n = 0$, which is referred to as the *trivial solution*. Some homogeneous systems also have other solutions, e.g., in the case $m = n = 1$ the system $0.\xi_1 = 0$ has every value of ξ_1 as a solution. We therefore wish to find in general the complete set of solutions of a homogeneous system. We will think of $x = (\xi_1, \ldots, \xi_n)$ as an element of the space of n-tuples R_n. The left-hand side

$$f_i(x) = \sum_{k=1}^{n} \alpha_{ik}\xi_k; \qquad f_i \in R_n^*$$

of each of the equations of the system is then a linear functional on R_n and therefore an element of R_n^*. The ith equation of the homogeneous system now means just that every solution vector x has to be orthogonal to f_i. Hence the set of all solutions of the homogeneous system is the dual subspace $L \subseteq R_n$ of the subspace $L^{\dagger} = L(f_1, \ldots, f_m)$ of R_n^* (see Example 6.2;3 and Theorem 6.2;6). The dimension of $L^{\dagger}$ is equal to the maximum number of linearly independent vectors in the set $\{f_1, \ldots, f_m\}$, i.e., equal to the maximum number of linearly independent rows in the matrix A, which is what we called the rank of A (see Problem 6.1;8). By Theorem 6.2;6, L therefore has the dimension $n - r(A)$ where $r(A)$ denotes the rank of A.

Theorem 1. The solutions of the homogeneous system

$$A\xi = 0$$

form a subspace L of R_n with dimension $n - r(A)$.

7.1.2 General Systems

Putting $\beta_i = 0$ for all $i = 1, \ldots, m$ in (1), we obtain the homogeneous system

$$A\eta = 0 \tag{2}$$

which we will refer to as the *homogeneous system associated* with the system (1). If $\zeta = \xi + \eta$ where ξ is a given solution of (1) and η is any solution of (2), then $A\zeta = A\xi + A\eta = -\beta$ and hence ζ is also a solution of (1). Similarly we can verify that the difference of any two solutions of (1) is a solution of (2) and hence we have the following result.

Theorem 2. If the general system (1) has a solution, then all solutions of (1) can be obtained by adding to any one of them all of the solutions of the associated homogeneous system (2).

Thus the solutions of (1) (if there are any) form a coset of the subspace L which corresponds to the solutions of (2). By Theorem 1, its dimension is $n - r(A)$. (The dimension of a coset means the dimension of the corresponding subspace.)

Naturally, the problem now is to find conditions for the general system (1) to have at least one solution. In order to answer this, we will think of the columns of A and the column β as elements $a_1, \ldots, a_n$, b of R_m. Then (1) can be written in the form

$$\sum_{k=1}^{n} \xi_k a_k + b = 0 \tag{3}$$

which means that $b \in L(a_1, \ldots, a_n) \subseteq R_m$. From this we have the following result.

Theorem 3. The general system (1) has a solution if and only if the rank of the matrix (A, β) is equal to the rank of A.

The matrix (A, β) is formed by adjoining the extra column β to A. For the proof, we only have to think of the ranks of the two matrices as column ranks. However, thinking of them as row ranks, we have the next theorem.

Theorem 4. The system (1) has a solution if and only if every linear relation between the rows of A is also satisfied by the β_i's.

132

We ask further, under what conditions for a given matrix A does the system (1) have a solution for all β_i, i.e., for all vectors $b \in R_m$. This is obviously the case when $L(a_1,\ldots,a_n) = R_m$, i.e., when there are m linearly independent vectors in the set $\{a_1,\ldots,a_n\}$.

Theorem 5. For a given matrix A, the system (1) has a solution for all possible β_i if and only if $r(A) = m$.

The case $m = n$ is of particular interest, i.e., when the number of equations is equal to the number of unknowns. The most important facts about this case are collected together in the following theorem.

Theorem 6. If $m = n$, then the following four statements are equivalent.
1. *The associated homogeneous system (2) has only the trivial solution.*
2. *The general system (1) has at least one solution for all β.*
3. *The general system (1) has exactly one solution for each β.*
4. *$\det A \neq 0$.*

Proof. From 1, it follows that $r(A) = n = m$ (Theorem 1) and hence 2 is satisfied (Theorem 5).

From 2, it follows that $r(A) = m = n$ (Theorem 5) and hence 3 is satisfied (Theorems 1 and 2).

From 3, it follows that $r(A) = m = n$ (Theorem 5) and hence 4 is satisfied (Theorem 4.2;7).

From 4, it follows that $r(A) = n$ (Theorem 4.2;7) and hence 1 is satisfied (Theorem 1).

If $m = n$ and $\det A \neq 0$, then the unique solution can be expressed explicitly by *Cramer's rule*. As in 4.3.1, we will denote the cofactor of the element α_{ik} in the matrix A by D_{ik}. Now if, for a fixed k, we multiply the ith equation by $(-1)^{i+k} D_{ik}$ $(i = 1,\ldots,n)$ and add all the resulting equations, we have

$$\xi_1 \sum_i (-1)^{i+k} \alpha_{i1} D_{ik} + \ldots + \xi_k \sum_i (-1)^{i+k} \alpha_{ik} D_{ik} + \ldots$$
$$\ldots + \sum_i (-1)^{i+k} \beta_i D_{ik} = 0.$$

By Theorem 4.3;1, the coefficient of ξ_1 is the determinant which is obtained from $\det A$ by replacing the kth column with the first and is therefore equal to 0 (unless $k = 1$). The same is true for the coefficients of the other unknowns except ξ_k whose coefficient is equal to $\det A$. Finally the last term is the determinant which is obtained from $\det A$ by replacing the kth column with β. If we denote this by $\det A_k$, then we have *Cramer's rule*,

$$\xi_k = -\frac{\det A_k}{\det A} \qquad (k = 1,\ldots,n).$$

133

This rule is unsuitable for the numerical calculation of the unknowns, because it involves an unnecessary amount of repetitious arithmetical effort.

Problems

1. Let $f(x, y)$ be a bilinear form on the vector spaces E and F. Let L_E be the subspace of E which consists of all those $x \in E$ for which $f(x, y) = 0$ for all $y \in F$ and let $L_F \subseteq F$ be defined in a similar manner. Show that

$$\dim E - \dim F = \dim L_E - \dim L_F.$$

2. Prove that the system of linear equations $A\xi = \beta$ has a solution if and only if β is orthogonal to every solution η of the transposed homogeneous system $A'\eta = 0$, (i.e., $\beta'\eta = 0$).

7.2 Numerical Solution of a System of Linear Equations

The *Gaussian Algorithm* described in this section is the most important method for the numerical solution of systems of linear equations. It can be carried out with the help of the exchange method.

Suppose we are given the system

$$\sum_{k=1}^{n} \alpha_{ik} \xi_k + \beta_i = 0 \qquad (i = 1, \ldots, m). \tag{1}$$

We write it in the form

$$\eta_i = \sum_{k=1}^{n} \alpha_{ik} \xi_k + \beta_i \tau = 0 \quad (i = 1, \ldots, m), \tag{2}$$

where we must remember that we have to put $\eta_i = 0$ $(i = 1, \ldots, m)$ and $\tau = 1$. Corresponding to these equations, there is the exchange tableau

$$
\begin{array}{c|ccc|c|}
 & \xi_1 & \cdots & \xi_n & \tau \\
\hline
\eta_1 = & & & & \\
\vdots & & A & & \beta \\
\vdots & & & & \\
\eta_m = & & & & \\
\hline
\end{array}
\tag{3}
$$

where we can consider the variables ξ_k, τ and η_i as linear functionals on R_{n+1}, i.e., as vectors in R_{n+1}^*. If the system is homogeneous, the τ column of the tableau can be ignored. Now we exchange as many of the η_i as possible with the ξ_k (but not with τ). The result (with a possible renumbering of the variables) is a tableau of the following form.

	η_1 $\cdots$ η_r	ξ_{r+1} $\cdots$ ξ_n	τ
$\xi_1 =$ $\vdots$ $\xi_r =$	B	C	D
$\eta_{r+1} =$ $\vdots$ $\eta_m =$	F	$G = 0$	H

$$(4)$$

where B, C, D, F, G and H are the matrices which are produced by the method. The matrix G consists of all zeros because, otherwise, it would be possible to exchange another ξ_k with an η_i. If $r=m$ or $r=n$, then F, G and H and respectively C and G do not appear.

Now, because the exchange method is reversible, the tableaux (3) and (4) represent equivalent sets of linear relationships between the variables, i.e., each set is a consequence of the other.

Thus a necessary and sufficient condition for the existence of a solution of (1) is that it should be possible to satisfy the relations represented by (4) with $\eta_i = 0$ $(i = 1, \ldots, m)$ and $\tau = 1$ ($\tau = 0$ for the associated homogeneous system). We distinguish the following cases.

1. $r = m$. ($\eta_{r+1}, \ldots, \eta_m$ do not appear on the left-hand side.)

1.1. $m < n$. The general solution is given by

$$\begin{pmatrix} \xi_1 \\ \vdots \\ \xi_m \end{pmatrix} = C \begin{pmatrix} \xi_{m+1} \\ \vdots \\ \xi_n \end{pmatrix} + D. \tag{5}$$

Thus there always exist solutions in this case. We can find them all by giving $\xi_{m+1}, \ldots, \xi_n$ arbitrary values and calculating $\xi_1, \ldots, \xi_m$ from (5). The dimension of the coset of solutions is $n - m$ and, by Theorems 1 and 2, it follows that $m = r(A)$ (see 6.2.7).

1.2. $r = m = n$. In this case we have

$$\begin{pmatrix} \xi_1 \\ \vdots \\ \xi_n \end{pmatrix} = D$$

and there is exactly one solution.

2. $r < m$. There exists a solution if and only if $H = 0$ which, for given α_{ik}, is a condition on the β_i. If this condition is satisfied, then the general solution is given by

$$\begin{pmatrix} \xi_1 \\ \vdots \\ \xi_r \end{pmatrix} = C \begin{pmatrix} \xi_{r+1} \\ \vdots \\ \xi_n \end{pmatrix} + D.$$

In particular, the solution is unique if $r = n$ ($< m$). The reader is advised to compare these results with the general theorems in 7.1.

Since the matrices B and F in (4) have no effect on the solutions, and since further all the pivots are in different columns, it is not necessary for the numerical solutions of the equations to carry each pivotal column into the later calculation.

Example 1

$$\xi_1 + \xi_2 + 2\xi_3 + 3\xi_4 = 1$$
$$\xi_1 \quad - \xi_3 + \xi_4 = 1$$
$$2\xi_1 + \xi_2 + \xi_3 + 4\xi_4 = 2$$

1st Tableau

	ξ_1	ξ_2	ξ_3	ξ_4	τ
$\eta_1 =$	1*	1	2	3	-1
$\eta_2 =$	1	0	-1	1	-1
$\eta_3 =$	2	1	1	4	-2
	*	-1	-2	-3	1

2nd Tableau

	ξ_2	ξ_3	ξ_4	τ
$\xi_1 =$	-1	-2	-3	1
$\eta_2 =$	-1*	-3	-2	0
$\eta_3 =$	-1	-3	-2	0
	*	-3	-2	0

3rd Tableau

	ξ_3	ξ_4	τ
$\xi_1 =$	1	-1	1
$\xi_2 =$	-3	-2	0
$\eta_3 =$	0	0	0**

No further exchange is possible. The system has a solution because of the zero indicated by **. The general solution is

$$\xi_1 = \xi_3 - \xi_4 + 1$$

$$\xi_2 = -3\xi_3 - 2\xi_4.$$

The dimension of the coset of solutions is therefore $2 = n - r(A)$. In fact $n = 4$ and $r(A) = 2$ because the third equation is the sum of the first two and the latter are linearly independent.

The Gaussian Algorithm is obtained from the present form of the exchange method by leaving out the pivotal rows from the subsequent calculation. This is possible because the coefficients in the cellar row of each tableau may be used to express the unknown in the top row which is to be exchanged in terms of the remaining ones. For example, if $\alpha_{i1} \neq 0$, it follows from

$$\eta_i = \sum_{k=1}^{n} \alpha_{ik} \xi_k + \beta_i \tau = 0, \quad \text{that} \quad \xi_1 = \sum_{k=2}^{n} \left(-\frac{\alpha_{ik}}{\alpha_{i1}} \right) \xi_k - \frac{\beta_i}{\alpha_{i1}} \tau.$$

The coefficients on the right-hand side are just the coefficients of the cellar row corresponding to the pivot α_{i1}.

In this form, the calculation for Example 1 appears as follows.

1st Tableau. As before. Using its cellar row, we can deduce that

$$\xi_1 = -\xi_2 - 2\xi_3 - 3\xi_4 + 1.$$

2nd Tableau

	ξ_2	ξ_3	ξ_4	τ
$\eta_2 =$	-1^*	-3	-2	0
$\eta_3 =$	-1	-3	-2	0
$\xi_2 =$	*	-3	-2	0

$$\xi_2 = -3\xi_3 - 2\xi_4$$

3rd Tableau

	ξ_3	ξ_4	τ
$\eta_3 =$	0	0	0

Hence ξ_3 and ξ_4 can take arbitrary values and

$$\xi_2 = -3\xi_3 - 2\xi_4$$
$$\xi_1 = -(-3\xi_3 - 2\xi_4) - 2\xi_3 - 3\xi_4 + 1 = \xi_3 - \xi_4 + 1.$$

Example 2

$$\xi_1 + 2\xi_2 \ - \xi_3 + \xi_4 - 6 = 0$$
$$\xi_1 - \ \xi_2 + 2\xi_3 - \xi_4 - 1 = 0$$
$$2\xi_1 + \ \xi_2 - \ \xi_3 - \xi_4 + 3 = 0$$
$$\xi_1 - \ \xi_2 - \ \xi_3 - \xi_4 + 8 = 0$$

1st Tableau

	ξ_1	ξ_2	ξ_3	ξ_4	τ
$\eta_1 =$	1*	2	−1	1	−6
$\eta_2 =$	1	−1	2	−1	−1
$\eta_3 =$	2	1	−1	−1	3
$\eta_4 =$	1	−1	−1	−1	8
	*	−2	1	−1	6

2nd Tableau

	ξ_1	ξ_3	ξ_4	τ
$\eta_2 =$	−3	3	−2	5
$\eta_3 =$	−3	1*	−3	15
$\eta_4 =$	−3	0	−2	14
$\xi_3 =$	3	*	3	−15

3rd Tableau

	ξ_2	ξ_4	τ
$\eta_2 =$	6	7	-40
$\eta_4 =$	-3	-2^*	14
$\xi_4 =$	$-\frac{3}{2}$	$*$	7

4th Tableau

	ξ_2	τ
$\eta_2 =$	$-\frac{9}{2}*$	9
$\xi_2 =$	$*$	2

It follows in sequence that

$$\xi_2 = 2$$
$$\xi_4 = -\tfrac{3}{2}\xi_2 + 7 = 4$$
$$\xi_3 = 3\xi_2 + 3\xi_4 - 15 = 3$$
$$\xi_1 = -2\xi_2 + \xi_3 - \xi_4 + 6 = 1.$$

Exercises

1. Solve the following systems of linear equations including in each case the associated homogeneous system.

(a)
$$\xi_1 - \xi_2 + 3\xi_3 + \xi_4 + 1 = 0$$
$$\xi_2 + 2\xi_3 - \xi_4 + 2 = 0$$
$$\xi_1 + \xi_2 - 2\xi_3 + 3\xi_4 + 3 = 0$$
$$2\xi_1 - 2\xi_2 + \xi_3 + 2\xi_4 + 4 = 0$$

(b)
$$\xi_1 + 2\xi_2 - 2\xi_3 + \xi_4 - \xi_5 + 1 = 0$$
$$2\xi_1 + 3\xi_2 + \xi_3 - \xi_4 + 2\xi_5 + 1 = 0$$
$$3\xi_1 + 5\xi_2 - \xi_3 + \xi_5 + 1 = 0$$
$$\xi_1 + \xi_2 + 3\xi_3 - 2\xi_4 + 3\xi_5 + 1 = 0$$

(c) $\quad \xi_1 - \xi_2 + 2\xi_3 + 3\xi_4 + \ \xi_5 - 1 = 0$

$\quad\quad 2\xi_1 - \xi_2 - \ \xi_3 - 2\xi_4 + \ \xi_5 - 1 = 0$

$\quad\quad \xi_1 + \xi_2 - 2\xi_3 + \ \xi_4 + 2\xi_5 - 1 = 0$

Solution. (a) Inhomogeneous: $\xi_1 = -5$, $\xi_2 = -\frac{7}{5}$, $\xi_3 = \frac{2}{5}$, $\xi_4 = \frac{7}{5}$

$\quad\quad\quad$ homogeneous: $\xi_1 = \xi_2 = \xi_3 = \xi_4 = 0$.

(b) Inhomogeneous: no solution.

$\quad\quad\quad$ homogeneous: $\xi_1 = -8\xi_3 + 5\xi_4 - 7\xi_5$

$\quad\quad\quad\quad\quad\quad\quad\quad \xi_2 = 5\xi_3 - 3\xi_4 + 4\xi_5.$

(c) Inhomogeneous: $\xi_1 = -2\xi_4 - \frac{3}{2}\xi_5 + 1$

$\quad\quad\quad\quad\quad\quad \xi_2 = -\frac{7}{3}\xi_4 - \frac{3}{2}\xi_5 + \frac{2}{3}$

$\quad\quad\quad\quad\quad\quad \xi_3 = -\frac{7}{3}\xi_4 - \frac{1}{2}\xi_5 + \frac{1}{3}$

$\quad\quad\quad$ homogeneous: $\xi_1 = -2\xi_4 - \frac{3}{2}\xi_5$

$\quad\quad\quad\quad\quad\quad\quad \xi_2 = -\frac{11}{3}\xi_4 - \frac{3}{2}\xi_5$

$\quad\quad\quad\quad\quad\quad\quad \xi_3 = -\frac{7}{3}\xi_4 - \frac{1}{2}\xi_5.$

2. The 4-dimensional vector spaces E and F form a dual pair of spaces with $\langle x, y \rangle = \sum \xi_k \eta_k$. The vectors a, $b \in E$ have components $(1, 2, -1, 3)$ and $(2, 1, 1, -3)$. Let $L = L(a, b) \subseteq E$. Find a basis for the dual subspace $L^\dagger \subseteq F$.

Solution. The vectors with components $(-1, 1, 1, 0)$ and $(3, -3, 0, 1)$ are an example of a basis for $L^\dagger$.

7.3 Positive Solutions of a System of Real Linear Equations

A solution $\xi_1, \ldots, \xi_n$ of the system of *real* linear equations

$$\sum_{k=1}^{n} \alpha_{ik} \xi_k + \beta_i = 0 \quad (i = 1, \ldots, m) \tag{1}$$

is said to be positive if

$$\xi_k \geqslant 0 \quad \text{for} \quad k = 1, \ldots, n. \tag{2}$$

We will now investigate the set of all positive solutions of a system of linear equations.

7.3.1 Homogeneous Systems

If the system (1) is homogeneous, then the positive solutions form a convex cone. We will show that this is actually a convex pyramid.

For any non-trivial positive solution $\xi_1, \ldots, \xi_n$, the quantity $\sigma = \sum\limits_{=1}^{n} \xi_k$
is strictly positive. Dividing the solution by σ, we obtain another solution
in which the sum of the unknowns is 1. Thus it is sufficient to find the solutions which satisfy the further condition

$$\sum_{k=1}^{n} \xi_k = 1. \tag{3}$$

Then all positive solutions can be found by multiplying the ones which
satisfy (3) by an arbitrary real number $\lambda \geqslant 0$.

The vectors $x = (\xi_1, \ldots, \xi_n) \in R_n$ which satisfy the conditions (2) and (3)
form a simplex S of dimension $n-1$ (see Problem 3.4;3). The solutions of
(1) form a subspace L of R_n when the system is homogeneous. By Theorem
3.4;6 the intersection $K = L \cap S$ is either a convex polyhedron or it is empty.
In the latter case, we have nothing more to do—it means that the trivial
solution is the only one which is positive. If $K = K(y_1, \ldots, y_r)$, then the set of
positive solutions is the positive hull $P = P(y_1, \ldots, y_r)$. Thus we have the
following result.

*Theorem 1. The positive solutions of a homogeneous system of linear equations
form a convex pyramid P. Note that it is possible for $P = \{0\}$.*

*Theorem 2. The intersection $Q = L \cap P$ of a subspace L with a convex pyramid
P is a convex pyramid.*

Proof. 1. Suppose $P = P(a_1, \ldots, a_r)$ is a convex pyramid and L is a subspace
of a finite-dimensional real vector space E. We will assume that $L(P) = E$.
This involves no significant restriction because otherwise we could replace
E by $L(P)$.

2. We define a linear mapping f of R_r onto E by

$$z = (\lambda_1, \ldots, \lambda_r) \in R_r \rightarrow f(z) = \sum_{k=1}^{r} \lambda_k a_k \in E.$$

The inverse image $\bar{L} = f^{-1}(L)$ is a subspace of R_r (see Problem 5.1;11). Let
$\bar{P} \subseteq R_r$ be the convex pyramid which is defined by $\lambda_k \geqslant 0$ ($k = 1, \ldots, r$) (see
Example 3.4;6). Then $f(\bar{P}) = P$.

3. By Theorem 1, $\bar{Q} = \bar{L} \cap \bar{P}$ is a convex pyramid in R. (By Theorem
6.2;2, $\bar{L}$ is the solution space of a finite system of linear equations.) Further
$Q = L \cap P = f(\bar{L} \cap \bar{P}) = f(\bar{Q})$ (see Problem 1.3;2). Now, since a linear mapping maps a convex pyramid into a convex pyramid (see Problem 5.1;8),
this completes the proof of the theorem.

Theorem 3. The solutions of a homogeneous system of linear equations, for which $\xi_k \geqslant 0$ for some given values of the index k, form a convex pyramid.

Proof. The set of these solutions is the intersection of the solution space of the system of equations with the convex pyramid defined by $\xi_k \geqslant 0$ (for the given k). Hence, by Theorem 2, it is a convex pyramid.

7.3.2 General Systems

We can make the general system (1) homogeneous by putting

$$\xi_k = \frac{\eta_k}{\eta_0} \qquad (k = 1, \ldots, n) \tag{4}$$

and multiplying (1) by η_0 to obtain

$$\sum_{k=1}^{n} \alpha_{ik} \eta_k + \beta_i \eta_0 = 0 \qquad (i = 1, \ldots, m). \tag{5}$$

The positive solutions of the homogeneous system (5) with $\eta_0 > 0$ give us by (4) the positive solutions of (1). By Theorem 1, we know that the positive solutions of (5) form a convex pyramid in R_{n+1}. Suppose that this is generated by the solutions $y_j = (\eta_{j0}, \ldots, \eta_{jn})$ $(j = 1, \ldots, r)$. Then every positive solution of (5) can be written in the form

$$y = \sum_{j=1}^{r} \lambda_j y_j, \quad \text{where} \quad \lambda_j \geqslant 0 \qquad (j = 1, \ldots, r),$$

i.e.,

$$\eta_k = \sum_{j=1}^{r} \lambda_j \eta_{jk} \qquad (k = 0, \ldots, n). \tag{6}$$

If all $\eta_{j0} = 0$ $(j = 1, \ldots, r)$, then (1) has no positive solution. In the other case, we divide the solutions y_j into two groups according to whether $\eta_{j0} \neq 0$ or $\eta_{j0} = 0$. With possible renumbering, assume that $\eta_{10}, \ldots, \eta_{s0} \neq 0$ and $\eta_{s+1, 0} = \ldots = \eta_{r0} = 0$. (Naturally s can be equal to r, i.e., the second group may not appear.) From (4) and (6), it now follows that the positive solutions of (1) and only these can be represented as follows.

$$\xi_k = \frac{\eta_k}{\eta_0} = \sum_{j=1}^{s} \left(\lambda_j \frac{\eta_{j0}}{\eta_0} \right) \frac{\eta_{jk}}{\eta_{j0}} + \sum_{j=s+1}^{r} \frac{\lambda_j}{\eta_0} \eta_{jk}$$

$$= \sum_{j=1}^{s} \mu_j \xi_{jk} + \sum_{j=s+1}^{r} \mu_j \eta_{jk} \qquad (k = 1, \ldots, n), \tag{7}$$

i.e.,

$$x = \sum_{j=1}^{s} \mu_j x_j + \sum_{j=s+1}^{r} \mu_j z_j, \tag{8}$$

where $\mu_j \geqslant 0$ $(j=1,\ldots,r)$ and $\sum\limits_{j=1}^{s} \mu_j = \sum\limits_{j=1}^{s} \lambda_j \dfrac{\eta_{j0}}{\eta_0} = 1$.

Theorem 4. The set P of positive solutions of the system (1) *is either empty or it is the sum of a convex polyhedron and a convex pyramid.*

Proof. By (8), $P = K(x_1,\ldots,x_s) + P(z_{s+1},\ldots,z_r)$, providing $P \neq \varnothing$. If $s = r$, then P is a convex polyhedron, i.e., the convex pyramid is equal to $\{0\}$.

Theorem 5. Theorem 4 is also valid, when the condition $\xi_k \geqslant 0$ is only required for some given indices k.

Proof. In view of Theorem 3, this proof is analogous to that of Theorem 4.

Problems

1. Show that the sum of a positive solution $\boldsymbol{\xi}$ of $A\boldsymbol{\xi} = \boldsymbol{\beta}$ and a positive solution $\boldsymbol{\eta}$ of $A\boldsymbol{\eta} = 0$ is a positive solution of $A\boldsymbol{\xi} = \boldsymbol{\beta}$.

2. Prove that the system $A\boldsymbol{\xi} = \boldsymbol{\beta}$ has a positive solution if and only if from $\boldsymbol{\eta}'A \geqslant 0$ it follows that $\boldsymbol{\eta}'\boldsymbol{\beta} \geqslant 0$.

7.4 Systems of Linear Inequalities

A system of linear inequalities has the form

$$\sum_{k=1}^{n} \alpha_{ik}\xi_k + \beta_i \geqslant 0 \qquad (i = 1,\ldots,m). \tag{1}$$

7.4.1 Homogeneous Systems

The system (1) is said to be homogeneous, when all the coefficients $\beta_i = 0$ $(i = 1,\ldots,m)$. We introduce m new variables $\eta_1, \ldots, \eta_m$ by defining

$$\eta_i = \sum_{k=1}^{n} \alpha_{ik}\xi_k \qquad (i = 1,\ldots,m).$$

The η_i are referred to as *slack variables*. With their help, the homogeneous system can be written as follows.

$$\sum_{k=1}^{n} \alpha_{ik}\xi_k - \eta_i = 0, \qquad \eta_i \geqslant 0 \qquad (i = 1,\ldots,m). \tag{2}$$

We now move into the vector space R_{m+n} with the vectors $(\xi_1,\ldots,\xi_n,\eta_1,\ldots,\eta_m)$. By Theorem 7.3;3, the solutions of (2) form a convex pyramid Q in R_{m+n}. We define a linear mapping of R_{m+n} onto R_n by $(\xi_1,\ldots,\xi_n,\eta_1,\ldots,\eta_m) \to (\xi_1,\ldots,\xi_n)$. Under this mapping, Q goes into a convex pyramid (see Problem 5.1;8) and we have the following result.

143

Theorem 1. The solutions of a homogeneous system of linear inequalities form a convex pyramid.

7.4.2 General Systems

For the investigation of the general system (1) we again introduce m slack variables $\eta_1, \ldots, \eta_m$ to obtain the following system (3) which, for $\xi_1, \ldots, \xi_n$, is equivalent to (1).

$$\sum_{k=1}^{n} \alpha_{ik}\xi_k - \eta_i + \beta_i = 0, \eta_i \geqslant 0 \qquad (i = 1, \ldots, m). \tag{3}$$

Here we are looking for the solutions $(\xi_1, \ldots, \xi_n, \eta_1, \ldots, \eta_m)$ of the system of equations (3) for which $\eta_i \geqslant 0$ $(i=1, \ldots, m)$. By Theorem 7.3;5, if there exist any solutions at all, then these solutions form the sum $P = P_1 + P_2$ of a convex polyhedron P_1 and a convex pyramid P_2. The transfer from R_{m+n} to R_n is achieved by the same linear mapping as in 7.4.1 and this maps P 1-1 onto the set K of solutions of (1). Hence we have the following theorem.

Theorem 2. The set of solutions of a system of linear inequalities (1) *is either empty or it is the sum* $K = K_1 + K_2$ *of a convex polyhedron* K_1 *and a convex pyramid* K_2. *Note that* K_2 *can be* $\{0\}$, *i.e.,* $K = K_1$.

We must now make several remarks in connection with this Theorem 2.

1. The polyhedron K_1 is not uniquely determined if $K_2 \neq \{0\}$. For instance, we can certainly extend K_1 by introducing any vector from $K \setminus K_1$ as an extra vertex vector. This does not alter K.

Suppose that $x_1, \ldots, x_r$ are the vertex vectors of K_1. We will say that one of these (x_1 say) is *superfluous* if $x_1 \in K(x_2, \ldots, x_r) + K_2$. We can then omit x_1 as a vertex vector (so making K_1 smaller) without changing K. If after the omission of x_1 there is a further superfluous vertex vector, then this can also be omitted, and so on. We may therefore assume that K_1 has no superfluous vertex vectors.

2. The pyramid K_2 consists of the solutions of the homogeneous system associated with (1). For, suppose $z \in K_2$, then, for any $x \in K_1$ and $\lambda > 0$, $x + \lambda z \in K$, or, in components,

$$\sum_{k=1}^{n} \alpha_{ik}(\xi_k + \lambda \zeta_k) + \beta_i$$

$$= \sum_{k=1}^{n} \alpha_{ik}\xi_k + \beta_i + \lambda \sum_{k=1}^{n} \alpha_{ik}\zeta_k \geqslant 0 \qquad (i = 1, \ldots, m). \tag{4}$$

From this, it follows that

$$\sum_{k=1}^{n} \alpha_{ik}\zeta_k \geqslant 0 \qquad (i = 1, \ldots, m). \tag{5}$$

Conversely, if (5) is satisfied and $x \in K_1$, then (4) follows for $\lambda \geqslant 0$. Hence $x + \lambda z \in K$, i.e., $z = (1/\lambda)(-x + k_1) + k_2$, where $k_1 \in K_1$, $k_2 \in K_2$ and are dependent on λ. As $\lambda \to \infty$ the components of $(1/\lambda)(-x + k_1)$ converge to 0 and therefore the components of k_2 converge to those of z. As a simple argument will show, it follows that $z \in K_2$.

3. Consequently, the subspace $L = K_2 \cap (-K_2)$ consists exactly of the solutions of the system of equations

$$\sum_{k=1}^{n} \alpha_{ik} \xi_k = 0 \qquad (i = 1, \ldots, m) \tag{6}$$

and therefore has the dimension $n - r(A)$ (Theorem 7.1;1), where $A = (\alpha_{ik})$.

4. Now, for a given (not superfluous) vertex vector $x_1 \in K_1$, suppose that exactly s of the inequalities are satisfied with the equality sign, i.e., assuming that they are the first s, suppose that

$$\sum_{k=1}^{n} \alpha_{ik} \xi_k + \beta_i = 0 \quad \text{for } i = 1, \ldots, s, \tag{7}$$

$$\sum_{k=1}^{n} \alpha_{ik} \xi_k + \beta_i > 0 \quad \text{for } i = s+1, \ldots, m. \tag{8}$$

Let M be the solution space of the system of equations

$$\sum_{k=1}^{n} \alpha_{ik} \xi_k = 0 \qquad (i = 1, \ldots, s). \tag{9}$$

If $d \in M$, then (7) is satisfied by $x = x_1 \pm \lambda d$ for all λ. Further, there exists a $\lambda_0 > 0$ such that, with $\lambda = \lambda_0$, (8) is also satisfied. Hence $x = x_1 \pm \lambda_0 d$ are solutions of (1). Hence

$$x_1 + \lambda_0 d = k_{11} + k_{21}$$

$$x_1 - \lambda_0 d = k_{12} + k_{22}, \quad \text{where} \quad k_{11}, k_{12} \in K_1 \text{ and } k_{21}, k_{22} \in K_2.$$

By adding these, it follows that

$$x_1 = k_1^* + k_2^*,$$

where $\quad k_1^* = \frac{1}{2}(k_{11} + k_{12}) \in K_1 \quad$ and $\quad k_2^* = \frac{1}{2}(k_{21} + k_{22}) \in K_2$.

But in this k_2^* must be zero because otherwise x_1 would be superfluous. Hence $x_1 = \frac{1}{2}(k_{11} + k_{12})$. Since x_1 is a vertex vector it follows that $k_{11} = k_{12} = x_1$. Hence $\lambda_0 d = k_{21} = -k_{22} \in K_2 \cap (-K_2) = L$. Therefore $d \in L$, and hence $M \subseteq L$. On the other hand, it is clear that $M \supseteq L$ and it follows that the rank of the system of equations (9) is equal to $r(A)$.

Therefore there are $r(A)$ of the s inequalities in (1) which are satisfied by x_1 with the equality sign, such that the corresponding left-hand sides of (9) are linearly independent. We will say, in this case, that the inequalities involved are also linearly independent.

Definition 1. *A solution* $x = (\xi_1, \ldots, \xi_n)$ *of the system* (*1*) *is said to be a 'basic solution' if it satisfies* $r(A)$ *linearly independent inequalities of the system with the equality sign.*

A basic solution is said to be *normal*, if exactly $r(A)$ inequalities are satisfied with the equality sign. A non-normal basic solution is said to be *degenerate*.

This proves the following theorem.

Theorem 3. *The convex polyhedron* K_1 (*Theorem 2*) *can be so chosen that its vertex vectors are basic solutions of the system* (*1*).

Example 1. Let $n = 2$, $r(A) = 2$ and consequently $m \geqslant 2$. If we replace the inequality sign by an equals sign in one of the inequalities we will obtain the equation of a line (assuming that not both of α_{i_1} and α_{i_2} are zero). Then the solutions of the corresponding inequality form one of the two half-planes bounded by the line. Thus K is the intersection of the halfplanes corresponding to the inequalities of (1).

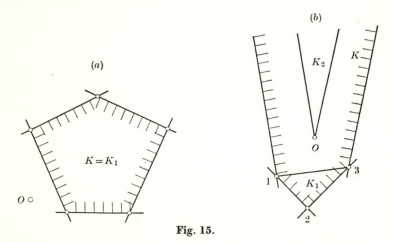

Fig. 15.

In Fig. 15a, $m = 5$ and $K_2 = \{0\}$. (The diagrams and the terminology naturally refer to the space G_2 which is isomorphic to R_2 (Example 2.1;1).) Every vertex vector of $K = K_1$ is a basic solution since it is the intersection of just two of the five lines.

In Fig. 15b, $m = 4$ and K_1 is the convex hull of the three numbered vertex vectors. All three are basic solutions. Here we could extend K_1 by introducing any vector $x \in K \setminus K_1$ as a new vertex vector. However, these would be superfluous and not basic solutions.

Example 2. Let $n=3$. Every inequality of (1) represents a halfspace which is bounded by the plane given by the corresponding equation. Thus K is the intersection of finitely many halfspaces.

1st case. $r(A)=3$ (hence $m \geqslant 3$). In this case, K_1 can be so chosen that each vertex is the unique intersection of three planes.

2nd case. $r(A)=2$ (hence $m \geqslant 2$). In this case there is a set of parallel lines S such that, for $x \in K$, the line of S which passes through x lies entirely in K.

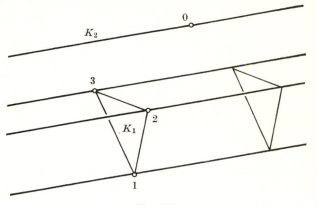

Fig. 16.

In Fig. 16, $m=3$ and K is a 3-sided prism. K_2 is the line of the set S which goes through 0. K_1 can be chosen to be the triangle with the numbered vertices. Every vertex vector lies in $r(A)=$ two faces. The vertex vectors are therefore basic solutions. More generally every point which lies on an edge of the prism K represents a basic solution.

In Fig. 17 ($m=4$), K_2 is a wedge which is bounded by two halfplanes whose common bonding line is the line from the set S which goes through 0. K_1 can again be chosen as a triangle.

3rd case. $r(A)=1$ (hence $m \geqslant 1$). In this case, K is either empty or it is a slice bounded by two parallel planes or a halfspace which is bounded by a plane. Every point of a boundary plane represents a basic solution.

4th case. $r(A)=0$. Now $K=R_3$ if all $\beta_i \geqslant 0$, otherwise $K=\varnothing$.

Example 3. The basic solutions in Figs. 15, 16, 17 were all normal. For instance, if $n=r(A)=3$, then a basic solution is degenerate when it is the intersection of at least four bounding planes. For example, if K is a regular

147

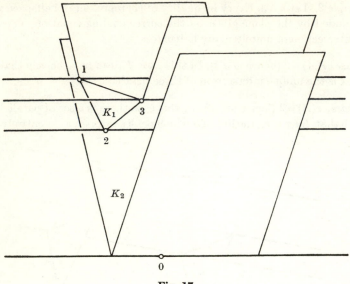

Fig. 17.

octahedron or icosahedron, then all vertices are degenerate basic solutions. On the other hand, if K is a cube, tetrahedron or a regular dodecahedron, then all basic solutions are normal.

CHAPTER 8

LINEAR PROGRAMMING

8.1 Linear Programmes

A *linear programme* is a problem of the following kind. We are given a system of linear inequalities

$$\rho_i = \sum_{k=1}^{n} \alpha_{ik}\xi_k + \beta_i \geqslant 0 \qquad (i = 1,\ldots,m) \tag{1}$$

and a linear functional

$$\theta = \sum_{k=1}^{n} \gamma_k \xi_k \tag{2}$$

which will be referred to as the *object function*. We wish to find those vectors $(\xi_1,\ldots,\xi_n) \in R_n$ which satisfy (1) and for which the object function takes its minimum value. In matrix notation, a linear programme can therefore be formulated as follows

$$\boldsymbol{\rho} = A\boldsymbol{\xi} + \boldsymbol{\beta} \geqslant 0 \tag{3}$$

$$\theta = \boldsymbol{\gamma}'\boldsymbol{\xi} = \min \tag{4}$$

(As before, we write $C \geqslant 0$ if all the elements of the matrix C are non-negative.)

The inequalities (3) are known as the *restrictions* of the programme. A vector $(\xi_1,\ldots,\xi_n) \in R_n$ which satisfies the restrictions is known as an *admissible* solution. An admissible solution which satisfies (4) is known as a *minimal* or *optimal* solution and the corresponding value of the object function is known as the *optimal value*.

The special case in which the restrictions include the inequalities

$$\xi_k \geqslant 0 \qquad (k = 1,\ldots,n), \qquad \text{i.e., } \boldsymbol{\xi} \geqslant 0, \tag{5}$$

is of particular importance. In this case the programme is said to be a *definite* linear programme.

There will also be cases in which the restrictions include only some of the inequalities (5), and then the variables which are not affected by (5) are known as *free variables* (see 8.4).

149

In practical applications, there will also be inequalities with $\leqslant$ in place of $\geqslant$. They may be changed into the latter form simply by changing the sign of both sides. If one of the restrictions is an equation, then it may be replaced by two inequalities ($\alpha = 0$ is replaced by $\alpha \geqslant 0$ and $-\alpha \geqslant 0$), so that we obtain the form (3) again. It is also possible that the object function has to be maximized. Since it is possible to replace $\theta = \max$ by $-\theta = \min$, this case is also covered by (4).

By Theorem 7.4;2, the set K of admissible solutions of a linear programme is either empty or it is the sum of a convex polyhedron K_1 and a convex pyramid K_2. If $K \neq \varnothing$, suppose that $a_1, \ldots, a_r$ are the vertex vectors of the polyhedron K_1 (by Theorem 7.4;3, we may assume that they are basic solutions) and that the pyramid K_2 is generated by $b_1, \ldots, b_s$. Then $x \in R_n$ is an admissible solution if and only if there are coefficients $\lambda_1, \ldots, \lambda_r$, $\mu_1, \ldots, \mu_s$, where $\lambda_k \geqslant 0$ $(k = 1, \ldots, r)$, $\sum_{k=1}^{r} \lambda_k = 1$, $\mu_k \geqslant 0$ $(k = 1, \ldots, s)$, such that

$$x = \sum_{k=1}^{r} \lambda_k a_k + \sum_{k=1}^{s} \mu_k b_k. \tag{6}$$

Denoting the object function by $g(x)$, we have

$$g(x) = \sum_{k=1}^{r} \lambda_k g(a_k) + \sum_{k=1}^{s} \mu_k g(b_k).$$

We distinguish the following three cases.

1. $g(b_k) > 0$ for $k = 1, \ldots, s$.

If x is to be a minimal solution then μ_k must be zero for all $k = 1, \ldots, s$. Let a_{k_0} be a vertex vector for which $g(a_k)$ takes the least value

$$\beta = g(a_{k_0}) = \min_{1 \leqslant k \leqslant r} g(a_k).$$

Then clearly $g(x) \geqslant \beta$ for every admissible solution x and $g(x) = \beta$ for $x = a_{k_0}$. In this case therefore, the basic solution $x = a_{k_0}$ is a minimal solution. Now it is possible that $g(a_k) = \beta$ for more of the vertex vectors a_k. These are also minimal solutions and the set of all minimal solutions is the convex polyhedron generated by these basic solutions.

2. $g(b_k) \geqslant 0$ for $k = 1, \ldots, s$ and, for at least one k, $g(b_k) = 0$.

A similar argument to that of case 1 shows that at least one minimal solution exists. The set of all minimal solutions is the sum of a convex polyhedron and a convex pyramid. The vertex vectors of the polyhedron are found as in case 1, and the pyramid is generated by those b_k for which $g(b_k) = 0$. In this case also there is at least one minimal solution which is a basic solution.

3. There is an index k_0, for which $g(b_{k_0}) < 0$. By choosing μ_{k_0} to be sufficiently large, we can construct admissible solutions for which the object function takes arbitrarily small values. In this case there is no minimal solution.

Collecting these results together, we obtain the following *Main Theorem of Linear Programming*.

Theorem 1. *The set K of admissible solutions of a linear programme is either empty or it is the sum of a convex polyhedron and a convex pyramid. If $K \neq \varnothing$, then one of the following three cases occurs.*

1. *There exists at least one minimal solution. The minimal solutions form a convex polyhedron P whose vertex vectors are basic solutions of the system (3).*

2. *There exists at least one minimal solution. The minimal solutions form the sum of a convex polyhedron P_1 and a convex pyramid $P_2 \neq \{0\}$. The vertex vectors of the polyhedron are basic solutions of (3). For $x \in P_2$, $g(x) = 0$.*

3. *There exist admissible solutions for which the object function takes arbitrarily small values. In this case, there is no minimal solution.*

The following three diagrams illustrate the three cases for $n = 2$. The arrows on the lines $g = $ const. indicate the direction in which the object function increases.

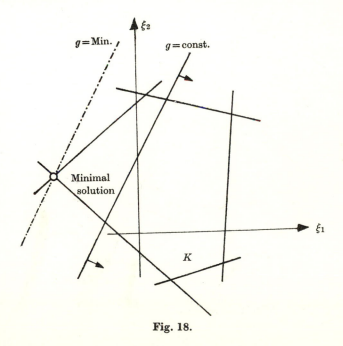

Fig. 18.

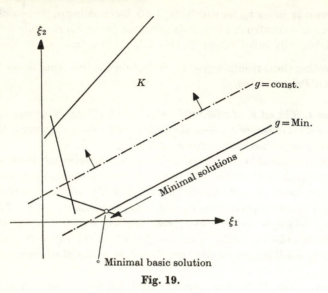

Fig. 19.

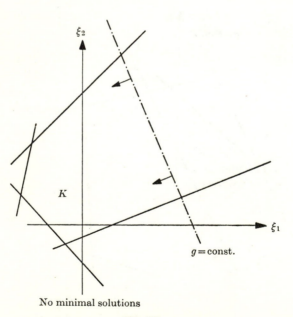

Fig. 20.

Problems

1. Show that the linear programme (3)/(4) has a minimal solution if and only if it has an admissible solution and γ' is a linear combination of the rows of A with non-negative coefficients.

2. Show that, if the rank of A is less than n, then the linear programme (3)/(4) either has no or infinitely many minimal solutions.

8.2 The Duality Law of Linear Programming

Consider the following two definite linear programmes.

$$\left. \begin{array}{c} A\xi \geqslant \beta \\ \xi \geqslant 0 \\ \theta_1 = \gamma'\xi = \min \end{array} \right\} \tag{1}$$

$$\left. \begin{array}{c} \eta' A \leqslant \gamma' \\ \eta' \geqslant 0' \\ \theta_2 = \eta'\beta = \max \end{array} \right\} \tag{2}$$

(2) is known as the *dual programme* of (1). (Note that the sign of β has been changed from that in the previous section.)

It is now natural to ask what is the dual programme of (2)? This is found by transforming (2) into the same form as (1) (i.e., using $\geqslant$ and min) and then constructing the dual in the same way as (2) was constructed from (1). It turns out that in fact this dual is the same programme as (1). In other words, the connection between a programme and its dual is a symmetric relationship.

Written out in full, the dual programmes (1) and (2) appear as follows,

$$\sum_{k=1}^{n} \alpha_{ik}\xi_k \geqslant \beta_i \qquad (i = 1, \ldots, m)$$
$$\xi_k \geqslant 0 \qquad (k = 1, \ldots, n)$$

$$\theta_1 = \sum_{k=1}^{n} \gamma_k \xi_k = \min$$

and

$$\sum_{i=1}^{m} \alpha_{ik}\eta_i \leqslant \gamma_k \qquad (k = 1, \ldots, n)$$
$$\eta_i \geqslant 0 \qquad (i = 1, \ldots, m)$$

$$\theta_2 = \sum_{i=1}^{m} \beta_i \eta_i = \max.$$

153

Theorem 1. If ξ and η' are admissible solutions of (1) and (2) respectively, then $\theta_1 = \gamma'\xi \geqslant \eta'\beta = \theta_2$.

Proof. Multiplying (1) and (2) by η' and ξ respectively, it follows that $\gamma'\xi \geqslant \eta' A\xi \geqslant \eta'\beta$.

Theorem 2. (Duality Theorem)

1. *If (1) has an optimal solution, then so has (2) and vice versa. The optimal values of the two object functions are equal.*

2. *If (1) [(2)] has admissible solutions with arbitrarily small [large] values of the object function, then (2) [(1)] has no admissible solutions.*

3. *If (1) [(2)] has no admissible solutions, then (2) [(1)] either has no admissible solutions or it has admissible solutions with arbitrarily large [small] values of the object function.*

Proof. 1. Suppose (1) has an optimal solution. Let the minimal value of the object function be σ_1. Then, it follows from

$$A\xi - \beta \geqslant 0, \quad \xi \geqslant 0, \quad \text{that} \quad \gamma'\xi - \sigma_1 \geqslant 0. \tag{3}$$

Using this we will now show further that,

if $\quad A\xi - \beta\zeta \geqslant 0, \quad \xi \geqslant 0, \quad \zeta \geqslant 0, \quad \text{then} \quad \gamma'\xi - \sigma_1\zeta \geqslant 0. \tag{4}$

Suppose first that $\zeta > 0$. We divide the conditions of (4) by ζ and obtain the conditions of (3) with $\zeta^{-1}\xi$ in place of ξ. By (3), it follows that

$$\gamma'\zeta^{-1}\xi - \sigma_1 \geqslant 0, \quad \text{i.e.,} \quad \gamma'\xi - \sigma_1\zeta \geqslant 0.$$

Suppose secondly that $\zeta = 0$. We must show that the conditions

$$A\xi_0 \geqslant 0, \quad \xi_0 \geqslant 0, \quad \gamma'\xi_0 < 0 \tag{5}$$

cannot be satisfied simultaneously. Let ξ_1 be an optimal solution of (1), i.e.,

$$A\xi_1 \geqslant \beta, \quad \xi_1 \geqslant 0 \quad \text{and} \quad \gamma'\xi_1 = \sigma_1,$$

then, if there were a ξ_0 satisfying (5), it would follow by addition that

$$A(\xi_0 + \xi_1) \geqslant \beta, \quad (\xi_0 + \xi_1) \geqslant 0 \quad \text{and} \quad \gamma'(\xi_0 + \xi_1) < \sigma_1.$$

This is a contradiction because σ_1 is the minimal value of the object function. This completes the proof of statement (4) and we will now write it in the form

if $\quad \begin{pmatrix} A & -\beta \\ I & 0 \\ 0 & 1 \end{pmatrix} \begin{pmatrix} \xi \\ \zeta \end{pmatrix} \geqslant 0, \quad \text{then} \quad (\gamma', -\sigma_1)\begin{pmatrix} \xi \\ \zeta \end{pmatrix} \geqslant 0.$

(The matrices in this are written in terms of submatrices as shown. In particular I is the $n \times n$ identity matrix.)

By Theorem 6.3;8, there is a matrix $(\boldsymbol{\eta}', \boldsymbol{\eta}_1', \boldsymbol{\eta}_2') \geqslant 0$, such that

$$(\boldsymbol{\eta}', \boldsymbol{\eta}_1', \boldsymbol{\eta}_2') \begin{pmatrix} A - \boldsymbol{\beta} \\ I \quad 0 \\ 0 \quad 1 \end{pmatrix} = (\boldsymbol{\gamma}', -\sigma_1),$$

i.e., $\qquad \boldsymbol{\eta}'A + \boldsymbol{\eta}_1' = \boldsymbol{\gamma}' \quad \text{and} \quad -\boldsymbol{\eta}'\boldsymbol{\beta} + \boldsymbol{\eta}_2' = -\sigma_1,$

i.e., $\qquad \boldsymbol{\eta}'A \leqslant \boldsymbol{\gamma}' \quad \text{and} \quad \boldsymbol{\eta}'\boldsymbol{\beta} \geqslant \sigma_1.$

Thus $\boldsymbol{\eta}'$ is an admissible solution of (2) and the corresponding value of the object function is at least σ_1. The first part of the theorem now follows because, by Theorem 1, the value of the object function of (2) is never greater than σ_1 at any of the admissible solutions of (2). (Using the fact that the relationship between dual programmes is symmetric.)

2. The second part of the Theorem is a direct consequence of Theorem 1.

3. The third part is a consequence of the first. Suppose for instance that (2) has admissible solutions, and suppose that the set of values of the object function on the set K of admissible solutions is bounded above. Then (2) has a maximal solution and therefore, by 1., (1) has a minimal solution.

Finally we see that it is possible for both programmes simultaneously to have no admissible solutions, for example, when $A = 0$, $\boldsymbol{\beta} > 0$ and $\boldsymbol{\gamma}' < 0$.

8.3 The Simplex Method for the Numerical Solution of Linear Programmes

Consider the following definite programme.

$$\left. \begin{aligned} \boldsymbol{\rho} = A\boldsymbol{\xi} + \boldsymbol{\beta} &\geqslant 0 \\ \boldsymbol{\xi} &\geqslant 0 \\ \boldsymbol{\gamma}'\boldsymbol{\xi} &= \min \end{aligned} \right\} \tag{1}$$

Our first task is to find the optimal basic solutions. It turns out that at the same time we will also obtain all the optimal solutions, i.e., those corresponding to section 2 of Theorem 8.1;1.

Because of the restrictions $\boldsymbol{\xi} \geqslant 0$, the rank of the system of inequalities in a definite programme is equal to n. Hence, by Theorem 7.4;3, there are n linearly independent variables in the set $\{\xi_1, \ldots, \xi_n, \rho_1, \ldots, \rho_m\}$ which take the value 0 in an admissible basic solution. The other variables and the object function can be expressed in terms of these linearly independent variables and the variable τ (which is to be set equal to 1) and this gives a tableau of the following kind.

$$
\begin{array}{c|ccc|c}
 & \xi & \rho & \tau \\
\hline
\rho = & & & \beta_1^* \\
 & & A^* & \vdots \\
\xi = & & & \beta_m^* \\
\hline
\theta = & \gamma_1^* & \cdots & \gamma_n^* & \delta^*
\end{array}
\tag{2}
$$

where the top row contains the n vanishing variables and τ, and the left-hand column contains the other variables and the object function θ. Since the tableau corresponds to an admissible solution $\beta_i^* \geqslant 0$ for all $i = 1, \ldots, m$, because these are the values of the variables on the left-hand side for the given basic solution.

Thus every admissible basic solution corresponds to a tableau of the form (2) with $\beta_i^* \geqslant 0$ ($i = 1, \ldots, m$). Conversely, we can construct an admissible basic solution from any tableau of this form by putting the variables in the top row equal to zero and calculating the others from the tableau.

Normal and Degenerate Basic Solutions

Tableau (2) corresponds to a normal basic solution if and only if $\beta_i^* > 0$ for all $i = 1, \ldots, m$. If $\beta_1^* = 0$, say, then, in addition to the n variables in the top row, the variable in the first row also takes the value 0 (and conversely). A tableau which corresponds to a normal (degenerate) basic solution will itself be referred to as normal (degenerate).

Optimal Basic Solutions

If $\gamma_k^* \geqslant 0$ ($k = 1, \ldots, n$), then tableau (2) represents an optimal basic solution and will then be said to be optimal itself. In every other admissible solution, at least one of the variables in the top row will take a strictly positive value. Hence the corresponding value of the object function is not smaller than that for the basic solution (2). If $\gamma_k^* > 0$ for all $k = 1, \ldots, n$, then (2) represents the unique optimal basic solution.

Conversely, if the tableau (2) is normal and optimal, then $\gamma_k^* \geqslant 0$ ($k = 1, \ldots, n$). For, if $\gamma_{k_0}^* < 0$ for some k_0, then, because $\beta_i^* > 0$ ($i = 1, \ldots, m$), we can choose a strictly positive value for the variable at the head of the k_0th column so that the variables on the left-hand side are still positive. In this way, we obtain a new admissible solution, which gives a smaller value of the object function.

On the other hand, some of the γ_k^* can be negative in a degenerate optimal tableau, e.g., when one of the β_i^*, say $\beta_{i_0}^*$, is zero and $\alpha_{i_0k}^* < 0$ $(k=1,\ldots,n)$.

The Existence of Many Optimal Solutions

In a normal optimal tableau (2), if $\gamma_{k_0}^* = 0$ for some index $k = k_0$, then there exist other optimal solutions apart from the one represented by the tableau. Since all the β_i^* are strictly positive, we can choose a strictly positive value for the variable which heads the column of $\gamma_{k_0}^*$ so that the variables on the left-hand side are still positive. The value of the object function is not altered by changing to this new admissible solution.

If the optimal tableau (2) is degenerate, then we can show that there exist further optimal solutions for example if some $\gamma_{k_0}^* = 0$ and if, from $\beta_i^* = 0$, it follows that $\alpha_{ik_0}^* \geqslant 0$ $(i=1,\ldots,m)$.

If $\gamma_{k_0}^* = 0$ and $\alpha_{ik_0}^* \geqslant 0$ $(i=1,\ldots,m)$, then the variable heading the column of $\gamma_{k_0}^*$ can be given an arbitrarily large positive value. Hence we obtain optimal solutions of the form $x = x_1 + x_2$, where x_1 belongs to the convex polyhedron P_1 generated by the optimal basic solutions and x_2 belongs to the convex pyramid P_2 (see Theorem 8.1;1, Case 2). This latter observation is naturally also true for degenerate optimal tableaux.

The *Simplex Method* which is due to G. B. Dantzig enables us to find the optimal tableaux, and hence the optimal solutions, of a linear programme. (See [10] p. 94.) It starts with the following tableau which comes directly from the given programme (1).

$$
\begin{array}{c|ccc|c}
 & \xi_1 & \cdots & \xi_n & \tau \\
\hline
\rho_1 = & & & & \beta_1 \\
\vdots & & A & & \vdots \\
\rho_m = & & & & \beta_m \\
\hline
\theta = & \gamma_1 & \cdots & \gamma_n & 0
\end{array}
\tag{3}
$$

where, for the time being, we will assume that $\beta_i \geqslant 0$ for all $i = 1, \ldots, n$, which means that $\xi_1 = \ldots = \xi_n = 0$ is an admissible solution (see 8.5 for the case when this condition is not satisfied).

The simplex method now enables us to reach the optimal tableaux by applying a finite sequence of exchange steps to (3). In this we attempt to carry out the individual exchange steps according to the following principles.

G1. Every tableau should represent an admissible basic solution, i.e., the coefficients in the τ column should be positive ($\geqslant 0$).

G2. If possible, degenerate tableaux should be avoided, i.e., the co-efficients in the τ column should actually be strictly positive (>0).

G3. If possible, the quantity δ in the bottom right-hand corner (which is equal to the value of the object function for the basic solution in question) should be strictly decreased by each exchange step and in any case should never be increased.

For the sake of brevity we will denote the coefficients of any of the tableaux in the sequence by α_{ik}, β_i, γ_k, δ. Let $\alpha_{i_0 k_0}$ be the pivot in this tableau for the exchange step which is to be carried out.

We distinguish the following three cases,

1. Normal tableaux which are not optimal,

2. Degenerate tableaux which are not optimal, and

3. Optimal tableaux.

Case 1. The tableau is normal but not optimal, i.e., $\beta_i > 0$ ($i=1,\ldots,m$) and not all $\gamma_k \geqslant 0$. Then

1. β_{i_0} goes into $-\beta_{i_0}/\alpha_{i_0 k_0}$ which should be $\geqslant 0$ and if possible >0. Because $\beta_{i_0} > 0$, it follows that $\alpha_{i_0 k_0}$ should be negative and then G2 will be satisfied for the new β_{i_0}.

2. δ goes into $\delta - \beta_{i_0}\gamma_{k_0}/\alpha_{i_0 k_0}$ which should be $\leqslant \delta$ and if possible $< \delta$. If $\alpha_{i_0 k_0} < 0$, it follows that γ_{k_0} should be $\leqslant 0$ and if possible <0.

3. $\beta_i (i \neq i_0)$ goes into $\beta_i - \beta_{i_0}\alpha_{ik_0}/\alpha_{i_0 k_0}$ which should be $\geqslant 0$ and if possible >0. This is automatically satisfied if $\alpha_{ik_0} \geqslant 0$. However, if $\alpha_{ik_0} < 0$, then we require $\beta_i/\alpha_{ik_0} \leqslant \beta_{i_0}/\alpha_{i_0 k_0}$, i.e. we determine the pivotal row i_0 from the pivotal column k_0 as follows. For each of the $i=1,\ldots,m$ where $\alpha_{ik_0} < 0$, we construct the *characteristic quotient* $\chi_i = \beta_i/\alpha_{ik_0}$. We then choose i_0 so that $\chi_{i_0} = \max_i \chi_i$. Now G2 can be satisfied if and only if, for $i_1 \neq i_0$, $\chi_{i_1} < \chi_{i_0}$, i.e., if the maximum value of the characteristic quotients is only taken at $i=i_0$.

Thus, in Case 1, we have the following method for the choice of the pivot.

1. We choose the pivotal column so that $\gamma_{k_0} < 0$.

2. We look to see if there is a coefficient in the pivotal column which is strictly negative.

2.1. If all α_{ik_0} are positive, then we can always obtain admissible solutions by giving the variable at the head of the pivotal column an arbitrary positive value, putting the rest of the top row of variables equal to zero and calculating those on the left-hand side from the tableau. Since $\gamma_{k_0} < 0$, the object function will take arbitrarily small values—the programme has admissible solutions but no optimal solution.

2.2. If some of the α_{ik_0} are strictly negative, then we construct the characteristic quotient for each of these and choose the pivotal row so that this has the greatest value.

With this method, G1 and the strong form of G3 will be satisfied. G2 will be satisfied if and only if χ_i takes its maximum value only for $i=i_0$. If it takes this value for other values of i, we can naturally try to avoid a degenerate tableau by looking for another column with $\gamma_k<0$ to be the pivotal column.

It is easy to see that the simplex method will lead to an optimal basic solution if there is one and if it is possible to avoid degenerate tableaux. Obviously, for the $m+n$ variables ξ_k and ρ_i, there are only finitely many tableaux. Now, since by the above method the quantity δ strictly decreases in each exchange step (providing no degenerate tableaux appear), no tableau can appear twice. We must therefore reach an optimal tableau after only a finite number of steps.

Example 1.

$$\rho_1 = \quad \xi_1-\xi_2+\ 3 \geqslant 0 \qquad \xi_1 \geqslant 0$$
$$\rho_2 = \quad\quad -\xi_2+\ 5 \geqslant 0 \qquad \xi_2 \geqslant 0$$
$$\rho_3 = -2\xi_1-\xi_2+20 \geqslant 0$$
$$\theta = -\ \xi_1-\xi_2 = \min$$

1st Tableau

	ξ_1	ξ_2	τ	χ
$\rho_1 =$	1	-1^*	3	-3
$\rho_2 =$	0	-1	5	-5
$\rho_3 =$	-2	-1	20	-20
$\theta =$	-1	-1	0	

Since both coefficients in the θ row are negative, we can choose either of the columns to be the pivotal column. We will use the second one. From the characteristic quotients written on the right-hand side, we see that the top one is greater than the other two. Thus $\alpha_{12}=-1$ is the pivot.

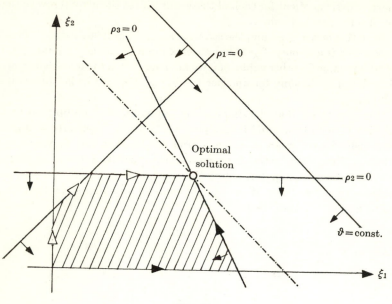

Fig. 21.

2nd Tableau

	ξ_1	ρ_1	τ	χ
$\xi_2 =$	1	-1	3	$*$
$\rho_2 =$	-1^*	1	2	-2
$\rho_3 =$	-3	1	17	$-\frac{17}{3}$
$\theta =$	-2	1	-3	

This corresponds to the basic solution $\xi_1 = 0$, $\xi_2 = 3$. The corresponding value of the object function is -3. Only the first column can be considered as a pivotal column. In view of the characteristic quotients $\alpha_{21} = -1$ is the pivot.

3rd Tableau

	ρ_2	ρ_1	τ
$\xi_2 =$	-1	0	5
$\xi_1 =$	-1	1	2
$\rho_3 =$	3	-2^*	11
$\theta =$	2	-1	-7

This corresponds to the basic solution $\xi_1 = 2$, $\xi_2 = 5$. θ has the value -7. The pivot must be chosen in the second column. Since the latter contains only one negative coefficient $\alpha_{32} = -2$, this must be the pivot.

4th Tableau

	ρ_2	ρ_3	τ
$\xi_2 =$	-1	0	5
$\xi_1 =$	$\frac{1}{2}$	$-\frac{1}{2}$	$\frac{15}{2}$
$\rho_1 =$	$\frac{3}{2}$	$-\frac{1}{2}$	$\frac{11}{2}$
$\theta =$	$\frac{1}{2}$	$\frac{1}{2}$	$-\frac{25}{2}$

The first two coefficients in the θ row are both positive and therefore this tableau gives an optimal solution, which is $\xi_1 = \frac{15}{2}$, $\xi_2 = 5$. The value of the object function is $-\frac{25}{2}$. The solution is not degenerate because $\beta_i > 0$ for all $i = 1, 2, 3$. It is the only optimal solution because $\gamma_k > 0$ for all $k = 1, 2$ (this can also be seen from Fig. 21).

If we had chosen the first column of the first tableau to be the pivotal column, then we would have passed through the basic solution $(10,0)$ to reach the optimal solution, i.e., in two instead of three steps (see Exercise 1). The two sequences of basic solutions which correspond to these two possible routes are marked by arrows in Fig. 21.

Example 2.

$$-11\xi_1+ \quad \xi_2- \quad \xi_3+110 \geqslant 0 \qquad \xi_1 \geqslant 0$$
$$\xi_1-11\xi_2- \quad \xi_3+110 \geqslant 0 \qquad \xi_2 \geqslant 0$$
$$-\xi_1 \qquad\quad -10\xi_3+110 \geqslant 0 \qquad \xi_3 \geqslant 0$$
$$-\xi_1- \quad \xi_2- \quad \xi_3+ \quad 25 \geqslant 0$$
$$\theta = -10\xi_1-10\xi_2-11\xi_3 = \min$$

1st Tableau

	ξ_1	ξ_2	ξ_3	τ
$\rho_1 =$	-11	1	-1	110
$\rho_2 =$	1	-11	-1	110
$\rho_3 =$	-1	0	-10^*	110
$\rho_4 =$	-1	-1	-1	25
$\theta =$	-10	-10	-11	0

2nd Tableau

	ξ_1	ξ_2	ρ_3	τ
$\rho_1 =$	$-\frac{109}{10}$	1	$\frac{1}{10}$	99
$\rho_2 =$	$\frac{11}{10}$	-11^*	$\frac{1}{10}$	99
$\xi_3 =$	$-\frac{1}{10}$	-1	$\frac{1}{10}$	11
$\rho_4 =$	$-\frac{9}{10}$	-1	$\frac{1}{10}$	14
$\theta =$	$-\frac{89}{10}$	-10	$\frac{11}{10}$	-121

3rd Tableau

	ξ_1	ρ_2	ρ_3	τ
$\rho_1 =$	$-\frac{108}{10}$	$-\frac{1}{11}$	$\frac{12}{110}$	108
$\xi_2 =$	$\frac{1}{10}$	$-\frac{1}{11}$	$\frac{1}{110}$	9
$\xi_3 =$	$-\frac{1}{10}$	0	$-\frac{1}{10}$	11
$\rho_4 =$	-1^*	$\frac{1}{11}$	$\frac{1}{11}$	5
$\theta =$	$-\frac{99}{10}$	$\frac{10}{11}$	$\frac{111}{110}$	-211

4th Tableau

	ρ_4	ρ_2	ρ_3	τ
$\rho_1 =$	$\frac{108}{10}$	$-\frac{118}{110}$	$-\frac{96}{110}$	54
$\xi_2 =$	$-\frac{1}{10}$	$-\frac{9}{110}$	$\frac{2}{110}$	$\frac{19}{2}$
$\xi_3 =$	$\frac{1}{10}$	$-\frac{1}{110}$	$-\frac{12}{110}$	$\frac{21}{2}$
$\xi_1 =$	-1	$\frac{1}{11}$	$\frac{1}{11}$	5
$\theta =$	$\frac{99}{10}$	$\frac{1}{110}$	$\frac{12}{110}$	$-\frac{521}{2}$

The fourth tableau contains the minimal solution $\xi_1 = 5$, $\xi_2 = \frac{19}{2}$, $\xi_2 = \frac{21}{2}$. The value of the object function is $-\frac{521}{2}$. The solution is not degenerate and it is the only minimal solution.

Case 2. The tableau is degenerate but not optimal. Thus $\beta_i \geqslant 0$ $(i = 1, \ldots, m)$, $\beta_{i_0} = 0$ for at least one i_0 and not all the γ_k are positive.

The method in this case is the same as in Case 1. However we must pay particular attention to the possibility that the coefficient β_{i_0} in the pivotal row is zero. Then two coefficients will be exchanged which both have the value zero in the basic solution represented by the tableau. In this case, the next tableau still represents the same basic solution. Correspondingly, the strong form of G3 will naturally not be satisfied. Such an exchange step is said to be *stationary*.

Because of this, we try to choose the pivotal column k_0 and the pivotal row i_0 so that the next β_{i_0} is strictly positive. If this is not possible, then the next exchange will be stationary again. In particular it can happen that, after a finite number of stationary steps, we will return to an earlier tableau. In this case, the method becomes cyclic and does not lead to the minimum. Such examples can actually be constructed. In practice, however, this danger is hardly significant.

Example 3.

$$
\begin{aligned}
\rho_1 &= & -\xi_2 + \xi_3 + 1 &\geqslant 0 & \xi_1 &\geqslant 0 \\
\rho_2 &= -\xi_1 & +\xi_3 + 1 &\geqslant 0 & \xi_2 &\geqslant 0 \\
\rho_3 &= & -\xi_2 - \xi_3 + 3 &\geqslant 0 & \xi_3 &\geqslant 0 \\
\rho_4 &= -\xi_1 & -\xi_3 + 3 &\geqslant 0 \\
\rho_5 &= \xi_1 & -\xi_3 + 1 &\geqslant 0 \\
\rho_6 &= & \xi_2 - \xi_3 + 1 &\geqslant 0
\end{aligned}
$$

$$\theta = -\xi_3 = \min$$

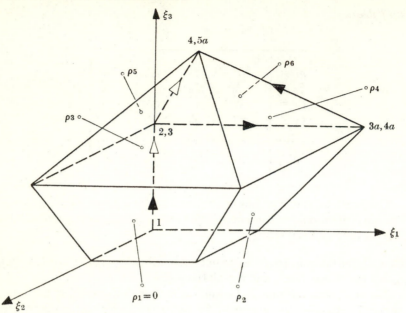

Fig. 22.

1st Tableau

	ξ_1	ξ_2	ξ_3	1
$\rho_1 =$	0	-1	1	1
$\rho_2 =$	-1	0	1	1
$\rho_3 =$	0	-1	-1	3
$\rho_4 =$	-1	0	-1	3
$\rho_5 =$	1	0	-1	1
$\rho_6 =$	0	1	-1^*	1
$\theta =$	0	0	-1	0

2nd Tableau

	ξ_1	ξ_2	ρ_6	1
$\rho_1 =$	0	0	-1	2
$\rho_2 =$	-1	1	-1	2
$\rho_3 =$	0	-2	1	2
$\rho_4 =$	-1	-1	1	2
$\rho_5 =$	1	-1^*	1	0
$\xi_3 =$	0	1	-1	1
$\theta =$	0	-1	1	-1

The second tableau is degenerate. In view of the rule $\gamma_{k_0} < 0$, only $\alpha_{52} = -1$ can be considered as a pivot. The next tableau still represents the same solution.

164

3rd Tableau

	ξ_1	ρ_5	ρ_6	τ
$\rho_1 =$	0	0	-1	2
$\rho_2 =$	0	-1	0	2
$\rho_3 =$	-2^*	2	-1	2
$\rho_4 =$	-2	1	0	2
$\xi_2 =$	1	-1	1	0
$\xi_3 =$	1	-1	0	1
$\theta =$	-1	1	0	-1

4th Tableau

	ρ_3	ρ_5	ρ_6	τ
$\rho_1 =$	0	0	1	2
$\rho_3 =$	0	-1	0	2
$\xi_1 =$	$-\frac{1}{2}$	1	$-\frac{1}{2}$	1
$\rho_4 =$	1	-1	1	0
$\xi_2 =$	$-\frac{1}{2}$	0	$\frac{1}{2}$	1
$\xi_3 =$	$-\frac{1}{2}$	0	$-\frac{1}{2}$	2
$\theta =$	$\frac{1}{2}$	0	$\frac{1}{2}$	-2

The fourth tableau gives the minimal solution $(1,1,2)$ with $\theta = -2$. This solution is degenerate because $\rho_3 = \rho_4 = \rho_5 = \rho_6 = 0$. Even though $\gamma_2 = 0$, there are no other optimal solutions because, if $\theta = -2$, then $\frac{1}{2}(\rho_3 + \rho_6) = 0$ and hence $\rho_3 = \rho_6 = 0$ and $\rho_4 = -\rho_5 = 0$.

If we drop the principle that $\gamma_{k_0} < 0$ and try to choose a pivot in the first column of the second tableau, say $\alpha_{21} = -1$, then we have

Tableau 3a

	ρ_2	ξ_2	ρ_6	τ
$\rho_1 =$	0	0	-1	2
$\xi_1 =$	-1	1	-1	2
$\rho_3 =$	0	-2	1	2
$\rho_4 =$	1	-2^*	2	0
$\rho_5 =$	-1	0	0	2
$\xi_3 =$	0	1	-1	1
$\theta =$	0	-1	1	-1

We come to a new degenerate solution, but without decreasing the object function.

Tableau 4a

	ρ_2	ρ_4	ρ_6	τ
$\rho_1 =$	0	0	-1	2
$\xi_1 =$	$-\frac{1}{2}$	$-\frac{1}{2}$	0	2
$\rho_3 =$	-1	1	-1	2
$\xi_2 =$	$\frac{1}{2}$	$-\frac{1}{2}$	1	0
$\rho_5 =$	-1^*	0	0	2
$\xi_3 =$	$\frac{1}{2}$	$-\frac{1}{2}$	0	1
$\theta =$	$-\frac{1}{2}$	$\frac{1}{2}$	0	-1

This tableau represents the same solution as 3a. The pivot has to be chosen in the first column and there are two possibilities for this (see Exercise 2).

Tableau 5a

	ρ_5	ρ_4	ρ_6	τ
$\rho_1 =$	0	0	-1	2
$\xi_1 =$	$\frac{1}{2}$	$-\frac{1}{2}$	0	1
$\rho_3 =$	-1	1	-1	0
$\xi_2 =$	$-\frac{1}{2}$	$-\frac{1}{2}$	1	1
$\rho_2 =$	-1	0	0	2
$\xi_3 =$	$-\frac{1}{2}$	$-\frac{1}{2}$	0	2
$\theta =$	$\frac{1}{2}$	$\frac{1}{2}$	0	-2

This tableau again represents the optimal solution. The two routes to the solution are again marked by arrows in Fig. 22.

Case 3. The tableau is optimal. We restrict the discussion to normal tableaux, i.e., we assume that $\beta_i > 0$ for all $i = 1, \ldots, m$. In this case, the condition of being optimal is equivalent to $\gamma_k \geqslant 0$ for all $k = 1, \ldots, n$. Further, we have already seen that the tableau represents the unique optimal solution if and only if $\gamma_k > 0$ for all $k = 1, \ldots, n$. If $\gamma_{k_0} = 0$ for some k_0 and all the corresponding α_{ik_0} are positive, then we can obtain optimal solutions by giving at least one variable an arbitrarily large value (Theorem 8.1;1, part 2). If $\gamma_{k_0} = 0$ and there is an i_0 such that $\alpha_{i_0 k_0} < 0$, then the same method can be applied as in Case 1. This produces a new basic solution which is again optimal because the value of the object function is not changed. We remark, without writing out the proof here, that all optimal basic solutions, and hence all optimal solutions, can be found in this way.

Example 4.

$$\rho_1 = -\xi_1 \qquad\qquad +5 \geqslant 0 \qquad \xi_1 \geqslant 0$$
$$\rho_2 = \qquad -\xi_2 \qquad +4 \geqslant 0 \qquad \xi_2 \geqslant 0$$
$$\rho_3 = \qquad\qquad -\xi_3 +3 \geqslant 0 \qquad \xi_3 \geqslant 0$$
$$\rho_4 = -\xi_1 -\xi_2 -\xi_3 +10 \geqslant 0$$
$$\theta = -\xi_1 -\xi_2 -\xi_3 = \min$$

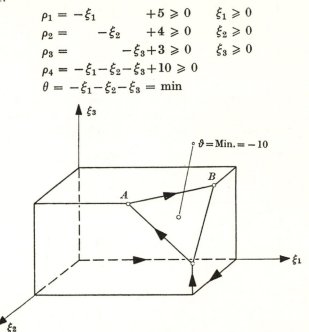

Fig. 23.

1st Tableau

	ξ_1	ξ_2	ξ_3	τ
$\rho_1 =$	-1^*	0	0	5
$\rho_2 =$	0	-1	0	4
$\rho_3 =$	0	0	-1	3
$\rho_4 =$	-1	-1	-1	10
$\theta =$	-1	-1	-1	0

2nd Tableau

	ρ_1	ξ_2	ξ_3	τ
$\xi_1 =$	-1	0	0	5
$\rho_2 =$	0	-1^*	0	4
$\rho_3 =$	0	0	-1	3
$\rho_4 =$	1	-1	-1	5
$\theta =$	1	-1	-1	-5

167

3rd Tableau

	ρ_1	ρ_2	ξ_3	τ
$\xi_1 =$	-1	0	0	5
$\xi_2 =$	0	-1	0	4
$\rho_3 =$	0	0	-1	3
$\rho_4 =$	1	1	-1^*	1
$\theta =$	1	1	-1	-9

4th Tableau

	ρ_1	ρ_2	ρ_4	τ
$\xi_1 =$	-1	0	0	5
$\xi_2 =$	0	-1	0	4
$\rho_3 =$	-1^*	-1	1	2
$\xi_3 =$	1	1	-1	1
$\theta =$	0	0	1	-10

The fourth tableau is optimal and gives the solution $(5, 4, 1)$. The vanishing coefficients γ_1 and γ_2 correspond to two further optimal basic solutions. For the next exchange, we choose the pivot to be in the first column.

5th Tableau

	ρ_3	ρ_2	ρ_4	τ
$\xi_1 =$	1	1	-1	3
$\xi_2 =$	0	-1	0	4
$\rho_1 =$	-1	-1^*	1	2
$\xi_3 =$	-1	0	0	3
$\theta =$	0	0	1	-10

6th Tableau

	ρ_3	ρ_1	ρ_4	τ
$\xi_1 =$	0	-1	0	5
$\xi_2 =$	1	1	1	2
$\rho_2 =$	-1	-1	1	2
$\xi_3 =$	-1	0	0	3
$\theta =$	0	0	1	-10

The last two tableaux correspond to the optimal basic solutions $(3, 4, 3)$ and $(5, 2, 3)$. The set of all optimal solutions is the 2-dimensional simplex which is generated by these three optimal basic solutions (see Fig. 23).

The Geometrical Meaning of the Simplex Method

The successive tableaux in the simplex method give a sequence of basic solutions which correspond to decreasing values of the object function, providing no degenerate tableaux appear. In the examples, we have seen that two successive basic solutions (vertices) are always connected by an edge of the convex polyhedron K_1. This is also the case in general. Because the successive tableaux have $n - 1$ variables in common in their top rows, the set of vectors, for which these variables take the value 0, is the edge which connects the vertices corresponding to the two tableaux. Thus the simplex method gives a sequence of vertices of the set K of admissible solutions and each two successive ones are connected by an edge of K. The value of the object function decreases from vertex to vertex. The last vertex of the sequence is an optimal solution if one exists.

Since we usually have many choices for the pivot, and hence for the next basic solution, it is natural to think of choosing the pivot so that the decrease in the object function is as large as possible. It is easy to make the appropriate adjustments to our rules and these are often used (see, e.g., [10] p. 98.). However, it is clear that although for many programmes this will reach the minimum more quickly, i.e., in fewer steps, there will still be some programmes for which there is no improvement.

Exercises

1. Choose a pivot in the first column of the initial tableau of Example 1.
2. Use the coefficient $\alpha_{31} = -1$ of Tableau 4a in Example 3 as the pivot for the further calculation.
3. Replace the object function in Example 4 by $\theta = -\xi_1 - \xi_2 - 2\xi_3$ and give a geometrical interpretation of the result with reference to Fig. 23.

Solution. There are two optimal solutions, $(3,4,3)$ and $(5,2,3)$, corresponding to the points A and B in Fig. 23.

4. Solve the following linear programmes.

(a)
$$\xi_1 - \xi_2 + 4 \geqslant 0, \qquad \xi_1 \geqslant 0$$
$$-\xi_1 - \xi_2 + 8 \geqslant 0, \qquad \xi_2 \geqslant 0$$
$$-2\xi_1 + 3\xi_2 + 6 \geqslant 0$$
$$\theta = -2\xi_1 + \xi_2 = \min$$

(b)
$$2\xi_1 - 3\xi_2 - 3\xi_3 + 18 \geqslant 0, \qquad \xi_1 \geqslant 0$$
$$-2\xi_1 + \xi_2 \qquad + 10 \geqslant 0, \qquad \xi_2 \geqslant 0$$
$$- \xi_3 + 4 \geqslant 0, \qquad \xi_3 \geqslant 0$$
$$\theta = -\xi_1 - \xi_2 - 3\xi_3 = \min$$

(c)
$$4\xi_1 - 3\xi_2 - \xi_3 - 2\xi_4 + 54 \geqslant 0, \qquad \xi_1 \geqslant 0$$
$$- \xi_1 \qquad\qquad + 18 \geqslant 0, \qquad \xi_2 \geqslant 0$$
$$\xi_1 - 3\xi_2 \qquad + 40 \geqslant 0, \qquad \xi_3 \geqslant 0$$
$$- \xi_2 - \xi_3 \qquad + 30 \geqslant 0, \qquad \xi_4 \geqslant 0$$
$$- \xi_3 - 2\xi_4 + 48 \geqslant 0$$
$$\theta = \xi_1 - \xi_2 - \xi_3 - \xi_4 = \min$$

Solutions. (a) $\xi_1 = 6$, $\xi_2 = 2$, $\theta = -10$

   (b) $\xi_1 = 9$, $\xi_2 = 8$, $\xi_3 = 4$, $\theta = -29$

   (c) $\xi_1 = 0$, $\xi_2 = 2$, $\xi_3 = 28$, $\xi_4 = 10$, $\theta = -40$.

8.4   The Treatment of Free Variables

In 8.3 we assumed that the programme was definite, i.e., that there were no free variables. We will now describe a variation of the simplex method which allows us to eliminate any free variables at the beginning of the calculation. We will restrict the discussion to normal tableaux, but it will not be difficult to see how we would deal with degenerate tableaux. Further, we will still assume that the zero-vector is an admissible solution, i.e., $\beta_i \geqslant 0$ for all $i = 1, \ldots, m$ (see 8.5 for the case when this is not so).

The free variables are initially in the top row of the tableau and, since we eliminate them (i.e., we do not carry them into the rest of the calculation when they have been exchanged with other variables), it follows that free variables will only ever appear in the top rows of tableaux. As before, we will keep to the principles G1, G2 and G3 in 8.3. Then, as a consequence of G1, it follows that all the coefficients in the τ-column of any of the tableaux must be positive (leaving out the row corresponding to the free variable which has been exchanged (see Example 1)).

Now we will assume that ξ_1 is free and therefore look for a pivot in the first column.

1. If $\alpha_{i1} = 0$ for all i, then we can always obtain admissible solutions by giving ξ_1 an arbitrary real value.

1.1. $\gamma_1 = 0$. Then θ is independent of ξ_1. Hence we need not consider ξ_1 at all in the solution.

1.2. $\gamma_1 \neq 0$. There exists no minimal solution.

2. Not all $\alpha_{i1} = 0$. Let $\alpha_{i_0 1} \neq 0$ be a possible pivot.

2.1. δ goes into $\delta - \beta_{i_0} \gamma_1 / \alpha_{i_0 1}$ which, to agree with G3 should be $\leqslant \delta$ and if possible $< \delta$.

2.1.1. $\gamma_1 < 0$. It follows that $\alpha_{i_0 1}$ should be < 0 and it is then possible to satisfy G3 in the strong form. (If all α_{i1} are positive, then there exists no minimal solution.)

2.1.2. $\gamma_1 > 0$. It follows that $\alpha_{i_0 1}$ should be > 0 and it is then possible to satisfy G3 in the strong form. (If all α_{i1} are negative, then there is no minimal solution.)

2.1.3. $\gamma_1 = 0$. It is only possible to satisfy G3 in the weak form.

2.2. For $i \neq i_0$, β_i goes into $\beta_i - \beta_{i_0} \alpha_{i1} / \alpha_{i_0 1}$ which, to agree with G1, should be $\geqslant 0$ and if possible > 0 to agree with G2.

2.2.1. $\gamma_1 < 0$. Then $\alpha_{i_0 1} < 0$ and G2 is satisfied whenever $\alpha_{i1} > 0$. If $\alpha_{i1} < 0$, then as before $\beta_i/\alpha_{i1} \leqslant \beta_{i_0}/\alpha_{i_0\,1}$ and so we choose i_0 such that $\chi_{i_0} = \max\limits_{i,\,\alpha_{i1}<0} \chi_i$

where $\chi_i = \beta_i/\alpha_{i1}$.

2.2.2. $\gamma_1 > 0$. Similarly, we choose i_0 such that $\chi_{i_0} = \min\limits_{i,\,\alpha_{i1}>0} \chi_i$.

2.2.3. $\gamma_1 = 0$. This case can be dealt with either as in 2.2.1 or 2.2.2.

Thus we have the following rule for choosing the pivot.

If $\gamma_1 > 0$ (< 0), then we choose the pivot from among the positive (negative) coefficients of the first column. The one which is actually chosen is decided by calculating the characteristic quotients $\chi_i = \beta_i/\alpha_{i1}$ for each of the positive (negative) coefficients α_{i1} and finding the index i_0 such that

$$\chi_{i_0} = \min\limits_{\alpha_{i1}>0} \chi_i \left(\max\limits_{i,\,\alpha_{i1}<0} \chi_i \right).$$ If $\gamma_1 = 0$, we can consider it to be either positive or negative. However, in this case, we do not decrease the object function. If all $\alpha_{i1} \leqslant 0 (\geqslant 0)$, then there exists no minimal solution.

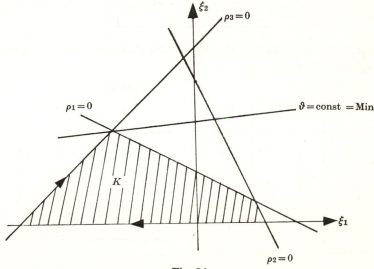

Fig. 24.

When the free variable ξ_1 has been exchanged onto the left-hand side of the tableau, the corresponding row will not be carried into the rest of the calculation and ξ_1 will only be calculated at the end. When all the free variables are eliminated, we continue with the usual simplex method as in 8.3. However, it is possible that not all of the free variables ξ_k can be exchanged. In this case either there exists no minimal solution or the free variables which are not exchanged take no part in the solution of the problem.

Example 1.

$$\rho_1 = -\ \xi_1 - 2\xi_2 + 7 \geqslant 0 \qquad \xi_1 \text{ free}$$
$$\rho_2 = -2\xi_1 - \ \xi_2 + 10 \geqslant 0 \qquad \xi_2 \geqslant 0$$
$$\rho_3 = \ \xi_1 - \ \xi_2 + 12 \geqslant 0$$
$$\theta = \ \xi_1 - 10\xi_2 = \min$$

1st Tableau

	ξ_1	ξ_2	1
$\rho_1 =$	-1	-2	7
$\rho_2 =$	-2	-1	10
$\rho_3 =$	1^*	-1	12
$\theta =$	1	-10	0

First, we want to eliminate the free variable ξ_1. Since $\gamma_1 > 0$, $\alpha_{31} = 1$ must be chosen as the pivot.

2nd Tableau

	ρ_3	ξ_2	1
$\rho_1 =$	-1	-3^*	19
$\rho_2 =$	-2	-3	34
$\xi_1 =$	1	1	-12
$\theta =$	1	-9	-12

3rd Tableau

	ρ_3	ρ_1	1
$\xi_2 =$	$-\frac{1}{3}$	$-\frac{1}{3}$	$\frac{19}{3}$
$\rho_2 =$	-1	1	15
$\theta =$	4	3	-69

From the third tableau, it follows that, for the optimal solution, $\xi_2 = \frac{19}{3}$, $\rho_1 = \rho_3 = 0$ and hence, from the second tableau, that $\xi_1 = \frac{19}{3} - 12 = -\frac{17}{3}$. The optimal value of the object function is -69.

8.5   General Linear Programmes

So far we have assumed in the simplex method that the zero vector is an admissible solution. We will now show how it is possible to deal with the problem when this condition is not satisfied. (Another possible method, which, however, is not applicable in all cases, will be described in 8.6.)

We start by considering the linear programme

$$A\xi+\beta \geqslant 0; \quad \begin{array}{l} \text{some of the } \xi_k \text{ are free} \\ \text{the others are } \geqslant 0 \end{array} \left.\vphantom{\begin{array}{l} a \\ a \end{array}}\right\} \tag{1}$$
$$\theta_1 = \theta_0+\omega\theta_2 = \gamma_0'\xi+\omega\gamma_2'\xi = \min$$

where ω is a constant. At the same time, we consider the programme (2) which has the same restrictions as (1) and the object function

$$\theta_2 = \gamma_2'\xi = \min. \tag{2}$$

Theorem 1. There exists a real number ω_0, such that for all $\omega > \omega_0$, every optimal solution of (1) is also an optimal solution of (2).

This statement of Theorem 1 corresponds to the intuitive idea that, for a large positive ω, in order to make θ_1 small we must first minimize θ_2 and then θ_0.

Proof. 1. An admissible solution

$$x = \sum_{k=1}^{r} \lambda_k a_k + \sum_{k=1}^{s} \mu_k b_k \tag{3}$$

(see 8.1;(6)) is optimal for an object function $\theta(x)$ if and only if

from $\lambda_k > 0$, it follows that a_k is an optimal solution $\quad (k = 1,\ldots,r)$

and from $\mu_k > 0$, it follows that $\theta(b_k) = 0 \quad (k = 1,\ldots,s)$.

2. Let K be the set of admissible solutions of the programmes (1) and (2) and let $x \in K$ be an admissible solution which is not minimal for θ_2. Then, for the representation (3) of x, either

there exists k_0, such that $\lambda_{k_0} > 0$ and a_{k_0} is not minimal for θ_2,

or

there exists k_0, such that $\mu_{k_0} > 0$ and $\theta_2(b_{k_0}) \neq 0$.

In the first case, there exists a $c_{k_0} \in K$ such that $\theta_2(c_{k_0}) < \theta_2(a_{k_0})$. From this, it follows that

$$\theta_1(c_{k_0}) = \theta_0(c_{k_0})+\omega\theta_2(c_{k_0}) < \theta_0(a_{k_0})+\omega\theta_2(a_{k_0}) = \theta_1(a_{k_0}),$$

$$\text{whenever} \quad \omega > \frac{\theta_0(c_{k_0}) - \theta_0(a_{k_0})}{\theta_2(a_{k_0}) - \theta_2(c_{k_0})}.$$

If ω satisfies this condition, then a_{k_0}, and hence x, is not optimal for θ_1.

In the second case

$$\theta_1(b_{k_0}) = \theta_0(b_{k_0}) + \omega\theta_2(b_{k_0})$$

$$\neq 0, \quad \text{when} \quad \omega \neq -\frac{\theta_0(b_{k_0})}{\theta_2(b_{k_0})}.$$

If ω satisfies this condition, then x is not optimal for θ_1.

3. Obviously, Theorem 1 will now be satisfied if we put ω_0 equal to the largest of the numbers

$$\frac{\theta_0(c_k) - \theta_0(a_k)}{\theta_2(a_k) - \theta_2(c_k)} \quad \text{and} \quad -\frac{\theta_0(b_k)}{\theta_2(b_k)},$$

where a_k runs through all the vectors in the set $\{a_1, \ldots, a_r\}$ for which θ_2 is not optimal, c_k satisfies the condition $\theta_2(c_k) < \theta_2(a_k)$ and b_k runs through those vectors in the set $\{b_1, \ldots, b_s\}$ for which $\theta_2(b_k) \neq 0$.

We will first carry out the method for a particular example.

Example 1.

$$\begin{aligned}
\rho_1 &= \xi_1 + 4\xi_2 - 8 \geqslant 0 & \xi_1 \geqslant 0 \\
\rho_2 &= 2\xi_1 + 3\xi_2 - 12 \geqslant 0 & \xi_2 \geqslant 0 \\
\rho_3 &= 2\xi_1 + \xi_2 - 6 \geqslant 0 & \\
\theta &= \xi_1 + \xi_2 = \min &
\end{aligned}\right\} \tag{4}$$

We add $\xi_3 + 12$ to each of the restrictions on the left-hand side, where ξ_3 is a new free variable, and we introduce a new restriction $\xi_3 + 12 \geqslant 0$.

We also replace the object function by $\theta + \omega(\xi_3 + 12) = \xi_1 + \xi_2 + \omega(\xi_3 + 12)$ where ω is a constant whose actual value will not need to be known. In this way we obtain the new programme

$$\begin{aligned}
\rho_1^* &= \xi_1 + 4\xi_2 + \xi_3 + 4 \geqslant 0 & \xi_1 \geqslant 0 \\
\rho_2^* &= 2\xi_1 + 3\xi_2 + \xi_3 \geqslant 0 & \xi_2 \geqslant 0 \\
\rho_3^* &= 2\xi_1 + \xi_2 + \xi_3 + 6 \geqslant 0 & \xi_3 \text{ free} \\
\rho_4^* &= \xi_3 + 12 \geqslant 0 & \\
\theta_1 &= \xi_1 + \xi_2 + \omega(\xi_3 + 12) = \min &
\end{aligned}\right\} \tag{5}$$

The programme (5) has the zero vector as an admissible solution (all $\beta_i \geqslant 0$) and can therefore be solved by the earlier methods. From Theorem 1, we also know that, for sufficiently large ω, every optimal solution of (5) is also optimal for the object function $\xi_3 + 12$ with the same restrictions. (The constant terms in the object functions obviously do not affect the solution.) Now however the minimum of $\xi_3 + 12$ is zero, providing (4) has any admissible solutions at all. If this is the case, then the solution of (5) will automatically give $\xi_3 + 12 = 0$ and the corresponding values of ξ_1 and ξ_2 will be an optimal solution of (4).

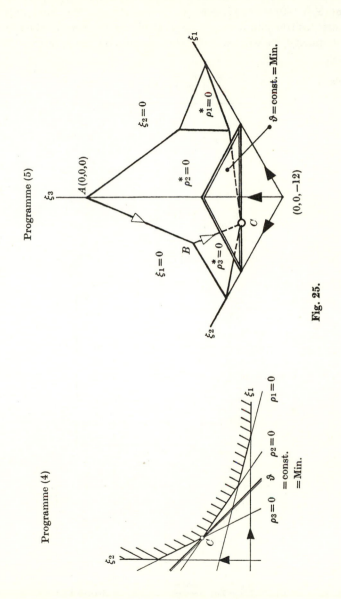

Fig. 25.

In order to carry out the calculation, it is not necessary to know the actual value of ω. We merely assume that ω is large enough to ensure that, for any constants $\alpha, \beta(>0)$ which appear in the calculation, the numbers $\alpha + \beta\omega$ and $\alpha - \beta\omega$ are positive and negative respectively irrespective of the sign of α.

For simplicity, we will leave out the asterisks on the ρ_i in the following tableaux.

1st Tableau

	ξ_1	ξ_2	ξ_3	τ
$\rho_1 =$	1	4	1	4
$\rho_2 =$	2	3	1*	0
$\rho_3 =$	2	1	1	6
$\rho_4 =$	0	0	1	12
$\theta_1 =$	1	1	ω	12ω

First we eliminate the free variable ξ_3.

2nd Tableau

	ξ_1	ξ_2	ρ_2	τ
$\rho_1 =$	-1	1	1	4
$\xi_3 =$	-2	-3	1	0
$\rho_3 =$	0	-2*	1	6
$\rho_4 =$	-2	-3	1	12
$\theta_1 =$	$1-2\omega$	$1-3\omega$	ω	12ω

3rd Tableau

	ξ_1	ρ_3	ρ_2	τ
$\rho_1 =$	-1	$-\frac{1}{2}$	$\frac{3}{2}$	7
$\xi_2 =$	0	$-\frac{1}{2}$	$\frac{1}{2}$	3
$\rho_4 =$	-2*	$\frac{3}{2}$	$-\frac{1}{2}$	3
$\theta_1 =$	$1-2\omega$	$\dfrac{3\omega-1}{2}$	$\dfrac{1-\omega}{3}$	$3(\omega+1)$

4th Tableau

	ρ_4	ρ_3	ρ_2	τ
$\rho_1 =$	$\frac{1}{2}$	$-\frac{5}{4}$	$\frac{7}{4}$	$\frac{11}{2}$
$\xi_2 =$	0	$-\frac{1}{2}$	$\frac{1}{2}$	3
$\xi_1 =$	$-\frac{1}{2}$	$\frac{3}{4}$	$-\frac{1}{4}$	$\frac{3}{2}$
$\theta_1 =$	$\dfrac{2\omega-1}{2}$	$\frac{1}{4}$	$\frac{1}{4}$	$\frac{9}{2}$

The fourth tableau is optimal. From it, we have $\xi_1=\frac{3}{2}$ and $\xi_2=3$. Hence, from the second tableau, we have $\xi_3=-3-9=-12$ and therefore $\xi_3+12=0$. This means that $\xi_1=\frac{3}{2}$, $\xi_2=3$ is an optimal solution of (4) and the minimum value of θ is $\frac{9}{2}$.

The programmes (4) and (5) are represented geometrically in Fig. 25. We note in particular that for (5) the planes $\theta_1=$ constant become steeper as ω decreases. If ω is too small, then the optimum is taken at point B instead of point C which corresponds to the optimum of (4).

This example shows that the method applied here is also applicable in general. This can be described in terms of the following rules for the solution of the general programme

$$
\left.
\begin{aligned}
&\sum_{k=1}^{n} \alpha_{ik}\xi_k+\beta_i \geqslant 0 \qquad (i = 1, \ldots, m)\\
&\text{some of the } \xi_k \text{ are free and some are} \geqslant 0\\
&\theta = \sum_{k=1}^{n} \gamma_k\xi_k = \min
\end{aligned}
\right\}
\tag{6}
$$

1. We construct the new programme

$$
\left.
\begin{aligned}
&\sum_{k=1}^{n} \alpha_{ik}\xi_k+\xi_{n+1}+\delta+\beta_i \geqslant 0 \qquad (i = 1, \ldots, m)\\
&\xi_{n+1}+\delta \geqslant 0\\
&\text{Conditions for } \xi_1, \ldots, \xi_n \text{ as in (6) and } \xi_{n+1} \text{ free}\\
&\theta_1 = \sum_{k=1}^{n} \gamma_k\xi_k+\omega(\xi_{n+1}+\delta) = \min
\end{aligned}
\right\}
\tag{7}
$$

where δ is chosen so that $\delta+\beta_i \geqslant 0$ for all $i=1, \ldots, m$.

2. We solve the programme (7), assuming that ω is so large that, for any constants α and $\beta(>0)$ which appear in the calculation, the numbers $\alpha + \beta\omega$ and $\alpha - \beta\omega$ are positive and negative respectively, irrespective of the sign of α.

3. If, in an optimal solution of (7), $\xi_{n+1} + \delta > 0$, then (6) has no admissible solution. On the other hand, if $\xi_{n+1} + \delta = 0$, then an optimal solution of (7) also gives an optimal solution of (6).

Exercise

Solve the linear programme

$$
\begin{aligned}
\xi_1 - \xi_2 - \ 2 &\geqslant 0, & \xi_1 &\geqslant 0 \\
\xi_1 + \xi_2 - \ 8 &\geqslant 0, & \xi_2 &\geqslant 0 \\
-2\xi_1 - \xi_2 + 20 &\geqslant 0 \\
\theta = 3\xi_1 + \xi_2 &= \min
\end{aligned}
$$

by introducing a new variable.

Solution. $\xi_1 = 5$, $\xi_2 = 3$, $\theta = 18$.

8.6   The Simplex Method and Duality

In this section we will show that the duality law of Linear Programming (Theorem 8.2;2) is contained in the duality law of the Exchange Method (see 3.3.2).

We consider the definite programme

$$
\left.
\begin{aligned}
\rho = A\boldsymbol{\xi} + \boldsymbol{\beta} &\geqslant 0 \\
\boldsymbol{\xi} &\geqslant 0 \\
\theta = \boldsymbol{\gamma}'\boldsymbol{\xi} &= \min
\end{aligned}
\right\}
\tag{1}
$$

where we will assume that $\boldsymbol{\beta} \geqslant 0$.

The dual programme, denoting the variables by η_k and σ_i, is

$$
\left.
\begin{aligned}
\boldsymbol{\sigma}' = -\boldsymbol{\eta}'A + \boldsymbol{\gamma}' &\geqslant 0 \\
\boldsymbol{\eta}' &\geqslant 0 \\
\tau = -\boldsymbol{\eta}'\boldsymbol{\beta} &= \max
\end{aligned}
\right\}
\tag{2}
$$

(Note that here τ now denotes one of the two object functions and is not to be set equal to 1.)

Suppose that the first and final tableaux for (1) are

	ξ_1 ... ξ_n	1
$\rho_1 =$		β_1
$\vdots$	A	$\vdots$
$\rho_m =$		β_m
$\theta =$	γ_1 ... γ_n	0

(3)

	ξ ρ	1
$\rho =$		β_1^*
	A^*	$\vdots$
$\xi =$		β_m^*
$\theta =$	γ_1^* ... γ_n^*	δ

(4)

where $\boldsymbol{\beta}^* \geqslant 0$, $\boldsymbol{\gamma}^{*\prime} \geqslant 0$, $\delta \leqslant 0$.

Now we interpret these two tableaux vertically, and write them as follows.

	$\sigma_1 =$ ... $\sigma_n =$	$\tau =$
$-\eta_1$		β_1
$\vdots$	A	$\vdots$
$-\eta_m$		β_m
1	γ_1 ... γ_n	0

(5)

	$\sigma =$ $\eta =$	$\tau =$
$-\eta$		β_1^*
	A^*	$\vdots$
$-\sigma$		β_m^*
1	γ_1^* ... γ_n^*	δ

(6)

(5) means that $\boldsymbol{\sigma}' = -\boldsymbol{\eta}'A + \boldsymbol{\gamma}'$ and $\tau = -\boldsymbol{\eta}'\boldsymbol{\beta}$.

It follows from tableau (6), which is equivalent to (5), that, when the left-hand variables are made equal to zero, then the variables in the top row take the values γ_k^* and hence they are positive or zero. Also τ takes the value $\delta \leqslant 0$. On the other hand, if the left-hand variables are given arbitrary positive values, then $\tau \leqslant \delta$. It follows that (6), and hence also (4), represents an optimal solution of the dual programme (2) of (1).

Thus the final tableau (4) for the programme (1) also represents an optimal solution of the dual programme (2). In this, the left-hand variables (in the tableau (6)) are made equal to zero, the variables in the top row are equal to γ_k^* and the maximum value of the object function is δ. This last result agrees with the duality law.

Example 1. The dual programme of Example 8.3;1 is

$$
\begin{aligned}
\sigma_1 &= -\eta_1 \qquad +2\eta_3-1 \geqslant 0 \\
\sigma_2 &= \quad \eta_1+\eta_2+\ \eta_3-1 \geqslant 0 \\
\eta_1 &\geqslant 0, \qquad \eta_2 \geqslant 0, \qquad \eta_3 \geqslant 0 \\
\tau &= -3\eta_1-5\eta_2-20\eta_3 = \max
\end{aligned}
$$

Its first tableau in the vertical form is

	$\sigma_1 =$	$\sigma_2 =$	$\tau =$
$-\eta_1$	1	-1	3
$-\eta_2$	0	-1	5
$-\eta_3$	-2	-1	20
1	-1	-1	0

and corresponds to the first tableau of Example 8.3;1. The final tableau can be taken straight from the earlier example without doing the calculation. It is

	$\eta_2 =$	$\eta_3 =$	$\tau =$
$-\sigma_2$	-1	0	5
$-\sigma_1$	$\frac{1}{2}$	$-\frac{1}{2}$	$\frac{15}{2}$
$-\eta_1$	$\frac{3}{2}$	$-\frac{1}{2}$	$\frac{11}{2}$
1	$\frac{1}{2}$	$\frac{1}{2}$	$-\frac{25}{2}$

Thus the corresponding optimal solution of (9) is $\eta_1=0$, $\eta_2=\eta_3=\frac{1}{2}$ and the optimal value of τ is $-\frac{25}{2}$.

In view of these results, it is now possible to solve the programme (1), in the case when not all the β_i are positive but all the γ_k are positive, without introducing an extra variable ξ_{n+1} as in 8.5. We first write the programme (1) in the form

$$\left.\begin{aligned}
\boldsymbol{\sigma}' &= -\boldsymbol{\eta}'(-A') + \boldsymbol{\beta}' \geq 0 \\
\boldsymbol{\eta}' &\geq 0 \\
\tau &= -\boldsymbol{\eta}'\boldsymbol{\gamma} = \max
\end{aligned}\right\} \tag{7}$$

construct the corresponding tableau (8) in the vertical interpretation and then apply the simplex method to this.

	$\sigma_1 =$	$\ldots$	$\sigma_m =$	$\tau =$
$-\eta_1$				γ_1
$\vdots$		$-A'$		$\vdots$
$-\eta_n$				γ_n
1	β_1	$\ldots$	β_m	0

$$(8)$$

Example 2.

$$\begin{aligned}
\sigma_1 &= \eta_1 + 4\eta_2 - 8 \geq 0 \qquad \eta_1 \geq 0 \\
\sigma_2 &= 2\eta_1 + 3\eta_2 - 12 \geq 0 \qquad \eta_2 \geq 0 \\
\sigma_3 &= 2\eta_1 + \eta_2 - 6 \geq 0 \\
-\tau &= \eta_1 + \eta_2 = \min
\end{aligned}$$

i.e.,

$$\tau = -\eta_1 - \eta_2 = \max.$$

This example is identical with Example 8.5;1 which we have already solved by introducing a free variable. Since the coefficients of the function $-\tau$ are positive, we can also solve the programme directly by writing the tableaux in the vertical form.

1st Tableau

	$\sigma_1 =$	$\sigma_2 =$	$\sigma_3 =$	$\tau =$
$-\eta_1$	-1	-2	-2	1
$-\eta_2$	-4^*	-3	-1	1
1	-8	-12	-6	0

2nd Tableau

	$\eta_2 =$	$\sigma_2 =$	$\sigma_3 =$	$\tau =$
$-\eta_1$	$\frac{1}{4}$	$-\frac{5}{4}$	$-\frac{7}{4}^*$	$\frac{3}{4}$
$-\sigma_1$	$-\frac{1}{4}$	$-\frac{3}{4}$	$-\frac{1}{4}$	$\frac{1}{4}$
1	2	-6	-4	-2

3rd Tableau

	$\eta_2 =$	$\sigma_2 =$	$\eta_1 =$	$\tau =$
$-\sigma_3$	$\frac{1}{7}$	$-\frac{5}{7}$	$-\frac{4}{7}$	$\frac{3}{7}$
$-\sigma_1$	$-\frac{2}{7}$	$-\frac{4}{7}*$	$\frac{1}{7}$	$\frac{1}{7}$
1	$\frac{10}{7}$	$-\frac{22}{7}$	$\frac{16}{7}$	$-\frac{26}{7}$

4th Tableau

	$\eta_2 =$	$\sigma_1 =$	$\eta_1 =$	$\tau =$
$-\sigma_3$	$\frac{2}{4}$	$\frac{5}{4}$	$-\frac{3}{4}$	$\frac{1}{4}$
$-\sigma_2$	$-\frac{2}{4}$	$-\frac{7}{4}$	$\frac{1}{4}$	$\frac{1}{4}$
1	3	$\frac{22}{4}$	$\frac{6}{4}$	$-\frac{18}{4}$

From the fourth tableau, we have the optimal solution $\eta_1 = \frac{3}{2}$, $\eta_2 = 3$. The maximum value of $+\tau$ is $-\frac{9}{2}$ and hence the minimal value of $-\tau$ is $+\frac{9}{2}$ which agrees with the earlier result.

Example 3.

$$\rho_1 = \quad \xi_1 + \xi_2 - 1 \geqslant 0$$
$$\rho_2 = \quad\quad -\xi_2 + 1 \geqslant 0$$
$$\theta = 2\xi_1 + \xi_2 = \min.$$

In the horizontal interpretation, this example can only be solved by introducing a free variable ξ_3. However, since the coefficients in θ are positive, we choose to use the tableaux in vertical form where we will denote the variables by ξ, ρ instead of η, σ.

1st Tableau

	$\rho_1 =$	$\rho_2 =$	$-\theta =$
$-\xi_1$	-1	0	2
$-\xi_2$	$-1*$	1	1
1	-1	1	0

2nd Tableau

	$\xi_2 =$	$\rho_2 =$	$-\theta =$
$-\xi_1$	1	-1	1
$-\rho_1$	-1	1	1
1	1	0	-1

The optimal solution is $\xi_1 = 0$, $\xi_2 = 1$ and the minimum value of θ is 1.

Exercises

1. Formulate the dual programmes of the three programmes in Exercise 8.3;4 and solve them.

Solutions. (a) $-\eta_1 + \eta_2 + 2\eta_3 - 2 \geqslant 0,$ $\eta_1 \geqslant 0$
            $\eta_1 + \eta_2 - 3\eta_3 + 1 \geqslant 0,$ $\eta_2 \geqslant 0$
                            $\eta_3 \geqslant 0$

$$\tau = -4\eta_1 - 8\eta_2 - 6\eta_3 = \max$$
$$\eta_1 = 0, \qquad \eta_2 = \tfrac{4}{5}, \qquad \eta_3 = \tfrac{3}{5}; \qquad \tau = -10$$

(b) $-2\eta_1 + 2\eta_2 \quad -1 \geqslant 0,$ $\eta_1 \geqslant 0$
       $3\eta_1 - \eta_2 \quad -1 \geqslant 0,$ $\eta_2 \geqslant 0$
       $3\eta_1 \quad +\eta_3 - 3 \geqslant 0,$ $\eta_3 \geqslant 0$

$$\tau = -18\eta_1 - 10\eta_2 - 4\eta_3 = \max$$
$$\eta_1 = \tfrac{3}{4}, \; \eta_2 = \tfrac{5}{4}, \; \eta_3 = \tfrac{3}{4}; \; \tau = -29$$

(c) $-4\eta_1 + \eta_2 - \eta_3 \qquad\qquad +1 \geqslant 0,$ $\eta_1 \geqslant 0$
       $3\eta_1 \qquad +3\eta_3 + \eta_4 \qquad -1 \geqslant 0,$ $\eta_2 \geqslant 0$
       $\eta_1 \qquad\qquad +\eta_4 + \eta_5 - 1 \geqslant 0,$ $\eta_3 \geqslant 0$
       $2\eta_1 \qquad\qquad\qquad +2\eta_5 - 1 \geqslant 0,$ $\eta_4 \geqslant 0$
                                        $\eta_5 \geqslant 0$

$$\tau = -54\eta_1 - 18\eta_2 - 40\eta_3 - 30\eta_4 - 48\eta_5 = \max$$
$$\eta_1 = \tfrac{1}{6}, \; \eta_2 = 0, \; \eta_3 = 0, \; \eta_4 = \tfrac{1}{2}, \; \eta_5 = \tfrac{1}{3}; \; \tau = -40$$

The solutions could also be read off from the final tableau of Exercise 8.3;4 by using the vertical interpretation.

2. Solve the linear programme of Exercise 8.5, without introducing a new variable, by using the vertical interpretation.

Solution. The initial tableau is

	$\rho_1 =$	$\rho_2 =$	$\rho_3 =$	$\tau =$
$-\xi_1$	-1	-1	2	3
$-\xi_2$	1	-1	1	1
1	-2	-8	20	0

CHAPTER 9

TCHEBYCHEV APPROXIMATIONS

9.1 Tchebychev's Method of Approximation

Consider the following real system of linear equations

$$\sum_{k=1}^{n} \alpha_{ik}\xi_k + \beta_i = 0 \quad (i = 1, \ldots, m), \qquad \text{i.e.,} \quad A\boldsymbol{\xi} + \boldsymbol{\beta} = 0 \qquad (1)$$

and for the moment suppose that (1) is the expression of some physical law, i.e., that the physical quantities $\alpha_{ik}, \beta_i, \xi_k$ are connected by the relationships given in (1). Further suppose that the coefficients α_{ik} and β_i can be measured by experiment and that the quantities ξ_k have then to be calculated using the equations (1). Since experimental measurements are always subject to error, it is possible that the system (1) will be inconsistent and have no solution even though physical considerations will show that there should be a solution. The problem then is to solve the system 'as closely as possible' in a sense which must be more precisely defined.

For arbitrary real values $\xi_1, \ldots, \xi_n$, we will put

$$\epsilon_i = \sum_{k=1}^{n} \alpha_{ik}\xi_k + \beta_i \, (i = 1, \ldots, m), \qquad \text{i.e.,} \quad \boldsymbol{\epsilon} = A\boldsymbol{\xi} + \boldsymbol{\beta}$$

and refer to these quantities as *residuals*. If $\boldsymbol{\xi}$ is a solution of the system, then $\boldsymbol{\epsilon} = 0$. Consequently, when the system cannot be solved exactly, we will try to find values of the unknowns $\xi_1, \ldots, \xi_n$ such that *the greatest of the absolute values of the residuals is as small as possible*. The solution of this problem is known as *Tchebychev's Method of Approximation*.

We remark here that, apart from Tchebychev's Method, there are several other methods of approximating to the solution and in particular there is the *method of least squares* which is due to Gauss and which will be met in 12.2. In fact the method of least squares is more suitable than Tchebychev's method for solving those problems in which the inconsistency of the equations (1) is due to statistical errors in the coefficients, as for example in the sort of physical problem considered at the beginning of this section.

(Cf. [20], p. 124). Tchebychev's method is more useful for the approximation of functions (see Example 4).

Now let σ be the largest of the quantities $|\epsilon_i|$ which we have to minimize. This can also be characterized as the least upper bound of the absolute values $|\epsilon_i|$, so that $|\epsilon_i| \leqslant \sigma$ for all $i = 1, \ldots, m$ and equality holds for at least one value of i. Thus we have to find values of the unknowns ξ_k which make σ as small as possible. Since $|\epsilon_i| \leqslant \sigma$ may be written in the form $\epsilon_i \leqslant \sigma$ and $-\epsilon_i \leqslant \sigma$, we obtain the following linear programme:

$$\left. \begin{array}{c} \epsilon_i = \sum_{k=1}^{n} \alpha_{ik} \xi_k + \beta_i \leqslant \sigma \\[2mm] -\epsilon_i = -\sum_{k=1}^{n} \alpha_{ik} \xi_k - \beta_i \leqslant \sigma \end{array} \right\} \quad i = 1, \ldots, m \\[4mm] \sigma = \min $$

$$\hspace{6cm}(2)$$

which has the free variables $\xi_1, \ldots, \xi_n$ and the object function $\theta = \sigma$ and clearly this could be solved directly using the techniques of Chapter 8. However, since the zero vector is not an admissible solution it pays to reformulate the programme by dividing through by σ and introducing the following notation.

$$\xi_k^* = \frac{\xi_k}{\sigma}, \qquad \xi_0^* = \frac{1}{\sigma}. \hspace{3cm} (3)$$

Then (2) may be rewritten in the form

$$\left. \begin{array}{ll} \rho_{i1} = \sum_{k=1}^{n} \alpha_{ik} \xi_k^* + \beta_i \xi_0^* + 1 \geqslant 0; & i = 1, \ldots, m \\[4mm] \rho_{i2} = -\sum_{k=1}^{n} \alpha_{ik} \xi_k^* - \beta_i \xi_0^* + 1 \geqslant 0; & i = 1, \ldots, m \end{array} \right\} \quad (4) \\[4mm] -\xi_0^* = \min $$

This is again a linear programme with the free variables $\xi_0^*, \xi_1^*, \ldots, \xi_n^*$. It has $\xi_0^* = \xi_1^* = \ldots = \xi_n^* = 0$ as an admissible solution and it is easy to find the simple rules by which it is derived from (1). In particular, we note that

$$\rho_{i1} + \rho_{i2} = 2 \quad \text{for } i = 1, \ldots, m. \hspace{2.5cm} (5)$$

The initial tableau for the simplex method is

	ξ_1^* $\cdots$ ξ_n^*	ξ_0^*	1
$\rho_{11} =$ $\vdots$ $\rho_{m1} =$	A	β_1 $\vdots$ β_m	1 $\vdots$ 1
$\rho_{12} =$ $\vdots$ $\rho_{m2} =$	$-A$	$-\beta_1$ $\vdots$ $-\beta_m$	1 $\vdots$ 1
$\theta =$	0 $\cdots$ 0	-1	0

$$(6)$$

During the calculation, it is helpful to keep the relations (5) in mind, because they must also be satisfied in each later tableau. This means,

1. If for some i, ρ_{i1} and ρ_{i2} are on the left-hand side, then the corresponding rows are identical except that they have opposite signs (apart from the last coefficients whose sum is equal to 2).

2. If ρ_{i1} is on the left-hand side and ρ_{i2} is in the top row, then the row corresponding to ρ_{i1} consists entirely of zeros except that the coefficient under ρ_{i2} is -1 and the last coefficient is 2.

By remembering these two rules, we can save ourselves a considerable amount of work in the calculation.

The final tableau of the method leads to the solution of the approximation problem as follows.

1. The minimal value of the object function is $\theta = -\xi_0^* = -1/\sigma$.
2. The values of the unknowns ξ_k are given by

$$\xi_k = \sigma \xi_k^*, \qquad k = 1, \ldots, n.$$

Example 1. We will first consider a trivial example, viz. the system

$$\xi_1 = 1 \qquad (7)$$
$$\xi_1 = 0$$

with only one unknown which is supposed to satisfy the incompatible equations (7). We see immediately that the Tchebychev method gives the solution $\xi_1 = \frac{1}{2}$. Both residuals then take the absolute value $\frac{1}{2}$, while, for every other value of ξ_1, at least one of them takes a larger absolute value. The linear programme corresponding to the system (7) is

$$\rho_{11} = \xi_1^* - \xi_0^* + 1 \geqslant 0$$
$$\rho_{12} = -\xi_1^* + \xi_0^* + 1 \geqslant 0$$
$$\rho_{21} = \xi_1^* \qquad + 1 \geqslant 0$$
$$\rho_{22} = -\xi_1^* \qquad + 1 \geqslant 0$$

$$\theta = -\xi_0^* = \min$$

1st Tableau

	ξ_1^*	ξ_0^*	1
$\rho_{11} =$	1	-1^*	1
$\rho_{12} =$	-1	1	1
$\rho_{21} =$	1	0	1
$\rho_{22} =$	-1	0	1
$\theta =$	0	-1	0

2nd Tableau (Elimination of ξ_0^*)

	ξ_1^*	ρ_{11}	1
$\xi_0^* =$	1	-1	1
$\rho_{12} =$	0	-1	2
$\rho_{21} =$	1	0	1
$\rho_{22} =$	-1^*	0	1
$\theta =$	-1	1	-1

3rd Tableau (Elimination of ξ_1^*)

	ρ_{22}	ρ_{11}	1
$\rho_{12} =$	0	-1	2
$\rho_{21} =$	-1	0	2
$\xi_1^* =$	-1	0	1
$\theta =$	1	1	-2

This tableau is already optimal and gives $\xi_1^* = 1$, $\rho_{11} = 0$, and, from the 2nd tableau it follows that $\xi_0^* = 2$. The minimum of θ is -2 and therefore $\sigma_{min} = \frac{1}{2}$ and hence $\xi_1 = \frac{1}{2}$. The calculation confirms the obvious result.

Example 2.

$$2\xi_1 \qquad -4 = 0$$
$$\xi_2 - 1 = 0 \qquad (8)$$
$$\xi_1 + \xi_2 - 2 = 0$$

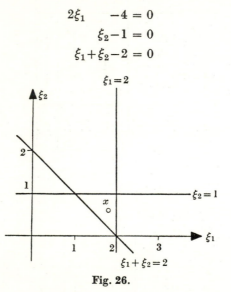

Fig. 26.

For any point (ξ_1, ξ_2) of the plane, $|\xi_1 + \xi_2 - 2|$ is the distance of the point from the line $\xi_1 + \xi_2 - 2 = 0$ multiplied by $\sqrt{2}$ and similarly $|2\xi_1 - 4|$ and $|\xi_2 - 1|$ are the distances from the lines $2\xi_1 - 4 = 0$ and $\xi_2 - 1 = 0$ multiplied by 2 and 1 respectively. The problem now is to find a point (ξ_1, ξ_2) which will make the largest of these distances (multiplied by the appropriate factor) as small as possible. (See Fig. 26, where the solution is denoted by x.)

The corresponding linear programme is

$$\rho_{11} = \quad 2\xi_1^* \quad\quad -4\xi_0^*+1 \geqslant 0$$
$$\rho_{12} = -2\xi_1^* \quad\quad +4\xi_0^*+1 \geqslant 0$$
$$\rho_{21} = \quad\quad \xi_2^*- \xi_0^*+1 \geqslant 0$$
$$\rho_{22} = \quad\quad -\xi_2^*+ \xi_0^*+1 \geqslant 0 \quad\quad (9)$$
$$\rho_{31} = \quad \xi_1^*+\xi_2^*-2\xi_0^*+1 \geqslant 0$$
$$\rho_{32} = - \quad \xi_1^*-\xi_2^*+2\xi_0^*+1 \geqslant 0$$

$$\theta = -\xi_0^* = \min$$

1st Tableau

	ξ_1^*	ξ_2^*	ξ_0^*	1
$\rho_{11} =$	2	0	-4^*	1
$\rho_{12} =$	-2	0	4	1
$\rho_{21} =$	0	1	-1	1
$\rho_{22} =$	0	-1	1	1
$\rho_{31} =$	1	1	-2	1
$\rho_{32} =$	-1	-1	2	1
$\theta =$	0	0	-1	0

2nd Tableau (elimination of ξ_0^*)

	ξ_1^*	ξ_2^*	ρ_{11}	1
$\xi_0^* =$	$\frac{1}{2}$	0	$-\frac{1}{4}$	$\frac{1}{4}$
$\rho_{12} =$	0	0	-1	2
$\rho_{21} =$	$-\frac{1}{2}^*$	1	$\frac{1}{4}$	$\frac{3}{4}$
$\rho_{22} =$	$\frac{1}{2}$	-1	$-\frac{1}{4}$	$\frac{5}{4}$
$\rho_{31} =$	0	1	$\frac{1}{2}$	$\frac{1}{2}$
$\rho_{32} =$	0	-1	$-\frac{1}{2}$	$\frac{3}{2}$
$\theta =$	$-\frac{1}{2}$	0	$\frac{1}{4}$	$-\frac{1}{4}$

3rd Tableau (elimination of ξ_1^*)

	ρ_{21}	ξ_2^*	ρ_{11}	1
$\rho_{12} =$	0	0	-1	2
$\xi_1^* =$	-2	2	$\frac{1}{2}$	$\frac{3}{2}$
$\rho_{22} =$	-1	0	0	2
$\rho_{31} =$	0	1	$\frac{1}{2}$	$\frac{1}{2}$
$\rho_{32} =$	0	-1^*	$-\frac{1}{2}$	$\frac{3}{2}$
$\theta =$	1	-1	0	-1

4th Tableau (elimination of ξ_2^*)

	ρ_{21}	ρ_{32}	ρ_{11}	1
$\rho_{12} =$	0	0	-1	2
$\rho_{22} =$	-1	0	0	2
$\rho_{31} =$	0	-1	0	2
$\xi_2^* =$	0	-1	$-\frac{1}{2}$	$\frac{3}{2}$
$\theta =$	1	1	$\frac{1}{2}$	$-\frac{5}{2}$

Since all the free variables are now eliminated, this is the point at which the simplex method should begin. However, since the first part of the last row consists of all positive numbers, we already have an optimal tableau. The optimal solution is

$$\xi_2^* = \tfrac{3}{2}, \qquad \rho_{11} = \rho_{21} = \rho_{32} = 0, \qquad \rho_{12} = \rho_{22} = \rho_{31} = 2.$$

From the third tableau it follows that $\xi_1^* = \tfrac{9}{2}$. The minimal value of θ is $-\tfrac{5}{2}$, hence $\sigma_{\min} = \tfrac{2}{5}$ and finally $\xi_1 = \tfrac{9}{5}$, $\xi_2 = \tfrac{3}{5}$. Substituting these values in (2), we obtain the residuals $\epsilon_1 = \epsilon_2 = \epsilon_3 = -\tfrac{2}{5}$. The fact that all three absolute values $|\epsilon_i|$ are equal to $\tfrac{2}{5}$ is the result of a more general rule, viz. the rank of the system of inequalities in programme (4) is equal to the rank of the matrix (A,β) and, in our example, this is equal to 3. Hence, in a normal basic solution, three of the inequalities (9) must be satisfied with the equality sign.

Example 3 (see Fig. 27).

$$\xi_1 + \xi_2 - 8 = 0$$
$$\xi_1 - \xi_2 - 2 = 0 \qquad (10)$$
$$\xi_1 + 2\xi_2 - 10 = 0$$

1st Tableau

	ξ_1^*	ξ_2^*	ξ_0^*	1
$\rho_{11} =$	1	1	-8	1
$\rho_{12} =$	-1	-1	8	1
$\rho_{21} =$	1	-1	-2	1
$\rho_{22} =$	-1	1	2	1
$\rho_{31} =$	1	2	-10	1
$\rho_{32} =$	-1	-2	10	1
$\theta =$	0	0	-1	0

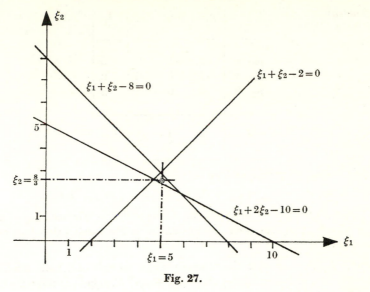

Fig. 27.

We find by elimination that

$$\xi_0^* = \tfrac{1}{10}\xi_1^* + \tfrac{2}{10}\xi_2^* - \tfrac{1}{10}\rho_{31} + \tfrac{1}{10}$$
$$\xi_1^* = -\tfrac{10}{8}\rho_{22} + \tfrac{14}{8}\xi_2^* - \tfrac{2}{8}\rho_{31} + \tfrac{12}{8}.$$

The *4th Tableau* then becomes

	ρ_{22}	ρ_{11}	ρ_{31}	1
$\xi_2^* =$	-1	-4	3	2
$\rho_{12} =$	0	-1	0	2
$\rho_{21} =$	-1	0	0	2
$\rho_{32} =$	0	0	-1^*	2
$\theta =$	$\tfrac{1}{2}$	$\tfrac{3}{2}$	-1	-1

At this point, we must start to use the simplex method.

5th Tableau

	ρ_{22}	ρ_{11}	ρ_{32}	1
$\rho_{12} =$	0	-1	0	2
$\rho_{21} =$	-1	0	0	2
$\rho_{31} =$	0	0	-1	2
$\theta =$	$\frac{1}{2}$	$\frac{3}{2}$	1	-3

This tableau is optimal and we find

$$\rho_{12} = \rho_{21} = \rho_{31} = 2, \qquad \rho_{11} = \rho_{22} = \rho_{32} = 0$$

and hence $\xi_2^* = 8$, $\xi_1^* = 15$. The minimal value of θ is -3, hence $\sigma_{\min} = \frac{1}{3}$ and $\xi_1 = 5$, $\xi_2 = \frac{8}{3}$. Again all three $|\epsilon_i|$ are equal to the same value $\frac{1}{3}$ as they should be.

Example 4. We start with the function $x(\tau) = \cos \pi\tau/2$ of a real variable τ and look for a polynomial $z(\tau) = \alpha_0 + \alpha_1 \tau + \alpha_2 \tau^2$ of degree at most 2 which most closely approximates to $x(\tau)$ at the points $\tau = 0$, ± 1, ± 2, i.e., for which the largest of the five absolute values $|z(\tau) - x(\tau)|$, $(\tau = 0, \pm 1, \pm 2)$ is as small as possible.

Since $x(\tau)$ is an even function, $z(\tau)$ must also be an even function and hence $\alpha_1 = 0$. Consequently, it is sufficient to consider only the values $\tau = 0, 1, 2$.

Now

$$
\begin{aligned}
z(0) - x(0) &= \alpha_0 & -1 \\
z(1) - x(1) &= \alpha_0 + \alpha_2 \\
z(2) - x(2) &= \alpha_0 + 4\alpha_2 + 1.
\end{aligned}
$$

Thus we have an approximation problem to solve for α_0 and α_2 using Tchebychev's method.

1st Tableau

	α_0^*	α_2^*	ξ_0^*	1
$\rho_{11} =$	1*	4	1	1
$\rho_{12} =$	-1	-4	-1	1
$\rho_{21} =$	1	1	0	1
$\rho_{22} =$	-1	-1	0	1
$\rho_{31} =$	1	0	-1	1
$\rho_{32} =$	-1	0	1	1
$\theta =$	0	0	-1	0

The solution is $\alpha_0 = \frac{3}{4}$, $\alpha_2 = -\frac{1}{2}$, $\sigma = \frac{1}{4}$. The required polynomial is therefore $z(\tau) = \frac{1}{4}(3 - 2\tau^2)$. It differs from $x(\tau)$ by the absolute value $\frac{1}{4}$ at each of the five points $\tau = 0$, ± 1, ± 2. Every other polynomial of degree at most 2 differs from $x(\tau)$ by more than $\frac{1}{4}$ at at least one of the five points.

Exercises

Use Tchebychev's method of approximation to solve the following systems of equations.

(a) $\xi_1 \quad = 5$
$\qquad \xi_2 = 7$
$\quad \xi_1 + \xi_2 = 0$
$\quad \xi_1 - \xi_2 = 1$

(b) $\quad 3\xi_1 + 3\xi_2 + 3\xi_3 = 90$
$\qquad \xi_1 + \xi_2 - \xi_3 = 20$
$\qquad \xi_1 - \xi_2 + \xi_3 = 20$
$\quad -\xi_1 + \xi_2 + \xi_3 = 20$

Solutions. (a) $\xi_1 = 1$, $\xi_2 = 3$.
$\qquad$ (b) $\xi_1 = \xi_2 = \xi_3 = 11$.

9.2 The Proof of Two Results used Earlier

9.2.1 Theorem 6.3;2

In the proof of Theorem 6.3;2, we assumed that

$$\theta = \min_{\lambda_k \geqslant 0} \max_{1 \leqslant i \leqslant n} |h_i(\lambda_1, \ldots, \lambda_r) - \xi_{i0}| \tag{1}$$

exists, where the h_i are linear functionals in the variables $\lambda_1, \ldots, \lambda_r$. We now see that this is a Tchebychev approximation problem for the system of linear equations

$$h_i(\lambda_1, \ldots, \lambda_r) - \xi_{i0} = 0 \qquad (i = 1, \ldots, n),$$

where however the variables are required to satisfy the further condition $\lambda_k \geqslant 0$ $(k = 1, \ldots, r)$. Thus (1) corresponds to a linear programme. This certainly has an optimal solution because $|h_i(\lambda_1, \ldots, \lambda_r) - \xi_{i0}| \geqslant 0$. Hence the quantity θ exists.

9.2.2 Theorem 6.3;9

In the statement of the Minimax Theorem 6.3;9, we assumed that the two quantities which appear in the theorem actually exist. Consider for instance

193

$\theta_0 = \min_{y \in Y} \max_{x \in X} f(x,y)$. For each $y \in Y$, $f(x,y)$ is a linear functional on E.

This takes its maximum value at a vertex vector of the polyhedron X. Therefore

$$\theta_0 = \min_{y \in Y} \max_i f(a_i, y),$$

where a_i runs through all the vertex vectors of X. Now, for $y \in Y$,

$$y = \sum_{k=1}^{s} \mu_k b_k \quad \text{where} \quad \sum_{k=1}^{s} \mu_k = 1, \qquad \mu_k \geqslant 0 \ (k = 1, \ldots, s), \qquad (2)$$

and $b_1, \ldots, b_s$ are the vertex vectors of Y.
Therefore

$$\theta_0 = \min_{\mu'} \max_i \left\{ \sum_{k=1}^{s} \mu_k f(a_i, b_k) \right\} \qquad (3)$$

where the minimum is taken over all $\mu' = (\mu_1, \ldots, \mu_s)$ which satisfy the conditions in (2). An argument similar to the one in 9.2.1 shows that θ_0 is the minimal value of the object function θ in the following linear programme.

$$\left. \begin{array}{l} \displaystyle\sum_{k=1}^{s} f(a_i, b_k) \mu_k \leqslant \sigma; \qquad i = 1, \ldots, r \\[2ex] \displaystyle\sum_{k=1}^{s} \mu_k = 1, \qquad \mu_k \geqslant 0; \qquad k = 1, \ldots, s \\[2ex] \theta = \sigma = \min \end{array} \right\} \qquad (4)$$

The minimal value exists, because an upper bound σ of a subset of the values $f(x,y)$ ($x \in X$, $y \in Y$) cannot be arbitrarily small.

Problems

1. Use the ideas in 9.2.2 to find a new proof of the Minimax Theorem for the case when $f(x,y) \geqslant 0$ for all $x \in X$; $y \in Y$. (Hint: The solutions of the linear programme (4) are still the same if the condition $\sum_k \mu_k = 1$ is replaced by $\sum_k \mu_k \geqslant 1$ and if the restriction $\sigma \geqslant 0$ is also included. Also the linear programme for $\max_{x \in X} \min_{y \in Y} f(x,y)$ is dual to (4).

2. Verify that the translation introduced in the proof in 9.2 never leads to a cyclic result.

CHAPTER 10

GAME THEORY

10.1 Two-Person Zero-Sum Games, Pure Strategies

Two players who, following Stiefel [16] p. 42, we will call 'Max' and 'Min' (the reason for these names will soon become obvious) are to play a game of the following kind.

An $m \times n$ matrix $A = (\alpha_{ik})$ is given. A move in the game consists of Max choosing a row and Min choosing a column of A. If row i and column k are chosen, then Min pays α_{ik} to Max. (If α_{ik} is negative, then Max pays $-\alpha_{ik}$ to Min.)

Example 1.

$$A = \begin{pmatrix} -1 & 2 & 3 \\ 3 & 0 & -2 \end{pmatrix}$$

If Max chooses $i = 1$ and Min chooses $k = 3$, then Min pays 3 to Max. If $i = 2$ and $k = 3$, then Max pays 2 to Min. If $i = k = 2$, then neither needs to pay.

The matrix A is referred to as the *pay-off matrix* of this *matrix game* which is an example of what is known as a *two-person* game. Since the gain for one player at each move is always equal to the loss for the other, the game is also an example of a *zero-sum* game. Finally, since each player has a finite number of possible choices (m and n respectively), the game is said to be *finite*. In the following we will give a brief description of the theory of finite two-person zero-sum games.

In fact, all finite two-person zero-sum games can be represented in the form of a matrix game of the kind just described. The m and n possible choices which Max and Min have available at each move are referred to as the *strategies* of the players—more accurately as the *pure strategies* to differentiate them from the *mixed strategies* which will be considered in 10.2.

The first aim of Game Theory is how to answer the following question for each player.

What is the optimal value of the pay-off (i.e., the greatest value for Max and the least for Min) that a player can guarantee in a move by a suitable choice of a pure strategy, irrespective of the other player's choice.

We will first answer this question for Max in our Example 1. If he chooses $i=1$, then at worst he will lose 1, viz., when Min chooses $k=1$. If he chooses $i=2$, then at worst he will lose 2, viz., when Min chooses $k=3$. Thus, if he chooses $i=1$, he guarantees that his payment will be at least -1 and he cannot guarantee more. Thus Max first looks in each row for the least element and then for the row in which this least element is as large as possible. The optimal value which Max can guarantee is therefore

$$\mu_{\text{Max}} = \max_i \min_k \alpha_{ik} = \alpha_{i_1 k_1}. \tag{1}$$

The subscript $i=i_1$, which gives this value is known as an *optimal pure strategy for Max*. (In some cases of course, there can be many of these.) In our example $i_1=1$ is the only optimal pure strategy for Max and $\mu_{\text{Max}}=-1$.

Now let us carry out the corresponding arguments for Min.

If Min chooses $k=1$, then the worst is $i=2$ with $\alpha_{21}=3$.
If Min chooses $k=2$, then the worst is $i=1$ with $\alpha_{12}=2$.
If Min chooses $k=3$, then the worst is $i=1$ with $\alpha_{13}=3$.

Therefore, if Min chooses $k=2$, then he will have to pay at most 2 and he cannot guarantee anything better.

Thus Min first looks in each column for the greatest element and then for the column in which the greatest element is as small as possible. The optimal value which he can guarantee is therefore

$$\mu_{\text{Min}} = \min_k \max_i \alpha_{ik} = \alpha_{i_2 k_2}. \tag{2}$$

A subscript $k=k_2$, which gives this value is known as an *optimal pure strategy for Min*. In Example 1, $k_2=2$ is the only optimal pure strategy for Min and $\mu_{\text{Min}}=2$.

Theorem 1. The optimal value which Max can guarantee is at most equal to that which Min can guarantee, i.e., $\mu_{\text{Max}} \leqslant \mu_{\text{Min}}$.

Proof. $\mu_{\text{Max}} = \alpha_{i_1 k_1} \leqslant \alpha_{i_1 k_2} \leqslant \alpha_{i_2 k_2} = \mu_{\text{Min}}$.

This expresses the already intuitive idea, that, in a zero-sum game, no player can guarantee a better (i.e., greater or smaller respectively) value than the other.

In some cases, it can happen that $\mu_{\text{Max}}=\mu_{\text{Min}}$, viz., when A has an element $\alpha_{i_0 k_0}$ which is both the smallest in its row and the largest in its column. Then $\mu_{\text{Max}}=\mu_{\text{Min}}=\alpha_{i_0 k_0}$. The pair (i_0, k_0) is then known as a *saddle point*. Thus, there is a common value which both players can guarantee if and only if A has a saddle point.

Example 2.

$$A = \begin{pmatrix} 3 & 2 & 4 \\ 0 & -1 & 2 \\ 2 & 0 & 4 \end{pmatrix}$$

The pair $(1,2)$ is a saddle point and hence $\mu_{\text{Max}} = \mu_{\text{Min}} = \alpha_{12} = 2$.

10.2 Mixed Strategies

The theory, which has not been very deep so far, becomes more interesting when we introduce a new kind of strategy known as a *mixed strategy*.

A mixed strategy for Max is a mixture of the rows of A or, more precisely, each subscript i is given a weight ξ_i such that $\sum_{i=1}^{m} \xi_i = 1$ and $\xi_i \geqslant 0$ for all $i = 1$, ..., m. We denote the corresponding mixed strategy by $x = (\xi_1, ..., \xi_m)$ and think of this as a vector in R_m.

Similarly a mixed strategy for Min consists in assigning a weight η_k to each column subscript k in such a way that $\sum_{k=1}^{n} \eta_k = 1$ and $\eta_k \geqslant 0$ $(k = 1, ..., n)$. We again denote this strategy by $y = (\eta_1, ..., \eta_n) \in R_n$.

Now, if Max and Min choose the mixed strategies x and y, then the amount which Min must pay to Max is defined to be

$$f(x,y) = \boldsymbol{\xi}' A \boldsymbol{\eta} = \sum_{i,k} \alpha_{ik} \xi_i \eta_k. \tag{1}$$

Example 1. In Example 10.1;1, $x = (\frac{1}{4}, \frac{3}{4})$ and $y = (\frac{1}{6}, \frac{2}{6}, \frac{3}{6})$ are possible mixed strategies for the two players. The corresponding pay-off is $\frac{1}{8}$.

Pure strategies can be realized as special cases of mixed strategies where, for instance, the pure strategy i for Max corresponds to the mixed strategy $x_i = (\delta_{1i}, ..., \delta_{mi}) = (0, ..., 0, 1, 0, ..., 0)$ and similarly for Min. (For the Kronecker delta, see 3.1.) For the mixed strategies x_i and y_k, we then have $f(x_i, y_k) = \alpha_{ik}$, i.e., the same value of the pay-off as for the two pure strategies.

The vectors corresponding to the mixed strategies of Max form an $(m-1)$-dimensional simplex S_1 in R_m which is the intersection of the hyperplane $\sum_i \xi_i = 1$ with the convex pyramid $\xi_i \geqslant 0$ $(i = 1, ..., m)$. Its vertex vectors are $(1, 0, ..., 0), ..., (0, 0, ..., 1)$ and therefore correspond to the pure strategies.

Similarly the vectors corresponding to the mixed strategies of Min form an $(n-1)$-dimensional simplex S_2 in R_n which again is the convex hull of the pure strategies. The pay-off function $f(x,y)$, extended to R_m and R_n by equation (1), is a bilinear form on R_m and R_n.

By introducing mixed strategies and the pay-off function (1), we have constructed a new game which is referred to as the *mixed extension* of the original game. This new game is no longer finite, except when $m=n=1$. Its pure strategies are the mixed strategies of the original game.

We now ask the same question for the mixed strategies as we did earlier for the pure strategies.

For each of the players, what is the optimal value of the pay-off which he can guarantee by a suitable choice of a mixed strategy irrespective of his opponent's choice?

A mixed strategy which guarantees this value for a player is known as an *optimal mixed strategy* for this player. If we denote the optimal values by μ_{Max}^* and μ_{Min}^*, then a similar argument to that for the pure strategies will show that

$$\mu_{\text{Max}}^* = \max_{x \in S_1} \min_{y \in S_2} f(x,y)$$

and

$$\mu_{\text{Min}}^* = \min_{y \in S_2} \max_{x \in S_1} f(x,y).$$

The following Fundamental Theorem of Game Theory follows immediately from the Minimax Theorem 6.3;9.

Theorem 1. $\mu_{\text{Max}}^* = \mu_{\text{Min}}^* = \mu$. *That is, in words; there is a value μ of the pay-off, which each of the players can guarantee by a suitable choice of a mixed strategy. Neither player can guarantee a better value than μ.*

The common value μ of μ_{Max}^* and μ_{Min}^* is known as the *value* of the game. The pay-off takes this value when each of the players chooses an optimal mixed strategy, x_0 and y_0 say, and then $\mu = f(x_0, y_0)$.

The calculation of μ and the optimal mixed strategies in a matrix game will be the content of section 10.3, but before that we will first consider an example.

Example 2.

$$A = \begin{pmatrix} 0 & 1 & 1 & 2 \\ 1 & 0 & 2 & 3 \\ 2 & 1 & 0 & 1 \end{pmatrix}$$

Suppose Max chooses the mixed strategy $x_0 = (\frac{3}{6}, \frac{1}{6}, \frac{2}{6})$. If Min now chooses $y_1 = (1,0,0,0)$ (i.e., the pure strategy $k=1$), then the pay-off is $f(x_0, y_1) = \frac{5}{6}$. The same value is obtained, if Min chooses $k=2$ or $k=3$. On the other hand for $k=4$, the value becomes $\frac{11}{6}$. If Min chooses any mixed strategy $y = (\eta_1, \eta_2, \eta_3, \eta_4)$, then

$$f(x_0, y) = \tfrac{5}{6}(\eta_1 + \eta_2 + \eta_3) + \tfrac{11}{6}\eta_4 \geqslant \tfrac{5}{6}.$$

Thus, by using the strategy x_0, Max ensures that the pay-off is at least $\frac{5}{6}$. It follows that $\mu = \mu^*_{Max} \geqslant \frac{5}{6}$. Now suppose that Min chooses $y_0 = (\frac{1}{6}, \frac{3}{6}, \frac{2}{6}, 0)$, then, as before, we see that $f(x, y_0) = \frac{5}{6}$ for any mixed strategy x for Max. Hence $\mu = \mu^*_{Min} \leqslant \frac{5}{6}$ and it follows that $\mu = \frac{5}{6}$ and x_0, y_0 are optimal mixed strategies for Max and Min.

At first sight it may seem surprising that, in this example, if Max chooses the optimal strategy x_0, then he guarantees the value μ but never any more so long as Min chooses any strategy in which $\eta_4 = 0$. On the other hand, by choosing y_0, Min always guarantees the value $\frac{5}{6}$ but never anything better whatever strategy Max chooses. This is an instance of a general fact which will be explained in Theorem 2.

We say that a pure strategy is *inessential* if it has the weight zero in every optimal mixed strategy. (Thus the pure strategy i for Max is inessential if $\xi_i = 0$ for every optimal mixed strategy $x = (\xi_1, \ldots, \xi_m)$.) In Example 3, Max has no inessential pure strategies, but for Min $k = 4$ is inessential.

Theorem 2. If Max (Min) chooses an optimal mixed strategy and Min (Max) chooses a mixed strategy in which every inessential pure strategy has the weight zero, then the payment is equal to the value μ of the game.

Proof. Let x_0 be the optimal mixed strategy chosen by Max. Suppose Min chooses the strategy $y \in S_2$ and that $f(x_0, y) > \mu$. We have to show that y contains an inessential pure strategy with a strictly positive weight. We will denote the pure strategies of Min by $y_1, \ldots, y_n$ and let

$$y = (\eta_1, \ldots, \eta_n) = \sum_{k=1}^{n} \eta_k y_k.$$

Then

$$f(x_0, y) = \sum_{k=1}^{n} \eta_k f(x_0, y_k) > \mu.$$

Therefore there exists a k_0 such that

$$f(x_0, y_{k_0}) > \mu \quad \text{and} \quad \eta_{k_0} > 0.$$

In this case y_{k_0} is inessential because, if y is a mixed strategy of Min in which y_{k_0} has a strictly positive weight, then it is easy to see that $f(x_0, y) > \mu$. Therefore y is not optimal.

The Practical Significance of Mixed Strategies

In the applications of Game Theory, mixed strategies are usually used in the following stochastic way. (The concepts of Probability Theory, which are used here may be found in [29].)

Let $x=(\xi_1,\ldots,\xi_m)$ and $y=(\eta_1,\ldots,\eta_n)$ be mixed strategies for Max and Min. We now think of the subscripts i, k of the pure strategies as being independent random variables which take the values $i=1,\ldots,m$ and $k=1,\ldots,n$ with the probabilities $\xi_1,\ldots,\xi_m$ and $\eta_1,\ldots,\eta_n$. Then the pay-off is also a random variable which corresponds to the matrix element α_{ik} with probability $\xi_i\eta_k$. The expected value of the pay-off is therefore

$$g(x,y) = \sum_{i,k} \alpha_{ik}\xi_i\eta_k = \boldsymbol{\xi}'A\boldsymbol{\eta} = f(x,y),$$

i.e., it is equal to the value of the pay-off for the mixed strategies x and y.

We now imagine that not just one but N moves are played. At each move. Max and Min each independently choose a pure strategy at random subject to the probabilities $\xi_1,\ldots,\xi_m$ and $\eta_1,\ldots,\eta_n$ given by the mixed strategies x and y. At each move the pay-off has the value of a matrix element α_{ik} (with a certain probability) and we can calculate the arithmetic mean of these values denoting it by $g_N(x,y)$. By the laws of Probability Theory, these means converge with probability 1 to the expected value as $N \to \infty$, i.e.,

$$\lim_{N\to\infty} g_N(x,y) = g(x,y) = f(x,y).$$

Example 3. In the game already considered (Example 2)

$$A = \begin{pmatrix} 0 & 1 & 1 & 2 \\ 1 & 0 & 2 & 3 \\ 2 & 1 & 0 & 1 \end{pmatrix}$$

Max has 3 and Min has 4 pure strategies available. Both players can now achieve their optimal mixed strategies,

$$x = (\tfrac{3}{6},\tfrac{1}{6},\tfrac{2}{6}) \quad \text{and} \quad y = (\tfrac{1}{6},\tfrac{3}{6},\tfrac{2}{6},0)$$

by the following stochastic method.

At each move, each player throws an ordinary die. Max chooses

$i = 1$ when he throws 1, 2 or 3
$i = 2$ when he throws 4

and $i = 3$ when he throws 5 or 6.

Min chooses

$k = 1$ when he throws 1
$k = 2$ when he throws 2, 3 or 4
$k = 3$ when he throws 5 or 6

and he never chooses $k=4$.

Thus at each move the fall of the die determines a pure strategy for each player. It corresponds to an element α_{ik} of A as the value of the pay-off.

If the arithmetic mean of a sufficiently large number of these pay-offs is now calculated, then it approaches arbitrarily close to the value of the game, i.e., to $\mu = \frac{5}{6}$.

The method of applying mixed strategies which we have described here must be used in practice with great care. As long as only a finite number of moves are played, there certainly is no guarantee that on average the players will reach the value of the game. The effective mean value will only approach within a given small distance ϵ of the value μ with a given probability when the number of moves is sufficiently large. However, if the number of moves is given, then the mean value will differ from μ by given amounts with given probabilities.

For example, if just one move is played in the game above with the given optimal strategies, then Max has the probability 10/36 that he will only achieve the value 0. The probability that the pay-off for Min is either 1 or 2 (i.e., worse than μ) is 26/36. If two moves are played, then Max has the probability 540/1296 of achieving on average only 0 or $\frac{1}{2}$ while the probability that Min will have an average pay-off of at least 1 is 756/1296.

In order to investigate this situation, it is necessary to make new approaches to the theory which are outside the scope of this book (see for example [26] chap. 13).

Problems

1. The pure strategy i_0 for Max is said to be *strictly dominated* if there is a mixed strategy $x = (\xi_1, \ldots, \xi_m)$ with $\xi_{i_0} = 0$ such that

$$\sum_{i=1}^{m} \xi_i \alpha_{ik} > \alpha_{i_0 k} \qquad (k = 1, \ldots, n). \tag{1}$$

Show that every strictly dominated pure strategy is inessential.

2. If the '$>$' sign in (1) (Problem 1) is replaced by '$\geqslant$', then the strategy i_0 is said to be dominated. Is every dominated pure strategy inessential?

3. Prove the assertion concerning saddle points which was stated at the end of 10.1.

10.3 The Evaluation of Games by the Simplex Method

If we add the same constant λ to all the elements of the matrix A, then the pay-off function $f(x, y)$ of the corresponding game becomes

$$f^*(x, y) = \xi'(A + \lambda D)\eta = \xi' A\eta + \lambda \xi' D\eta = f(x, y) + \lambda,$$

where D is the $m \times n$ matrix all of whose elements are equal to 1, i.e., $f(x,y)$ is simply increased by the constant λ. Hence the value of the game is also increased by λ and the optimal strategies are the same in both cases. Thus, if we know the value and the optimal strategies for the game A, then we also know them for the game $A + \lambda D$ and conversely.

By choosing λ sufficiently large, we can ensure that the value $\mu + \lambda$ of $A + \lambda D$ is strictly positive. In particular this will be the case when all the elements of $A + \lambda D$ are strictly positive. Because the pay-off function will then take only strictly positive values and hence the value of the game will also be strictly positive. Thus there will be no essential loss of generality for the numerical evaluation of a game if we assume that its value is strictly positive.

Thus we will consider a game with the matrix A and the value $\mu > 0$. An optimal strategy $y = (\eta_1, \ldots, \eta_n)$ for Min satisfies the inequalities

$$A\boldsymbol{\eta} \leqslant \boldsymbol{\mu}, \tag{1}$$

where $\boldsymbol{\mu}$ is the $m \times 1$ matrix all of whose elements are equal to μ. Indeed μ is the least real number for which there exists $\boldsymbol{\eta}$ such that both (1) and

$$\sum_{k=1}^{n} \eta_k = 1; \qquad \eta_k \geqslant 0 \qquad (k = 1, \ldots, n) \tag{2}$$

are satisfied. If there were a smaller number $\mu_1 < \mu$ with this property, then

$$f(x,y) = \boldsymbol{\xi}' A \boldsymbol{\eta} \leqslant \boldsymbol{\xi}' \boldsymbol{\mu}_1 = \mu_1$$

for every mixed strategy of Max, so that the value of the game would be less than or equal to μ_1.

The relations (1) and (2) together with $\mu = \min$ constitute a linear programme, the solution of which will give the optimal strategies for Min and the value μ. As in 9.1, it is useful to modify the programme by dividing by μ. Putting $\eta_k^* = \eta_k/\mu$, we have

$$\left. \begin{aligned} \boldsymbol{\sigma} = -A\boldsymbol{\eta}^* + 1 &\geqslant 0 \\ \boldsymbol{\eta}^* &\geqslant 0 \\ -1'\boldsymbol{\eta}^* = -\sum_{k=1}^{n} \eta_k^* = -\frac{1}{\mu} &= \min \end{aligned} \right\} \tag{3}$$

The initial tableau of the simplex method for the definite programme (3) is

	η_1^*	$\cdots$	η_n^*	1
$\sigma_1 =$				1
$\vdots$		$-A$		$\vdots$
$\sigma_m =$				1
$\theta =$	-1	$\cdots$	-1	0

$$(4)$$

Suppose that the final tableau is

	η^*		σ	1
$\sigma =$				β_1
		A^*		$\vdots$
$\eta^* =$				β_m
$\theta =$	γ_1	$\cdots$	γ_n	δ

where $\beta_i \geqslant 0$ $(i=1,\ldots,m)$, $\gamma_k \geqslant 0$ $(k=1,\ldots,n)$ and $\delta < 0$. Then the value of the game is $\mu = -1/\delta$, and we obtain an optimal strategy for Min by putting the η_k^* in the top row equal to zero and those on the left-hand side equal to the corresponding β_k. Finally $\eta_k = \mu \eta_k^*$ $(k=1,\ldots,n)$. The possible existence of further optimal strategies may be decided from the final tableau by using the method described in 8.3.

The corresponding arguments for Max show that the optimal strategies for him are given by the solutions of the definite programme

$$
\left.
\begin{aligned}
\boldsymbol{\rho}' &= \boldsymbol{\xi}^{*\prime} A - \mathbf{1}' \geqslant 0 \\
\boldsymbol{\xi}^{*\prime} &\geqslant 0 \\
\boldsymbol{\xi}^{*\prime}(-1) &= - \sum_{i=1}^{m} \xi_i^* = \max
\end{aligned}
\right\}
\qquad (5)
$$

where $\xi_i^* = \xi_i/\mu$ $(i=1,\ldots,m)$. But this is the dual programme of (3) and will therefore be solved by the same initial and final tableaux as (3) providing we write them as follows and interpret them vertically.

	$\rho_1 =$	$\cdots$	$\rho_n =$	$\theta =$
$-\xi_1^*$				1
$\vdots$		$-A$		$\vdots$
$-\xi_m^*$				1
1	-1	$\cdots$	-1	0

	$\rho =$	$\xi^* =$	$\theta =$
$-\xi^*$			β_1
	A^*		$\vdots$
$-\rho$			β_m
1	γ_1	$\cdots \ \gamma_n$	δ

We obtain an optimal strategy for Max by putting the ξ_i^* on the left-hand side equal to zero and those in the top row equal to the corresponding γ_i and finally calculating $\xi_i = \mu\xi_i^*$ $(i=1,\ldots,m)$. It is clear that the optimal strategies of both players can be obtained simultaneously by carrying out the following procedure.

1. We set out the tableau (4), but writing ξ_k^* on the left-hand side instead of σ_k $(k=1,\ldots,m)$. (But note that, with this interpretation, the tableau no longer expresses valid linear relations.)

2. We apply the normal simplex method to this tableau. From the final tableau, we obtain

2.1. the value of the game $\mu = -1/\delta$

2.2. an optimal strategy for Max, by putting the ξ_i^* on the left-hand side equal to 0 and those in the top row equal to the corresponding γ_k and finally putting $\xi_i = \mu\xi_i^*$ $(i=1,\ldots,m)$

2.3. an optimal strategy for Min, by putting the η_k^* in the top row equal to 0 and those on the left-hand side equal to the corresponding β_k and finally putting $\eta_k = \mu\eta_k^*$ $(k=1,\ldots,n)$.

Example 1. Consider again the Example 10.2;2

$$A = \begin{pmatrix} 0 & 1 & 1 & 2 \\ 1 & 0 & 2 & 3 \\ 2 & 1 & 0 & 1 \end{pmatrix}$$

whose value μ is strictly positive.

The initial tableau is

	η_1^*	η_2^*	η_3^*	η_4^*	1
ξ_1^*	0	-1	-1	-2	1
ξ_2^*	-1	0	-2	-3	1
ξ_3^*	-2	-1	0	-1	1
1	-1	-1	-1	-1	0

After carrying out the simplex method, the final tableau is

	ξ_1^*	ξ_2^*	ξ_3^*	η_4^*	1
η_1^*					$\frac{1}{5}$
η_2^*		A^*			$\frac{3}{5}$
η_3^*					$\frac{2}{5}$
1	$\frac{3}{5}$	$\frac{1}{5}$	$\frac{2}{5}$	$\frac{6}{5}$	$-\frac{6}{5}$

Hence $\mu = -\dfrac{1}{\delta} = \dfrac{5}{6}$.

An optimal mixed strategy for Min is

$$y = \tfrac{5}{6}(\tfrac{1}{5}, \tfrac{3}{5}, \tfrac{2}{5}, 0) = (\tfrac{1}{6}, \tfrac{3}{6}, \tfrac{2}{6}, 0)$$

and for Max is

$$x = \tfrac{5}{6}(\tfrac{3}{5}, \tfrac{1}{5}, \tfrac{2}{5}) = (\tfrac{3}{6}, \tfrac{1}{6}, \tfrac{2}{6}).$$

At the same time, the tableau also shows that these are the only optimal mixed strategies.

Example 2.

$$A = \begin{pmatrix} 2 & 4 \\ 6 & 3 \\ 1 & 3 \end{pmatrix}$$

Since all $\alpha_{ik} > 0$, then μ is also strictly positive.

The initial tableau is

	η_1^*	η_2^*	1
ξ_1^*	-2	-4	1
ξ_2^*	-6	-3	1
ξ_3^*	-1	-3	1
1	-1	-1	0

and the final tableau is

	ξ_1^*	ξ_2^*	1
η_2^*			$\frac{2}{9}$
η_1^*		A^*	$\frac{1}{18}$
ξ_3^*			$\frac{5}{18}$
1	$\frac{1}{6}$	$\frac{1}{9}$	$-\frac{5}{18}$

Hence $\mu = 18/5$.

The optimal mixed strategies are

$$\tfrac{18}{5}(\tfrac{1}{6}, \tfrac{1}{9}, 0) = (\tfrac{3}{5}, \tfrac{2}{5}, 0) \quad \text{for Max}$$

and

$$\tfrac{18}{5}(\tfrac{1}{18}, \tfrac{2}{9}) = (\tfrac{1}{5}, \tfrac{4}{5}) \quad \text{for Min.}$$

It again follows from the final tableau that each player has only one optimal mixed strategy.

Exercises

Calculate the optimal mixed strategies x for Max and y for Min and the value μ of the following matrix games.

$$(a) \; A = \begin{pmatrix} 1 & 0 & 3 \\ 2 & 3 & 1 \end{pmatrix} \qquad (b) \; A = \begin{pmatrix} 2 & 1 & 3 \\ 0 & 4 & 2 \\ 3 & 2 & 1 \end{pmatrix}$$

Solutions. (a) $x = (\tfrac{1}{3}, \tfrac{2}{3})$, $y = (\tfrac{2}{3}, 0, \tfrac{1}{3})$; $\mu = \tfrac{5}{3}$

(b) $x = (\tfrac{2}{5}, \tfrac{1}{5}, \tfrac{2}{5})$, $y = (\tfrac{1}{3}, \tfrac{1}{3}, \tfrac{1}{3})$; $\mu = 2$.

CHAPTER 11

FORMS OF THE SECOND DEGREE

11.1 Quadratic Forms on Real Vector Spaces

In this first section we will restrict the discussion to real vector spaces. The corresponding discussion for complex vector spaces will be carried out in 11.2.

11.1.1 Bilinear Forms

The concept of a bilinear form was introduced in Definition 6.2;2. We will now consider the special case in which the spaces of the two variables are both the same. Definition 6.2;2 then becomes

Definition 1. A 'bilinear form' $f(x,y)$ on a real vector space E is a real-valued function of two variables x, $y \in E$ for which

$$f(\alpha_1 x_1 + \alpha_2 x_2, y) = \alpha_1 f(x_1, y) + \alpha_2 f(x_2, y)$$

and
$$f(x, \beta_1 y_1 + \beta_2 y_2) = \beta_1 f(x, y_1) + \beta_2 f(x, y_2)$$

for all x, x_1, x_2, y, y_1, $y_2 \in E$ and all scalars α_1, α_2, β_1, β_2.

Example 1. Let E be the vector space C (see Example 2.1;7) and let $k(\sigma, \tau)$ be a real-valued continuous function of the real variables σ, τ in the range $-1 \leqslant \sigma, \tau \leqslant +1$. Then

$$f(x,y) = \int\limits_{-1}^{+1} \int\limits_{-1}^{+1} k(\sigma, \tau) x(\sigma) y(\tau) \, d\sigma \, d\tau$$

is a bilinear form on C.

Definition 2. A bilinear form $f(x,y)$ on E is said to be 'symmetric' if $f(x, y) = f(y, x)$ for all $x, y \in E$.

The bilinear form in Example 1 is symmetric if $k(\sigma, \tau) = k(\tau, \sigma)$, i.e., if k is symmetric. It is also true that f can only be symmetric if k is symmetric, but we will omit the proof which can easily be carried out using some simple results from analysis.

Now suppose that E is finite-dimensional and that $\{e_1, \ldots, e_n\}$ is a basis of E. If ξ_i and η_k are the components of x and y respectively, then for any bilinear form $f(x, y)$

$$f(x, y) = \sum_{i=1}^{n} \sum_{k=1}^{n} \alpha_{ik} \xi_i \eta_k = \boldsymbol{\xi}' A \boldsymbol{\eta} \tag{1}$$

where $\alpha_{ik} = f(e_i, e_k)$ $(i, k = 1, \ldots, n)$.

Conversely, for any real $n \times n$ matrix A, (1) always represents a bilinear form f on E. For a given basis the bilinear form f and the matrix A are each uniquely determined by the other.

Theorem 1. The bilinear form f is symmetric if and only if the corresponding matrix A is symmetric.

Proof. 1. Suppose $A = A'$. Then $f(y, x) = \boldsymbol{\eta}' A \boldsymbol{\xi} = \boldsymbol{\xi}' A' \boldsymbol{\eta} = f(x, y)$ (any 1×1 matrix is equal to its transpose).

2. Suppose f is symmetric. Then

$$\boldsymbol{\xi}' A \boldsymbol{\eta} = f(x, y) = f(y, x) = \boldsymbol{\eta}' A \boldsymbol{\xi} = \boldsymbol{\xi}' A' \boldsymbol{\eta}$$

for arbitrary ξ_i and η_k. Therefore $A = A'$.

If we carry out the change of basis in E which is given by $\boldsymbol{\xi} = S \boldsymbol{\xi}^*$, then $f(x, y) = \boldsymbol{\xi}' A \boldsymbol{\eta} = \boldsymbol{\xi}^{*'} (S' A S) \boldsymbol{\eta}^*$, i.e., A goes into $A^* = S' A S$.

11.1.2 Quadratic Forms

Definition 3. Let E be a real vector space. 'A quadratic form' on E is a function $q(x)$ on E which can be represented in the form $q(x) = f(x, x)$, where $f(x, y)$ is a symmetric bilinear form. $f(x, y)$ is referred to as the 'polar form' of $q(x)$. (See Theorem 2.)

The condition of symmetry for f does not entail any real restriction because the quadratic form which arises from $f(x, y)$ is the same as that arising from the symmetric bilinear form $\frac{1}{2}[f(x, y) + f(y, x)]$.

Theorem 2. The polar form of a quadratic form $q(x)$ is uniquely determined by $q(x)$.

Proof. $q(x+y) - q(x-y) = f(x+y, x+y) - f(x-y, x-y) = 4f(x, y)$.

Example 2. For each function $k(\sigma, \tau)$ satisfying the conditions of Example 1, we have a quadratic form $q(x)$ on the vector space C given by

$$q(x) = \int_{-1}^{+1} \int_{-1}^{+1} k(\sigma, \tau) x(\sigma) x(\tau) \, d\sigma \, d\tau.$$

Without losing any of the generality of the example, we may assume that $k(\sigma, \tau)$ is symmetric.

Definition 4. A quadratic form $q(x)$ on a real vector space E is said to be 'positive definite' if it never takes negative values and if it only takes the value 0 for $x = 0$.

Example 3. A positive definite quadratic form on C is given by $q(x) = \int_{-1}^{+1} [x(\sigma)]^2 \, d\sigma$. Its polar form is $f(x, y) = \int_{-1}^{+1} x(\sigma) y(\sigma) \, d\sigma$.

If the space E has finite dimension n, then every quadratic form can be written in the form

$$q(x) = \boldsymbol{\xi}' A \boldsymbol{\xi} = \sum_{i=1}^{n} \sum_{k=1}^{n} \alpha_{ik} \xi_i \xi_k \tag{2}$$

for some symmetric real matrix A. Conversely, for any real matrix A, this expression defines a quadratic form f on E. For a given basis, the quadratic forms and the symmetric real matrices each uniquely determine the other.

For example, a positive definite form can be constructed from the matrix $A = I$ and this gives $q(x) = \sum_{k=1}^{n} \xi_k^2$.

11.1.3 The Reduction of Quadratic Forms

If we put $A\boldsymbol{\xi} = \boldsymbol{\eta}$, then the quadratic form (2) becomes

$$q(x) = \boldsymbol{\xi}' A \boldsymbol{\xi} = \boldsymbol{\xi}' \boldsymbol{\eta} = \boldsymbol{\eta}' \boldsymbol{\xi} = \sum_{k=1}^{n} \xi_k \eta_k.$$

As in 5.6;(3), suppose that the matrix A is partitioned into the four sub-matrices A_1, B, C, D. If A is a square symmetric matrix, then so are A_1 and D and further $C = B'$. We then obtain the tableau

	ξ^1	ξ^2
$\eta^1 =$	A_1	B
$\eta^2 =$	B'	D

$$(3)$$

where we have written ξ^1 in place of $\xi_1, \ldots, \xi_r$, etc.

We will now assume that A_1 is non-singular, i.e., that A_1^{-1} exists. Then $\eta_1, \ldots, \eta_r$ can be exchanged with $\xi_1, \ldots, \xi_r$ and, as in 5.6;(6), this will give the tableau

	η^1	ξ^2
$\xi^1 =$	P	Q
$\eta^2 =$	R	S

(4)

where $P = A_1^{-1}$, $Q = -A_1^{-1}B$, $R = B'A_1^{-1}$, $S = D - B'A_1^{-1}B$. Therefore S is symmetric ($S' = D' - B'(A_1^{-1})B = S$), P is symmetric ($P' = (A_1^{-1})' = (A_1')^{-1} = A_1^{-1} = P$) and $R' = -Q$.

Tableau (4) expresses the relations

$$\xi^1 = P\eta^1 + Q\xi^2$$
$$\eta^2 = R\eta^1 + S\xi^2 \text{ (or } \eta^{2'} = \eta^{1'}R' + \xi^{2'}S').$$

From all of these, it follows that

$$q(x) = \eta'\xi = \eta^{1'}\xi^1 + \eta^{2'}\xi^2 \tag{5}$$
$$= \eta^{1'}[P\eta^1 + Q\xi^2] + [\eta^{1'}R' + \xi^{2'}S']\xi^2$$
$$= \eta^{1'}P\eta^1 + \xi^{2'}S\xi^2$$
$$= \sum_{i,k=1}^{r} \pi_{ik}\eta_i\eta_k + \sum_{i,k=r+1}^{n} \sigma_{ik}\xi_i\xi_k$$

where $P = (\pi_{ik})$ and $S = (\sigma_{ik})$.

Thus $q(x)$ is the sum of a quadratic form in the variables $\eta_1, \ldots, \eta_r$ and a quadratic form in the variables $\xi_{r+1}, \ldots, \xi_n$. That is, by introducing the variables $\eta_1, \ldots, \eta_r$, $q(x)$ is divided into two independent parts and is said to be *reduced*. Since A_1^{-1} exists, the change from the original variables $\xi_1, \ldots, \xi_n$ to the new variables $\eta_1, \ldots, \eta_r, \xi_{r+1}, \ldots, \xi_n$ corresponds to a change of basis, so that the reduction of the quadratic form (2) to the form (5) is the result of a change of basis.

In particular the above method can be applied to the special case $r=1$ which means that the variable ξ_1 is replaced by η_1. This is only possible if $\alpha_{11} \neq 0$. In this way, a quadratic form $\lambda_1\eta_1^2$ in η_1 is split off from $q(x)$, where $\lambda_1 = 1/\alpha_{11}$, $(P = A_1^{-1})$.

Thus
$$q(x) = \lambda_1 \eta_1^2 + q^*(x)$$

where $q^*(x)$ is a quadratic form in $\xi_2, \ldots, \xi_n$. The form $q^*(x)$ can now be dealt with in the same way, providing its coefficient matrix has a non-zero diagonal element, and then $q(x)$ takes the form

$$q(x) = \lambda_1 \eta_1^2 + \lambda_2 \eta_2^{*2} + q^{**}(x).$$

Providing there are always non-zero diagonal elements available, we can continue in this way until

$$q(x) = \sum_{k=1}^{n} \lambda_k \zeta_k^2 \tag{6}$$

where the new variables are now uniformly denoted by ζ_k. The coefficients λ_k are equal to the reciprocals of the pivots. When put into the form (6), $q(x)$ is said to be *completely reduced* or in *canonical form*. Since each individual step in the reduction is the result of a change of basis, then so is the passage from (2) to (6). The new variables $\zeta_1, \ldots, \zeta_n$ are the components of x with respect to the new basis.

So far, the ability to carry out the complete reduction has depended on the existence at each step of a non-zero diagonal element. However we can remove this restriction as follows. Suppose for instance that in (2) every $\alpha_{ii} = 0$ $(i = 1, \ldots, n)$. Then, either $q(x) = 0$ for all x and $q(x)$ is already completely reduced, or there is an element $\alpha_{ik} \neq 0$ where $i \neq k$. Suppose that $\alpha_{12} \neq 0$. We make the following change of basis.

$$\xi_1 = \xi_1^* + \xi_2^*$$
$$\xi_2 = \xi_1^* - \xi_2^*$$
$$\xi_k = \xi_k^* \qquad (k = 3, \ldots, n)$$

Then $\xi_1 \xi_2 = \xi_1^{*2} - \xi_2^{*2}$. Thus, in the new variables ξ_k^*, $q(x)$ contains two perfect squares which it is easy to see are not cancelled by any of the others. Now a variable can be split off using the method described above. Thus we have the following general result.

Theorem 3. The quadratic form (2) can be completely reduced (i.e., put into the form (6)) by a suitable change of basis (or by a real linear transformation of the variables).

We remark that the reduction method described here is not the only one possible. For instance, the basis with respect to which $q(x)$ is completely reduced is in no way uniquely determined, nor are the coefficients λ_k in (6) uniquely determined by $q(x)$. (Nevertheless see Theorem 6.)

211

In the special case of a positive definite quadratic form (Definition 4), all diagonal elements are strictly positive after each exchange step. To show this, it is sufficient to show that no diagonal element of the matrix A is negative or 0 if the form represented by (2) is positive definite. But, for instance, if $\alpha_{11} \leqslant 0$, then $q(x) \leqslant 0$ for $\xi_1 = 1$, $\xi_2 = \ldots = \xi_n = 0$, which is a contradiction of the definition of positive definite.

Thus, if $q(x)$ is positive definite, then it is always possible at each exchange step to use the next element of the main diagonal as the pivot. If we denote these pivots by $\delta_1, \ldots, \delta_n$, then, by 4.3.2,

$$\delta_1 = \alpha_{11} = \Delta_1, \quad \delta_2 = \frac{\Delta_2}{\Delta_1}, \ldots, \delta_n = \frac{\Delta_n}{\Delta_{n-1}}$$

where
$$\Delta_r = \begin{vmatrix} \alpha_{11} & \cdots & \alpha_{1r} \\ \cdots & \cdots & \cdots \\ \alpha_{r1} & \cdots & \alpha_{rr} \end{vmatrix} \quad (r = 1, \ldots, n),$$

i.e., Δ_r is the determinant formed from the intersection of the first r rows and columns of A. Hence the reduced form is

$$q(x) = \frac{1}{\delta_1} \zeta_1^2 + \ldots + \frac{1}{\delta_n} \zeta_n^2$$

$$= \frac{1}{\Delta_1} \zeta_1^2 + \frac{\Delta_1}{\Delta_2} \zeta_2^2 + \ldots + \frac{\Delta_{n-1}}{\Delta_n} \zeta_n^2. \tag{7}$$

Since q is positive definite, the coefficients must be strictly positive. Conversely, if all the coefficients are strictly positive, then q is obviously positive definite. Thus we have

Theorem 4. *The quadratic form $q(x)$ with the symmetric matrix A is positive definite if and only if the determinants $\Delta_1, \ldots, \Delta_n$ are strictly positive.*

For example, the quadratic form $q(x) = \alpha_{11} \xi_1^2 + 2\alpha_{12} \xi_1 \xi_2 + \alpha_{22} \xi_2^2$ in two variables is positive definite if and only if

$$\alpha_{11} > 0 \quad \text{and} \quad \begin{vmatrix} \alpha_{11} & \alpha_{12} \\ \alpha_{12} & \alpha_{22} \end{vmatrix} = \alpha_{11} \alpha_{22} - \alpha_{12}^2 > 0.$$

Example 4. Let $q(x) = \xi_1^2 + 2\xi_2^2 + \xi_3^2 + 4\xi_1\xi_2 + 8\xi_1\xi_3 + 2\xi_2\xi_3 = \xi' A \xi$,

where
$$A = \begin{pmatrix} 1 & 2 & 4 \\ 2 & 2 & 1 \\ 4 & 1 & 1 \end{pmatrix}$$

1st Tableau

	ξ_1	ξ_2	ξ_3
$\zeta_1 = \eta_1 =$	1*	2	4
$\eta_2 =$	2	2	1
$\eta_3 =$	4	1	1

2nd Tableau

	ξ_2	ξ_3
$\zeta_2 = \eta_2' =$	-2*	-7
$\eta_3' =$	-7	-15

Note that η_2' and η_3' are not identical with the original η_2 and η_3.

3rd Tableau

	ξ_3
$\zeta_3 = \eta_3'' =$	$\frac{19}{2}$

In the calculations it is possible to make use of the fact that all the tableaux are symmetric. For instance, in the 2nd tableau, only one of the coefficients, -7, needs to be calculated—the other follows by symmetry.

Thus $q(x)$ is put into the reduced form

$$q(x) = \zeta_1^2 - \tfrac{1}{2}\zeta_2^2 + \tfrac{2}{19}\zeta_3^2 \qquad (8)$$

by the transformation

$$\zeta_1 = \xi_1 + 2\xi_2 + 4\xi_3$$
$$\zeta_2 = \qquad -2\xi_2 - 7\xi_3 \qquad (9)$$
$$\zeta_3 = \qquad\qquad \tfrac{19}{2}\xi_3$$

The coefficients of ζ_k^2 are the reciprocals of the pivots. From (8) it follows that q is not positive definite.

The equations (9) are particularly easy to solve for the ξ_k because the matrix of coefficients is triangular (i.e., all elements below the main diagonal are equal to zero). We first solve the 3rd equation for ξ_3, then the 2nd for ξ_2, and finally the 1st for ξ_1 to obtain

$$\xi_1 = \zeta_1 + \zeta_2 + \tfrac{6}{19}\zeta_3$$
$$\xi_3 = \qquad -\tfrac{1}{2}\zeta_2 - \tfrac{7}{19}\zeta_3 \qquad (10)$$
$$\xi_3 = \qquad\qquad \tfrac{2}{19}\zeta_3$$

We can easily verify (8) by substituting (10) in the original representation of $q(x)$.

The equations (10) could also be found by doing the reduction with the full tableaux (see Example 12.1;4).

It is also easy to find the basis vectors f_1, f_2, f_3 to which the components $\zeta_1, \zeta_2, \zeta_3$ are referred. For, if we write (9) in the form $\boldsymbol{\zeta} = S\boldsymbol{\xi}$, then $\mathbf{f} = S^* \mathbf{e} = (S^{-1})' \mathbf{e}$ where $\{e_1, e_2, e_3\}$ is the original basis (Theorem 5.4;2). Now S^{-1} can be found from (10) and we have

$$f_1 = e_1$$
$$f_2 = e_1 - \tfrac{1}{2}e_2$$
$$f_3 = \tfrac{6}{19}e_1 - \tfrac{7}{19}e_2 + \tfrac{2}{19}e_3.$$

The method can obviously be applied in the way just shown (using the diagonal elements in their natural order as pivots) whenever $\Delta_1, \dots, \Delta_n$ are not zero and in particular for positive definite forms. The matrix S in (9) is always then a triangular matrix. Since the matrix of the reduced form is diagonal (i.e., all elements off the main diagonal are zero) and in view of the remark after the proof of Theorem 1, we have the following result.

Theorem 5. If, for a real symmetric matrix A, the determinants $\Delta_1, \dots, \Delta_n$ are not zero, then there exists a non-singular real triangular matrix S such that $A^* = S'AS$ is diagonal. (See Problem 2.)

11.1.4 The Inertia Theorem

If the reduced quadratic form $q(x) = \sum\limits_{k=1}^{n} \lambda_k \zeta_k^2$ is transformed by substituting

$$\zeta_k^* = \sqrt{|\lambda_k|}\, \zeta_k, \qquad \text{if } \lambda_k \neq 0$$

and
$$\zeta_k^* = \zeta_k, \qquad \text{if } \lambda_k = 0,$$

then $q(x)$ becomes

$$q(x) = \sum_{k=1}^{n} \mu_k \zeta_k^{*2} \quad \text{where} \quad \mu_k = \pm 1 \text{ or } 0 \ (k = 1, \dots, n). \tag{11}$$

Of course, without carrying out the reduction, it is not possible to say which μ_k are equal to $+1$, -1 or 0. However we do have the following *Inertia Theorem* which is due to Sylvester.

Theorem 6. If the quadratic form $q(x)$ is represented in the form (11), the numbers of positive and negative terms are uniquely determined by $q(x)$. In particular, these numbers are independent of the method of reduction and the transformation of the variables used to reach the form (11).

Proof. If the theorem were false, then there would be a quadratic form with two representations

$$q(x) = \xi_1^2 + \ldots + \xi_p^2 - \xi_{p+1}^2 - \ldots - \xi_r^2$$
$$= \eta_1^2 + \ldots + \eta_q^2 - \eta_{q+1}^2 - \ldots - \eta_s^2,$$

where $p > q$. It could happen that no negative terms appear in either or both of these representations. On the other hand, we can assume that some positive terms actually appear in the first representation. Suppose that the components ξ_k refer to the basis $\{e_1, \ldots, e_n\}$, and η_k refer to the basis $\{f_1, \ldots f_n\}$. Let $L_1 = L(e_1, \ldots, e_p)$ and $L_2 = L(f_{q+1}, \ldots, f_n)$. By Theorem 3.2;8 there is an $x_0 \in L_1 \cap L_2$, $x_0 \neq 0$. For this vector x_0, $\xi_{p+1} = \ldots = \xi_n = \eta_1 = \ldots = \eta_q = 0$, while not all of $\xi_1, \ldots, \xi_p$ are zero. From the first representation of q, it follows that $q(x_0) > 0$ and from the second that $q(x_0) \leqslant 0$. Thus we have reached a contradiction.

Exercise

Reduce the following two quadratic forms to canonical form and state the corresponding transformation of the variables. Are these forms positive definite?

(a) $q(x) = \xi_1^2 + 3\xi_2^2 + 5\xi_3^2 - 2\xi_1\xi_2 + 4\xi_1\xi_3 - 6\xi_2\xi_3$

(b) $q(x) = 2\xi_1^2 + 3\xi_2^2 + 8\xi_3^2 + 2\xi_1\xi_2 - 8\xi_1\xi_3 + 6\xi_2\xi_3.$

Solution. (a) $q(x) = \zeta_1^2 + \tfrac{1}{2}\zeta_2^2 + 2\zeta_3^2$

where $\zeta_1 = \xi_1 - \xi_2 + 2\xi_3$

$\zeta_2 = \quad 2\xi_2 - \xi_3$

$\zeta_3 = \qquad \tfrac{1}{2}\xi_3$

$q(x)$ is positive definite.

(b) $q(x) = \tfrac{1}{2}\zeta_1^2 + \tfrac{2}{5}\zeta_2^2 - \tfrac{1}{10}\zeta_3^2$

where $\zeta_1 = 2\xi_1 + \xi_2 - 4\xi_3$

$\zeta_2 = \quad \tfrac{5}{2}\xi_2 + 5\xi_3$

$\zeta_3 = \qquad -10\xi_3$

$q(x)$ is not positive definite.

Problems

1. Let A and S be $n \times n$ matrices. Show that, if A is symmetric, then so is $S'AS$. (Cf. Theorem 1.)

2. Show that Theorem 5 is not true for the symmetric matrix $A = \begin{pmatrix} 0 & 1 \\ 1 & 0 \end{pmatrix}$.

3. Prove that, if A is a real $n \times n$ matrix and $\det A \neq 0$, then the quadratic form $q(x) = \boldsymbol{\xi}' A A' \boldsymbol{\xi}$ is positive definite.

4. Suppose that, for the real $n \times n$ matrix A, the determinants $\varDelta_1, \dots, \varDelta_n$ (cf. Theorem 4) are all strictly positive. Prove that all the diagonal elements α_{kk} are also strictly positive.

5. Suppose that the quadratic form $q(x)$ has the reduced form (6) with respect to a given basis. What is the polar form of $q(x)$ with respect to the same basis?

6. Prove that, if A is a real symmetric matrix, then there is a real number λ such that the quadratic form $\boldsymbol{\xi}'(A + \lambda I)\boldsymbol{\xi}$ is positive definite.

7. If A is a real symmetric matrix, show that $\boldsymbol{\xi}' A \boldsymbol{\xi}$ is positive definite if and only if $\boldsymbol{\xi}' A^{-1} \boldsymbol{\xi}$ is positive definite.

8. Prove that the symmetric matrix A of the quadratic form $q(x) = \boldsymbol{\xi}' A \boldsymbol{\xi}$ with respect to a given basis is uniquely determined by q.

9. Let $f(x,y) = \boldsymbol{\xi}' A \boldsymbol{\eta}$ be a bilinear form on the vector space E. Prove that $f(x,x) = 0$ for all $x \in E$ if and only if $f(x,y) = -f(y,x)$ for all $x, y \in E$, i.e., if $A = -A'$.

11.2 Hermitian Forms on Complex Vector Spaces

Part of the discussion in 11.1 can be carried over directly to the case of complex vector spaces. However, we meet with difficulties for example when we come to the definition of positive definite quadratic forms. Apart from this, it is preferable on formal grounds to proceed somewhat differently in the complex case as will be described in the present section. In this, we will not repeat proofs which are simple generalizations of those in 11.1. If α is a complex number, we will denote the complex conjugate of α by $\bar{\alpha}$.

11.2.1 Quasi-bilinear Forms

Definition 1. A 'quasi-bilinear form' on a complex vector space E is a complex-valued function $f(x,y)$ of two variables $x, y \in E$ which is linear in x and which is dependent on y according to the rules

$$f(x, y_1 + y_2) = f(x, y_1) + f(x, y_2)$$

and
$$f(x, \alpha y) = \bar{\alpha} f(x, y)$$

for all $x, y, y_1, y_2 \in E$ and all scalars α.

Example 1. We use the same notation as in Example 11.1;1, except that now $x(\sigma)$, $y(\tau)$ and $k(\sigma, \tau)$ are complex-valued functions of the real variables σ, τ. Then

$$f(x, y) = \int_{-1}^{+1} \int_{-1}^{+1} k(\sigma, \tau) x(\sigma) \overline{y}(\tau) \, d\sigma \, d\tau$$

is a quasi-bilinear form.

216

If E is finite-dimensional, the representation 11.1;(1) corresponds to the following representation of the quasi-bilinear form $f(x,y)$

$$f(x,y) = \sum_{i,k=1}^{n} \alpha_{ik} \xi_i \bar{\eta}_k = \boldsymbol{\xi}' A \bar{\boldsymbol{\eta}}, \quad \text{where} \quad \alpha_{ik} = f(e_i, e_k) \tag{1}$$

and where $\bar{\boldsymbol{\eta}}$ means the matrix whose elements are the complex conjugates of the elements of $\boldsymbol{\eta}$. For a given basis, f and A again determine each other uniquely.

Under a change of basis $\boldsymbol{\xi} = S\boldsymbol{\xi}^*$, A goes into $A^* = S' A \bar{S}$.

Definition 2. The quasi-bilinear form is said to be 'Hermitian' if

$$f(y,x) = \bar{f}(x,y).$$

(We write $\bar{f}(x,y)$ instead of $\overline{f(x,y)}$.)

We note that the property $f(x, \alpha y) = \bar{\alpha} f(x,y)$ of Definition 1 is a consequence of the linearity with respect to x and the condition $f(y,x) = \bar{f}(x,y)$. ($f(x, \alpha y) = \bar{f}(\alpha y, x) = \bar{\alpha} \bar{f}(y,x) = \bar{\alpha} f(x,y)$.) The quasi-bilinear form in Example 1 is Hermitian if and only if $k(\tau, \sigma) = \bar{k}(\sigma, \tau)$.

Theorem 1. If the complex vector space E is finite-dimensional, then $f(x,y) = \boldsymbol{\xi}' A \bar{\boldsymbol{\eta}}$ is Hermitian if and only if $A' = \bar{A}$. A matrix A with this last property is also said to be Hermitian.

The simple proof of this is the content of Problem 5.

11.2.2 Quadratic and Hermitian Forms

Definition 3. A complex-valued function $q(x)$ on a complex vector space E is said to be a 'quadratic form' if there is a quasi-bilinear form $f(x,y)$ (the 'polar form' of q) such that $q(x) = f(x,x)$.

Theorem 2. A quadratic form and its polar form on a complex vector space determine each other uniquely.

Proof. $f(x,y) = \frac{1}{4}[q(x+y) + iq(x+iy) - q(x-y) - iq(x-iy)]$.

Definition 4. A quadratic form on a complex vector space is said to be 'Hermitian' if its polar form is Hermitian. We then speak briefly of a 'Hermitian' form.

Thus, in view of Example 1, a Hermitian form is given by

$$q(x) = \int_{-1}^{+1} k(\sigma, \tau) x(\sigma) \bar{x}(\tau) \, d\sigma \, d\tau$$

whenever $k(\sigma, \tau) = \bar{k}(\tau, \sigma)$.

If E is finite-dimensional, then Hermitian forms may be written in the form

$$q(x) = \boldsymbol{\xi}' A \bar{\boldsymbol{\xi}} \tag{2}$$

where A is Hermitian. Conversely (2) always gives a Hermitian form whenever A is Hermitian. For a given basis, q and A determine each other uniquely.

Theorem 3. A quadratic form $q(x)$ on a complex vector space is Hermitian if and only if it only takes real values.

Proof. 1. If q is Hermitian, then $q(x) = f(x,x) = \bar{f}(x,x) = \bar{q}(x)$.

  2. If q only takes real values, then in particular $q(x \pm y)$ and $q(x \pm iy)$ are real. Now if, in the proof of Theorem 2, we interchange x and y, $q(x+y)$ and $q(x-y)$ do not change while $q(x+iy)$ and $q(x-iy)$ interchange. Therefore $f(y,x) = \bar{f}(x,y)$.

11.2.3   The Reduction of Hermitian Forms

In this section we will formulate the theorems corresponding to the results of 11.1.3. We will omit the proofs which are simple generalizations of those in 11.1.3.

Theorem 4. Given a Hermitian form $q(x)$ on a finite-dimensional complex vector space, then there is a basis with respect to which $q(x)$ takes the form

$$q(x) = \sum_{k=1}^{n} \lambda_k \zeta_k \bar{\zeta}_k = \sum_{k=1}^{n} \lambda_k |\zeta_k|^2 \tag{3}$$

where the coefficients λ_k are real.

A Hermitian form is said to be *positive definite* if it never takes negative values and if it only takes the value 0 for $x = 0$. Clearly the Hermitian form (3) is positive definite if and only if all the coefficients λ_k are strictly positive.

  The determinants $\Delta_1, \ldots, \Delta_n$ (defined as in 11.1.3) are real for a Hermitian matrix A (see Problem 4).

Theorem 5. If $q(x)$ is a Hermitian form and the determinants $\Delta_1, \ldots, \Delta_n$ are not zero, then there is a basis such that

$$q(x) = \frac{1}{\Delta_1} |\zeta_1|^2 + \frac{\Delta_1}{\Delta_2} |\zeta_2|^2 + \ldots + \frac{\Delta_{n-1}}{\Delta_n} |\zeta_n|^2.$$

Theorem 6. A Hermitian form is positive definite if and only if $\Delta_1, \ldots, \Delta_n$ are strictly positive.

Theorem 7. (*'Inertia Theorem'*) *If a Hermitian form is written in two different ways in the form* (3), *then the numbers of positive, negative and zero coefficients are the same in both forms.*

Problems

1. Show that $\bar{A}A'$ is Hermitian for all matrices A.

2. Show that, if A is a square matrix and $\det A \neq 0$, then $\boldsymbol{\xi}'\,\bar{A}A'\,\boldsymbol{\xi}$ is positive definite.

3. Show that, if A is Hermitian, then so is A^{-1} (providing A^{-1} exists).

4. Prove that, if A is Hermitian, then $\det A$ is real.

5. Prove Theorem 11.2;1.

6. Let $f(x,y) = \boldsymbol{\xi}'A\boldsymbol{\eta}$ be a quasi-bilinear form on a complex vector space E. Prove that $f(x,x) = 0$ for all $x \in E$ if and only if $f(x,y) = 0$ for all $x, y \in E$, i.e., if $A = 0$.

7. If A is a Hermitian matrix, show that $S'A\bar{S}$ is also Hermitian, where S is an arbitrary matrix for which the product is defined.

8. Adapt the reduction method described in 11.1.3 so that it can be applied to Hermitian forms and use it to prove Theorems 4, 5 and 6.

CHAPTER 12

EUCLIDEAN AND UNITARY VECTOR SPACES

As in the previous chapter, we will first consider the real case (Euclidean spaces) in sections 12.1 to 12.3 and then the complex case (Unitary spaces) in section 12.4.

12.1 Euclidean Vector Spaces

12.1.1 The Concept of a Euclidean Vector Space

Definition 1. A Euclidean vector space is a real vector space together with a positive definite quadratic form which is known as the 'fundamental form' of the space.

If $q(x)$ is the fundamental form of the Euclidean space E, then we refer to the real number $\|x\| = +\sqrt{\{q(x)\}}$ as the *length* or the *norm* of the vector $x \in E$. The zero vector is the only vector which has norm 0. For all scalars α and all $x \in E$, $\|\alpha x\| = |\alpha| \, \|x\|$. A vector $x \in E$ is said to be *normalized* if $\|x\| = 1$. If $y \in E$ and $y \neq 0$, then $x = \|y\|^{-1} y$ is normalized.

The polar form of the fundamental form $q(x)$, which we will now write briefly as (x, y) is known as the *scalar product* of the vectors x and y. Thus the scalar product is a symmetric bilinear form on E.

Example 1. The vectors of the 3-dimensional space of Euclidean geometry form a Euclidean vector space when $\|x\|$ is defined to be the length of the vector x in the usual sense. It is from this example that the name 'Euclidean vector space' is derived. (See Problem 1).

Example 2. The real n-dimensional space R_n is a Euclidean vector space when

$$\|x\|^2 = \sum_{k=1}^{n} \xi_k^2, \qquad \text{and hence } (x, y) = \sum_{k=1}^{n} \xi_k \eta_k.$$

Example 3. The vector space C (Example 2.1;7) is a Euclidean vector space when

$$\|x(\tau)\|^2 = \int_{-1}^{+1} [x(\tau)]^2 \, d\tau$$

220

(see Example 11.1;3). The scalar product is $(x,y) = \int\limits_{-1}^{+1} x(\tau)y(\tau)\,\mathrm{d}\tau$.

Theorem 1. (Cauchy–Schwarz Inequality.) For all vectors x, y of a Euclidean vector space

$$|(x,y)| \leqslant \|x\|\,\|y\|.$$

Equality holds if and only if x and y are linearly dependent.

Proof. 1. If x and y are linearly dependent, $y = \lambda x$ say, then both sides are equal to $|\lambda|\,\|x\|^2$.

2. If x and y are linearly independent, then we think of

$$q(\lambda x + \mu y) = (\lambda x + \mu y, \lambda x + \mu y) = \|x\|^2\lambda^2 + 2(x,y)\lambda\mu + \|y\|^2\mu^2$$

as a quadratic form in the variables λ, μ. This is positive definite, and therefore

$$\Delta_2 = \begin{vmatrix} \|x\|^2 & (x,y) \\ (x,y) & \|y\|^2 \end{vmatrix} = \|x\|^2\|y\|^2 - (x,y)^2 > 0.$$

(see Theorem 11.1;4).

In view of the Cauchy–Schwarz inequality, it is possible to define the angle α between two vectors x and $y (\neq 0)$ by the formula $\cos\alpha = (x,y)/\|x\|\,\|y\|$.

Theorem 2. (Triangle Inequality.) For all vectors x,y of a Euclidean vector space

$$\big|\,\|x\| - \|y\|\,\big| \leqslant \|x+y\| \leqslant \|x\| + \|y\|.$$

Proof. By Theorem 1,

$$\|x+y\|^2 = \|x\|^2 + 2(x,y) + \|y\|^2 \leqslant (\|x\| + \|y\|)^2$$

and the second inequality follows from this (remembering that both $\|x+y\|$ and $\|x\| + \|y\|$ are positive or zero). The first inequality follows from the second because

$$\|x\| = \|(x+y) - y\| \leqslant \|x+y\| + \|y\|.$$

The vectors x and y are said to be *orthogonal* if $(x,y) = 0$. Since (x,y) is a symmetric bilinear form, orthogonality is a symmetric relation between vectors.

The zero vector is the only vector which is orthogonal to all vectors $x \in E$. This means that every Euclidean vector space E forms a dual pair of spaces with itself using the scalar product (x,y) (see Definition 6.2;3). If E is finite-dimensional, the definitions and theorems of sections 6.2.3 to 6.2.6 can be brought directly into the present work.

In particular the dual space, or as we will now say the *orthogonal comple-ment*, $L^\dagger$ of a subspace L of E consists of all those vectors $y \in E$ which are orthogonal to every vector $x \in L$. By Theorem 6.2;6, $L^{\dagger\dagger} = L$ (for finite-dimensional spaces) and the sum of the dimensions of L and $L^\dagger$ is equal to the dimension of E. On the other hand, $L \cap L^\dagger = \{0\}$, because 0 is orthogonal to itself and, if $\|x\|^2 = 0$, then $x = 0$. From these we have the following result. (Cf. Theorems 2.4;10 and 3.2;6.)

Theorem 3. Let E be a finite-dimensional Euclidean vector space, L a subspace of E and $L^\dagger$ its orthogonal complement. Then $E = L \oplus L^\dagger$, i.e., for each $x \in E$ there is a unique representation $x = x_1 + x_2$ where $x_1 \in L$ and $x_2 \in L^\dagger$.

Theorem 4. In a Euclidean vector space E, the vectors $a_1, \ldots, a_r \in E$ are linearly dependent if and only if Gram's determinant

$$G = \mathrm{Det}\Big((a_i, a_k)\Big)$$

is equal to zero (see Problem 2).

The elements of Gram's determinant are the scalar products of the vectors $a_1, \ldots, a_n$ taken in pairs.

Proof. If $\sum\limits_{i=1}^{r} \lambda_i a_i = 0$, then

$$(\sum_i \lambda_i a_i, a_k) = \sum_i \lambda_i(a_i, a_k) = 0 \quad \text{for } k = 1, \ldots, r. \tag{1}$$

Conversely (1) means that $\sum\limits_i \lambda_i a_i \in [L(a_1, \ldots, a_r)]^\dagger$. On the other hand $\sum\limits_i \lambda_i a_i \in L(a_1, \ldots, a_r)$. Hence by Theorem 3, it follows from (1) that $\sum\limits_i \lambda_i a_i = 0$. Thus a linear relation between $a_1, \ldots, a_r$ is equivalent to the same relation between the rows of Gram's determinant and the theorem follows from Theorem 4.2;7.

12.1.2 Orthogonalization

Definition 2. A subset A of a Euclidean vector space E is said to be 'orthogonal' if any two of its elements are orthogonal. It is said to be 'orthonormal' if, in addition, its elements are normalized.

Theorem 5. Any orthogonal set A which does not contain the zero vector is linearly independent. Thus in particular an orthonormal set is linearly independent.

Proof. If $\sum\limits_{x \in A} \lambda_x x = 0$, then, for any $y \in A$, it follows from the conditions of the theorem that $\sum\limits_{x \in A} \lambda_x(x, y) = \lambda_y(y, y) = 0$, and hence $\lambda_y = 0$.

Theorem 6. *Let E be a Euclidean vector space and let A be a finite or countably infinite subset of E. Then the subspace $L(A)$ of E has an orthonormal basis. In particular every finite-dimensional Euclidean vector space has an orthonormal basis.*

Proof. Since $L(A)$ has a finite or countably infinite basis (Theorem 3.1;10), we will not lose any generality by assuming that the set A is linearly independent. We will prove the theorem by using the *Schmidt Orthogonalization Method* which will also produce a technique for constructing an orthonormal basis for $L(A)$.

Suppose that the vectors of the linearly independent set A are a_1, a_2, a_3, ... We first construct an orthogonal basis e_1, e_2, e_3, ... by setting

$$\left. \begin{aligned} e_1 &= a_1 \\ e_2 &= \lambda_{21} e_1 + a_2 \\ e_3 &= \lambda_{31} e_1 + \lambda_{32} e_2 + a_3, \end{aligned} \right\} \tag{2}$$

etc., where the coefficients are to be determined as follows. Since $(e_1, e_2) = 0$, it follows that

$$(e_1, \lambda_{21} e_1 + a_2) = \lambda_{21}\|e_1\|^2 + (e_1, a_2) = 0$$

and, since $\|e_1\| = \|a_1\| \neq 0$, λ_{21} can be determined from this equation. Since $(e_1, e_3) = 0$, it follows similarly that

$$(e_1, \lambda_{31} e_1 + \lambda_{32} e_2 + a_3) = \lambda_{31}\|e_1\|^2 + \lambda_{32}(e_1, e_2) + (e_1, a_3)$$
$$= \lambda_{31}\|e_1\|^2 + (e_1, a_3) = 0$$

and λ_{31} can be found from this. The condition $(e_2, e_3) = 0$, similarly leads to the equation $\lambda_{32}\|e_2\|^2 + (e_2, a_3) = 0$ for λ_{32}. (Since a_1 and a_2 may be written as linear combinations of e_1 and e_2 from (2), it follows that e_1 and e_2 are linearly independent and hence that $\|e_2\| \neq 0$.)

It is clear that this Schmidt Orthogonalization Method can be continued in such a way that a new vector e_k is assigned to each vector a_k. In view of the method of construction, any two of these vectors e_k will be orthogonal. The set e_1, e_2, e_3, ... is a basis of $L(A)$ because for any r, the vectors $a_1, ..., a_r$ may be written as linear combinations of $e_1, ..., e_r$ from (2). We can now obtain an orthonormal basis simply by normalizing the vectors e_k, i.e., by replacing e_k with $\|e_k\|^{-1} e_k$.

We remark that the proof also shows that the vectors e_k are uniquely

223

determined by the formulae (2) and the condition of orthonormality. It is easy to see that the vectors e_k are uniquely determined except for scalar factors, if we require any two to be orthogonal and each e_k to be a linear combination of $a_1, \ldots, a_k$ $(k=1,2,3,\ldots)$.

With a view to producing a concise technique for calculation, we will now show that the Schmidt Method can be carried out by using the Exchange Method. We first prove the following result.

Theorem 7. A basis B of a Euclidean vector space E is orthogonal if and only if the fundamental form is completely reduced with respect to this basis, i.e.,

$$\|x\|^2 = q(x) = \sum_{e \in B} \lambda_e \zeta_e^2. \tag{3}$$

Further, B is orthonormal if and only if $\lambda_e = 1$ for all $e \in B$.

Proof. 1. For any basis B, we have

$$q(x) = (x, x) = \sum_{e \in B} \sum_{f \in B} (e, f)\, \zeta_e \zeta_f. \tag{4}$$

Now, if B is orthogonal, then $(e, f) = 0$ for $e \neq f$ and hence we have (3) with $\lambda_e = (e, e)$. Further, if B is orthonormal, then $\lambda_e = (e, e) = 1$ for all $e \in B$.

2. From (3) and (4), it follows that $(e, f) = 0$ for $e \neq f$ and hence that B is orthogonal (see Problem 11.1;8). Further $\lambda_e = (e, e)$ for all $e \in B$ so that B is orthonormal when all $\lambda_e = 1$.

Thus, in view of Theorem 7, we can obtain an orthogonal basis by reducing the fundamental form. The Reduction Method of 11.1.3 can be easily extended to a countably infinite number of variables simply by allowing tableaux with a countably infinite number of rows and columns. Except for scalar factors, this method will give the same orthogonal basis as the Schmidt Method, because again each e_k is a linear combination of $a_1, \ldots, a_k$ $(k=1,2,3,\ldots)$. The norms $\|e_k\|$, which are used to normalize the basis vectors e_k, also appear in the method, because, from (3), we have $\|e_k\| = +\sqrt{\lambda_k}$. But $\lambda_k = 1/\delta_k$ where δ_k is the kth pivot. Thus the vector e_k can be normalized by multiplying by $+\sqrt{\delta_k}$.

Example 4. The space P_3 (Example 2.1;5) is Euclidean with the fundamental form

$$\|x\|^2 = q(x) = \int_{-1}^{+1} [x(\tau)]^2 \, d\tau$$

since, if a polynomial $x(\tau)$ vanishes on $-1 \leqslant \tau \leqslant +1$, it vanishes for all τ and therefore $x = 0$. A basis of P_2 is given by $a_0(\tau) = 1$; $a_1(\tau) = \tau$; $a_2(\tau) = \tau^2$;

$a_3(\tau) = \tau^3$ (see Example 3.1;5). With respect to this basis, we have by (4) that

$$q(x) = \sum_{i,\,k=0}^{3} (a_i, a_k)\, \xi_i \xi_k$$

where

$$(a_i, a_k) = \int_{-1}^{+1} a_i(\tau)\, a_k(\tau)\, \mathrm{d}\tau$$

$$= \int_{-1}^{+1} \tau^{i+k}\, \mathrm{d}\tau$$

$$= \begin{cases} 0 & \text{if } (i+k) \text{ is odd} \\[2mm] \dfrac{2}{i+k+1} & \text{if } (i+k) \text{ is even.} \end{cases}$$

Hence the initial tableau for the reduction of the fundamental form is

	ξ_0	ξ_1	ξ_2	ξ_3
$\zeta_0 = \eta_0$	2*	0	$\frac{2}{3}$	0
η_1	0	$\frac{2}{3}$	0	$\frac{2}{5}$
η_2	$\frac{2}{3}$	0	$\frac{2}{5}$	0
η_3	0	$\frac{2}{5}$	0	$\frac{2}{7}$

Using the elements in the main diagonal in their natural order as the pivots, we reach the following final tableau after four exchange steps.

	ζ_0	ζ_1	ζ_2	ζ_3
ξ_0	$\frac{1}{2}$	0	$-\frac{15}{8}$	0
ξ_1	0	$\frac{3}{2}$	0	$-\frac{105}{8}$
ξ_2	0	0	$\frac{45}{8}$	0
ξ_3	0	0	0	$\frac{175}{8}$

Here the pivots have the values 2, 2/3, 8/45, 8/175.

Thus the orthogonal basis is

$$e_0 = \tfrac{1}{2}a_0 = \tfrac{1}{2}$$

$$e_1 = \tfrac{3}{2}a_1 = \tfrac{3}{2}\tau$$

$$e_2 = -\tfrac{15}{8}(a_0 - 3a_2) = -\tfrac{15}{8}(1 - 3\tau^2)$$

$$e_3 = -\tfrac{35}{8}(3a_1 - 5a_3) = -\tfrac{35}{8}(3\tau - 5\tau^3).$$

To normalize these polynomials it is necessary to multiply them by $\sqrt{2}$, $\sqrt{(2/3)}$, $\sqrt{(8/45)}$, $\sqrt{(8/175)}$ respectively. This gives

$$\bar{e}_0 = \frac{1}{\sqrt{2}} \qquad \bar{e}_2 = \sqrt{(\tfrac{5}{8})}\,(3\tau^2 - 1)$$

$$\bar{e}_1 = \sqrt{(\tfrac{3}{2})}\,\tau \qquad \bar{e}_3 = \sqrt{(\tfrac{7}{8})}\,(5\tau^3 - 3\tau)$$

which are the first four normalized Legendre polynomials. (We note, however, that the Legendre polynomials are not usually normalized in this way but by the condition that they should take the value $+1$ for $\tau = +1$.) All the Legendre polynomials can be obtained using this method. However for this special purpose there are much simpler methods. (Cf. [18], p. 190.)

12.1.3   Orthogonal Transformations and Matrices

Let E be an n-dimensional Euclidean vector space and let $\xi_1, \ldots, \xi_n$ and $\eta_1, \ldots, \eta_n$ be the components of a vector $x \in E$ with respect to two orthonormal bases. Then, by Theorem 7,

$$\|x\|^2 = q(x) = \sum_{k=1}^{n} \eta_k^2 = \sum_{k=1}^{n} \xi_k^2 = \boldsymbol{\xi}'\boldsymbol{\xi} = \boldsymbol{\eta}'\boldsymbol{\eta}.$$

Now, if

$$\boldsymbol{\eta} = S\boldsymbol{\xi}, \tag{5}$$

(cf. Theorem 5.4; 2), then $\boldsymbol{\xi}'\boldsymbol{\xi} = \boldsymbol{\xi}'S'S\boldsymbol{\xi}$ and hence, in view of the symmetry of $S'S$,

$$S'S = I, \text{ i.e., } S^{-1} = S', \text{ i.e., } S^* = S \tag{6}$$

(see 5.4;(5)).

Definition 3. A real square matrix S which has the property (6) is said to be 'orthogonal'. The linear transformation (5) of the vector components which corresponds to S is also said to be 'orthogonal'.

Now suppose that (5) represents a change of basis in which the ξ-basis is orthonormal. If S is orthogonal, then

$$\boldsymbol{\eta}'\boldsymbol{\eta} = \boldsymbol{\xi}'S'S\boldsymbol{\xi} = \boldsymbol{\xi}'\boldsymbol{\xi} = q(x)$$

so that the η-basis is also orthonormal (Theorem 7).

If S is orthogonal, then, by Theorem 5.4;2, (5) corresponds to the transformation $f = S^* e = Se$ of the vector components. Thus the vector components are transformed in the same way as the basis vectors. Putting all this together, we have

Theorem 8. A change of basis between orthonormal bases is represented by an orthogonal matrix. Conversely, every orthogonal matrix can be expressed as the representation of a change of basis between orthonormal bases. In the transfer from one orthonormal basis to another, the vector components undergo the same transformation as the basis vectors.

From (6) it is clear that an orthogonal matrix always has an inverse which is also orthogonal $((S^{-1})' = (S')^{-1} = (S^{-1})^{-1})$. Further, the product of two orthogonal matrices S_1, S_2 is also orthogonal $((S_1 S_2)^{-1} = S_2^{-1} S_1^{-1} = S_2' S_1' = (S_1 S_2)')$. Since the identity matrix I is orthogonal, we have the following theorem.

Theorem 9. The set of all $n \times n$ orthogonal matrices is a group with the usual matrix multiplication and is known as the orthogonal group of degree n.

If S is orthogonal, then

$$(\det S)^2 = \det S \det S' = \det SS' = \det I = 1$$

and therefore $\det S = \pm 1$. Both possibilities occur. For example, when $n = 1$, $S_1 = (+1)$ and $S_2 = (-1)$ are orthogonal. The first matrix has the determinant $+1$ and the second has determinant -1. The set of all orthogonal matrices of degree n which have determinant $+1$ is clearly also a group which is a subgroup of the orthogonal group.

If we interpret (5) (referred to an orthonormal basis) as representing an endomorphism f of E (cf. 5.3), then it follows from the orthogonality of S that f is an automorphism. Also, for all $x_1, x_2 \in E$,

$$(f(x_1), f(x_2)) = \eta_1' \eta_2 = \xi_1' S' S \xi_2 = \xi_1' \xi_2 = (x_1, x_2),$$

i.e., the scalar product of two vectors and hence the norm of a vector is invariant under the automorphism f. An automorphism with this property is called a *rotation* of the Euclidean vector space E. If $\det S = +1$, we refer to f as a proper rotation and if $\det S = -1$, we refer to f as a reflection. It is easy to verify that the matrix S in (5) is orthogonal if (5) represents a rotation of E. Conversely we now have the following theorem.

Theorem 10. The set of all rotations of a Euclidean vector space of dimension n and the set of all proper rotations are groups (with the multiplication of mappings). The first is isomorphic to the orthogonal group of degree n and is a subgroup of the group $A(E)$. (Theorem 5.3;1.)

A rotation of a Euclidean vector space E is often referred to simply as an automorphism of E because all the structural properties of E, i.e., the linear relationships, the norms and scalar products of vectors, are preserved.

Exercises

1. Use Gram's determinant to prove that the vectors $a_k(\tau) = \tau^k$ $(k = 0, 1, 2, 3)$ in the vector space C (Example 2.1;7) are linearly independent.

Solution. Make C Euclidean as in Example 3, and Gram's determinant is equal to $256/23625 \neq 0$.

2. (a) Prove that the vector space C (Example 2.1;7) becomes Euclidean with the scalar product $(x, y) = \int_{-1}^{+1} \tau^2 x(\tau) y(\tau) \, d\tau$.

(b) In this Euclidean space, orthogonalize the sequence of vectors $a_0 = 1$, $a_1 = \tau$, $a_2 = \tau^2$, $a_3 = \tau^3$.

Solution. If the fundamental form $q(x) = \sum_{i,k=0}^{3} (a_i, a_k) \xi_i \xi_k$ (on the subspace $P_3 \subseteq C$) is reduced by the exchange method, then the corresponding change of the basis vectors is

$$e_0 = \tfrac{3}{2} a_0 = \tfrac{3}{2}$$

$$e_1 = \tfrac{5}{2} a_1 = \tfrac{5}{2}\tau$$

$$e_2 = -\tfrac{105}{8} a_0 + \tfrac{175}{8} a_2 = \tfrac{35}{8}(5\tau^2 - 3)$$

$$e_3 = -\tfrac{315}{8} a_1 + \tfrac{441}{8} a_3 = \tfrac{63}{8}(7\tau^3 - 5\tau)$$

3. If the polynomials $x(\tau)$ are considered only for the values $\tau = -2, -1, 0, 1, 2$, then they form a 5-dimensional vector space E. (Two polynomials are considered to be equal if and only if they take the same values for the five given values of τ (see Example 12.2;2).) The set of vectors $a_1(\tau) = \tau^i$ $(i = 0, \ldots, 4)$ is a basis of E, and E is Euclidean with the fundamental form $q(x) = \sum_{k=-2}^{2} [x(k)]^2$. Orthogonalize the given basis.

Solution. $e_0 = \tfrac{1}{5}$, $e_1 = \tfrac{1}{10}\tau$, $e_2 = \tfrac{1}{14}(\tau^2 - 2)$,

$$e_3 = \tfrac{1}{72}(5\tau^3 - 17\tau), \quad e_4 = \tfrac{1}{288}(35\tau^4 - 155\tau^2 + 72).$$

The norms of these polynomials are $\sqrt{(\frac{1}{5})}$, $\sqrt{(\frac{1}{10})}$, $\sqrt{(\frac{1}{14})}$, $\sqrt{(\frac{5}{72})}$, $\sqrt{(\frac{35}{288})}$.

Problems

1. Prove the assertion made in Example 1.

2. Prove that Gram's determinant (Theorem 4) is strictly positive if $a_1, \ldots, a_r$ are linearly independent. (Hint: Consider the quadratic form
$$\sum_{i,k=1}^{r} (a_i, a_k) \xi_i \xi_k .)$$

3. Use Gram's determinant to show that the polynomials $a_0(\tau) = 1$, ..., $a_4(\tau) = \tau^4$ are a basis of P_4 (Example 2.1;5).

4. Prove that, if E is a Euclidean vector space, $\{e_1, \ldots, e_n\}$ is a basis of E and $\alpha_1, \ldots, \alpha_n$ are real numbers, then there is exactly one vector $x \in E$ such that $(x, e_k) = \alpha_k$ for $k = 1, \ldots, n$.

5. Show that the set L of all odd functions in the Euclidean vector space C (Example 3) (i.e., those functions $x(\tau)$ for which $x(-\tau) = -x(\tau)$ in $-1 \leqslant \tau \leqslant +1$) is a subspace of C. Which functions are in L†?

6. Let E be a Euclidean vector space and let $a \in E$. For which vectors $x \in E$ are $(x+a)$ and $(x-a)$ orthogonal?

7. Prove that
$$\left[\sum_{k=1}^{n} \xi_k \eta_k \right]^2 \leqslant \left[\sum_{k=1}^{n} \xi_k^2 \right] \left[\sum_{k=1}^{n} \eta_k^2 \right]$$

for all real numbers $\xi_1, \ldots, \xi_n, \eta_1, \ldots, \eta_n$.

8. Prove that
$$\left[\int_{-1}^{+1} x(\tau) y(\tau) \, d\tau \right]^2 \leqslant \left[\int_{-1}^{+1} [x(\tau)]^2 \, d\tau \right] \left[\int_{-1}^{+1} [y(\tau)]^2 \, d\tau \right] \text{ for all } x, y \in C.$$

9. Prove that Pythagoras's theorem,

'$\|x \pm y\|^2 = \|x\|^2 + \|y\|^2$ if x and y are orthogonal,'

is true in any Euclidean vector space.

10. Show that a 2×2 orthogonal matrix is either of the form
$$\begin{pmatrix} \cos \alpha \sin \alpha \\ -\sin \alpha \cos \alpha \end{pmatrix} \text{ or of the form } \begin{pmatrix} \cos \alpha & \sin \alpha \\ \sin \alpha & -\cos \alpha \end{pmatrix}.$$

Which of these represent proper rotations?

11. Show that, if f is a reflection of a Euclidean plane (i.e., a 2-dimensional Euclidean space), then $f^2 = e$.

12. Suppose f is a rotation of a finite-dimensional Euclidean vector space E. What is the dual endomorphism of f?

13. Which rotations of a Euclidean vector space—in particular a Euclidean plane—are symmetric endomorphisms? (Cf. Definition 1.3.2;1.)

14. Show that, if $\{e_1,\ldots,e_n\}$ and $\{f_1,\ldots,f_n\}$ are orthonormal bases of the Euclidean vector space E, then there is exactly one rotation t of E for which $t(e_k)=f_k$ $(k=1,\ldots,n)$.

15. Prove that, given a basis of a real vector space, then there is a fundamental form which makes this basis orthonormal.

16. Prove that a subspace L of a Euclidean vector space E is itself a Euclidean vector space when $\|x\|$ $(x \in L)$ is defined to be equal to the norm of x in E.

12.2 Approximation in Euclidean Vector Spaces, The Method of Least Squares

12.2.1 The General Approximation Problem

Suppose that we are given a finite set of vectors a_1, ..., a_r in a Euclidean vector space E and suppose that, for an arbitrary vector $z \in E$, we wish to find a vector $x_0 \in L=L(a_1,\ldots,a_r)$ for which $\|z-x_0\|$ is as small as possible, i.e., $\|z-x_0\|=\min\limits_{x\in L}\|z-x\|$.

Theorem 1. There is exactly one vector $x_0 \in L$ for which

$$\|z-x_0\| = \min_{x\in L}\|z-x\|.$$

This is the vector x_0 which is uniquely determined by $x_0 \in L$ and $z-x_0 \in L^\dagger$.

Proof. By Theorem 12.1;3 applied to the subspace $L(L,z)$, there is just one vector $x_0 \in L$ such that $z-x_0 \in L^\dagger$. Now suppose that $x \in L$ and $x\neq x_0$. Then

$$\|z-x\|^2 = \|(z-x_0)+(x_0-x)\|^2$$
$$= \|z-x_0\|^2+2(z-x_0,x_0-x)+\|x_0-x\|^2.$$

Since $z-x_0 \in L^\dagger$ and $x_0-x \in L$, the middle term vanishes. Further, since $x_0-x\neq 0$, it follows that $\|z-x\|^2 > \|z-x_0\|^2$ and the theorem is proved.

In the applications of the theorem it is important to be able to calculate the coefficients λ_k in the representation $x_0= \sum\limits_{k=1}^{r} \lambda_k a_k$.

Since $(z-x_0,a_i)=0$ for all $i=1$, ..., r, it follows that

$$\sum_{k=1}^{r} (a_k,a_i)\lambda_k = (z,a_i) \qquad (i = 1,\ldots,r) \tag{1}$$

is a system of r linear equations for the r coefficients λ_k. Every solution

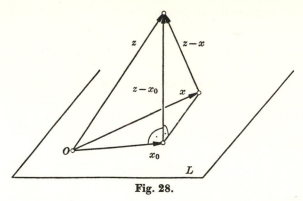

Fig. 28.

$(\lambda_1, \ldots, \lambda_r)$ of this system provides a representation of x_0. The $\lambda_1, \ldots, \lambda_r$ are uniquely determined if and only if $a_1, \ldots, a_r$ are linearly independent. It is only in this latter case that the determinant of the system (1) is not equal to zero (Theorem 12.1;4).

Definition 1. The vector x_0 which is uniquely determined as in Theorem 1 is said to be the 'best approximation to' z as a linear combination of the vectors $a_1, \ldots, a_r$.

The following examples will illustrate three important types of problem to which the theory may be applied.

Example 1. Let E be the Euclidean vector space C (Example 12.1;3) and let $a_k \in C$ be the polynomial $a_k(\tau) = \tau^k$ $(k = 0, \ldots, r)$. Then, for a given continuous function $z \in C$, the method described above will produce a 'best approximation in mean squares' for z on the interval $-1 \leqslant \tau \leqslant +1$ by a polynomial x_0 whose degree is at most r. The terminology stems from the fact that, for a given $z \in C$, the integral

$$\int\limits_{-1}^{+1} (z(\tau) - x(\tau))^2 \, d\tau \; = \; \|z - x\|^2$$

takes a smaller value for $x = x_0$ than for any other polynomial of degree at most r.

The procedure for the solution is as follows:

1. For each $i, k = 0, \ldots, r$, we calculate the quantities

$$\alpha_{ik} = (a_i, a_k) = \int\limits_{-1}^{+1} \tau^{i+k} \, d\tau = \begin{cases} 0 & \text{if } (i+k) \text{ is odd} \\ \dfrac{2}{i+k+1} & \text{if } (i+k) \text{ is even.} \end{cases}$$

231

2. We further calculate the quantities

$$\beta_i = (z, a_i) = \int\limits_{-1}^{+1} z(\tau)\,\tau^i\,\mathrm{d}\tau \qquad (i = 0, \ldots, r).$$

3. We then solve the system of linear equations

$$\sum_{k=0}^{r} \alpha_{ik}\lambda_k = \beta_i \qquad (i = 0, \ldots, r)$$

for the unknowns λ_k.

4. Finally $\qquad\qquad x_0(\tau) = \sum\limits_{k=0}^{r} \lambda_k \tau^k.$

Example 2. Suppose that the physical quantity ξ is dependent on another quantity τ, i.e., $\xi = \xi(\tau)$, although the exact nature of the function $\xi(\tau)$ is not actually known. In order to gain some insight into this functional relationship, suppose that we measure the quantity ξ for each of n distinct values $\tau_1, \ldots, \tau_n$ of τ. Let us denote these measurements by $\zeta_1, \ldots, \zeta_n$. Because of experimental errors the value ζ_p will not be exactly equal to $\xi(\tau_p)$ in general. In order to find at least an approximate representation of ξ as a function of τ, we think of $z = (\zeta_1, \ldots, \zeta_n)$ as an element of the n-dimensional space R_n which is made Euclidean by introducing the norm

$$\|z\|^2 = \sum_{p=1}^{n} \zeta_p^2.$$

We then introduce a vector space L of functions $x(\tau)$ and look for an $x_0 \in L$ such that the sum of the squares of the errors is minimal. That is,

$$\sum_{p=1}^{n} (\zeta_p - x_0(\tau_p))^2 = \min_{x \in L} \sum_{p=1}^{n} (\zeta_p - x(\tau_p))^2. \tag{2}$$

This problem can also be reduced to the general approximation problem. The correspondence $x \in L \to \bar{x} = (x(\tau_1), \ldots, x(\tau_n)) \in R_n$ is a linear mapping of L onto a subspace $\bar{L}$ of R_n, and (2) can be written in the form

$$\|z - \bar{x}_0\| = \min_{\bar{x} \in \bar{L}} \|z - \bar{x}\|. \tag{3}$$

Also $\bar{L}$ is isomorphic to the quotient space L/N, where N is the subspace of L which consists of all $x \in L$ such that $x(\tau_p) = 0$ for all $p = 1, \ldots, n$ (see Theorem 5.1;9).

The vector $\bar{x}_0$ is uniquely determined by (3) and x_0 is then determined by $\bar{x}_0$ apart from a term in N.

It is usual to take L to be P_r, i.e., the space of polynomials with degree at most r. If we choose $r \leqslant n-1$, then x_0 is uniquely determined, because a polynomial $x(\tau)$ of degree at most $n-1$ which takes the value 0 at $\tau = \tau_1, \ldots,$

τ_n must take the value 0 for all τ. In fact there is no point in choosing $r \geqslant n$ because it is already possible to satisfy the condition $x_0(\tau_p) = \zeta_p (p = 1, \ldots, n)$ for $r = n - 1$. We will therefore assume in the following that $r \leqslant n - 1$.

The calculation runs as follows.

1. We put $a_k(\tau) = \tau^k \ (k = 0, \ldots, r)$. Since $r \leqslant n - 1$, the corresponding images $\bar{a}_k \in \bar{L} \subset R_n$ are linearly independent and a spanning set for $\bar{L}$.

2. We calculate the quantities

$$\alpha_{ik} = (\bar{a}_i, \bar{a}_k) = \sum_{p=1}^{n} \tau_p^{i+k} \qquad (i, k = 0, \ldots, r).$$

3. We also calculate

$$\beta_i = (z, \bar{a}_i) = \sum_{p=1}^{n} \zeta_p \tau_p^{i} \qquad (i = 0, \ldots, r).$$

4. We solve the system of equations

$$\sum_{k=0}^{r} \alpha_{ik} \lambda_k = \beta_i \qquad (i = 0, \ldots, r).$$

5. The solution is the polynomial

$$x_0(\tau) = \sum_{k=0}^{r} \lambda_k \tau^k.$$

This form of the procedure is known as approximation by a polynomial of degree r by the method of least squares.

Example 3. In 9.1, we approximated to the solution of a system of linear equations which did not have an exact solution by using Tchebychev's Method. Now we will meet another kind of approximation which is known as the method of least squares and is due to Gauss. It has the advantage over the Tchebychev Method of being simpler from the point of view of calculating technique. It also gives the expected solution according to the laws of Probability Theory when the system does not have an exact solution because of errors in the coefficients which are distributed statistically. (cf. [29].)

Suppose therefore that we have the system

$$\sum_{k=1}^{n} \alpha_{pk} \xi_k + \beta_p = 0 \qquad (p = 1, \ldots, m),$$

i.e.,

$$A\xi + \beta = 0. \qquad (4)$$

As in 9.1 we introduce the residuals

$$\epsilon_p = \sum_{k=1}^{n} \alpha_{pk} \xi_k + \beta_p \qquad (p = 1, \ldots, m). \qquad (5)$$

However here we try to find unknowns ξ_k which minimize the sum of the squares of the residuals

$$\delta^2 = \sum_{p=1}^{m} \epsilon_p^2 = \sum_{p=1}^{m} \left(\sum_{k=1}^{n} \alpha_{pk}\xi_k + \beta_p \right)^2.$$

In analogy with 7.1;(3), (5) can be written as an equation in the m-dimensional space R_m

$$e = \sum_{k=1}^{n} \xi_k a_k + b.$$

The space R_m is made Euclidean by setting $\|z\|^2 = q(z) = \sum_{p=1}^{m} \zeta_p^2.$
The problem now is to find $\xi_1, \ldots, \xi_n$ such that

$$\|e\| = \|b + \sum_{k=1}^{n} \xi_k a_k\|$$

is minimal. According to the general rules we proceed as follows.

1. We calculate $\gamma_{ik} = (a_i, a_k) = \sum_{p=1}^{m} \alpha_{pi}\alpha_{pk}$ $(i, k = 1, \ldots, n)$, i.e., $C = (\gamma_{ik}) = A'A$.

2. We calculate $\delta_i = (b, a_i) = \sum_{p=1}^{m} \alpha_{pk}\beta_p$ $(k = 1, \ldots, n)$, i.e., $\boldsymbol{\delta} = A'\boldsymbol{\beta}$.

3. We solve the system of equations

$$C\boldsymbol{\xi} + \boldsymbol{\delta} = 0, \qquad \text{i.e., } A'(A\boldsymbol{\xi} + \boldsymbol{\beta}) = 0,$$

i.e., a system of n equations for the n unknowns $\xi_1, \ldots, \xi_n$.

These equations are known as the *normal equations*. They are obtained from (4) by multiplying on the left by A'. The unknowns are uniquely determined by the normal equations (and hence by the condition $\delta^2 = \min$) if and only if $a_1, \ldots, a_n$ are linearly independent, i.e., if A has rank n. (Theorems 7.1;6 and 12.1;4.)

12.2.2 Approximation with Orthogonal Systems

The method of approximation described in 12.2.1 becomes particularly easy when the vectors $a_1, \ldots, a_r$ form an orthogonal set, in which case we will denote them by $e_1, \ldots, e_n$. Since (a_i, a_k) is now equal to $(e_i, e_k) = \delta_{ik}\|e_k\|^2$, the system of equations 12.2;(1) becomes $\sum_{k=1}^{r} \|e_k\|^2 \delta_{ik}\lambda_k = \|e_i\|^2\lambda_i = (z, e_i)$, and therefore

$$\lambda_i = \frac{(z, e_i)}{\|e_i\|^2} = \frac{(z, e_i)}{(e_i, e_i)} \qquad (i = 1, \ldots, r). \tag{6}$$

If the set $\{e_1, \ldots, e_r\}$ is actually orthonormal, then

$$\lambda_i = (z, e_i) \qquad (i = 1, \ldots, r) \tag{7}$$

and the best approximation is

$$x_0 = \sum_{i=1}^{r} (z, e_i)\, e_i.$$

The coefficient λ_i is called the *Fourier coefficient of z for the vector e_i*. It is determined by z and e_i alone and is independent of the other vectors in the orthogonal set.

Example 4. To find the best approximation in mean squares for the function $z(\tau) = \tau^4 + \tau$ on the interval $-1 \leqslant \tau \leqslant +1$ by a polynomial of degree at most 2. With the Legendre polynomials $\bar{e}_0, \bar{e}_1, \bar{e}_2$ (see 12.1.2) we obtain the Fourier coefficients

$$\lambda_0 = (z, \bar{e}_0) = \int_{-1}^{+1} \frac{1}{\sqrt{2}} (\tau^4 + \tau)\, \mathrm{d}\tau = \frac{1}{\sqrt{2}} \cdot \frac{2}{5}$$

$$\lambda_1 = (z, \bar{e}_1) = \int_{-1}^{+1} \sqrt{\tfrac{3}{2}}\, \tau (\tau^4 + \tau)\, \mathrm{d}\tau = \sqrt{\tfrac{3}{2}} \cdot \tfrac{2}{3}$$

$$\lambda_2 = (z, \bar{e}_2) = \int_{-1}^{+1} \sqrt{\tfrac{5}{8}}(3\tau^2 - 1)\, (\tau^4 + \tau)\, \mathrm{d}\tau = \sqrt{\tfrac{5}{8}} \cdot \tfrac{16}{35}$$

and hence the approximation

$$x_0(\tau) = \lambda_0\, \bar{e}_0(\tau) + \lambda_1\, \bar{e}_1(\tau) + \lambda_2\, \bar{e}_2(\tau)$$

$$= -\tfrac{3}{35} + \tau + \tfrac{6}{7}\tau^2$$

If we decide to approximate the function by a polynomial of degree at most 3, the work we have just done will not be wasted since all we need to do is to add $\lambda_3 \bar{e}_3$.

Bessel's Inequality

Let $\{e_1, e_2, e_3, \ldots\}$ be an orthonormal set in a Euclidean vector space E. The best approximation of a vector $z \in E$ by a linear combination of the first r vectors $e_1, \ldots, e_r$ is then

$$x_{0r} = \sum_{k=1}^{r} \lambda_k e_k$$

where the λ_k are the Fourier coefficients of z for $e_1, \ldots, e_r$. The closeness of the approximation is measured by the 'defect'

$$\delta_r^2 = \|z - x_{0r}\|^2 = (z - \sum_i \lambda_i e_i, z - \sum_k \lambda_k e_k)$$

$$= (z, z) - 2 \sum_k \lambda_k (z, e_k) + \sum_{i,k} \lambda_i \lambda_k (e_i, e_k)$$

$$= \|z\|^2 - \sum_{k=1}^r \lambda_k^2 \tag{8}$$

Since the defect is non-negative it follows that

$$\sum_{k=1}^r \lambda_k^2 \leqslant \|z\|^2.$$

Since this is valid for every natural number r when the orthonormal set is infinite, the series $\sum_{k=1}^\infty \lambda_k^2$ converges and we have

$$\sum_{k=1}^\infty \lambda_k^2 \leqslant \|z\|^2 \quad \text{(Bessel's Inequality).} \tag{9}$$

(If the set is finite, the ∞ is replaced by the number of vectors in the set.)

Definition 2. The orthonormal set $\{e_1, e_2, e_3, \ldots\}$ is said to be 'complete in E' if

$$\sum_{k=1}^\infty \lambda_k^2 = \|z\|^2 \quad \text{(Parseval's Equation)} \tag{10}$$

for all $z \in E$.

(Again, the ∞ is to be replaced by the number of vectors in the set when it is finite.)

Completeness means that, by (8), every vector $z \in E$ can be arbitrarily closely approximated (i.e., with arbitrarily small defect) by the partial sums of its *Fourier series* $\sum_{k=1}^\infty \lambda_k e_k$. If E is finite-dimensional, then an orthonormal set is complete if and only if it is a basis of E.

Exercises

1. (a) Approximate to the function $z = e^\tau$ by a polynomial of degree at most 2 in the interval $-1 \leqslant \tau \leqslant +1$ by using the method of least squares (see Example 12.2;1).

(b) Do the same exercise as in (a) but using the orthonormal system introduced in Example 12.1;4.

(c) Do the same exercise for the fundamental form $\|x(\tau)\|^2 = \int\limits_{-1}^{+1} \tau^2[x(\tau)]^2 \, d\tau$ with the orthogonal system calculated in Exercise 12.1;2.

Solutions. (a), (b) $\quad x_0 = \dfrac{3(11-e^2)}{4e} + \dfrac{3}{e}\tau + \dfrac{15(e^2-7)}{4e}\tau^2$

$$= 0{\cdot}9963 + 1{\cdot}1036\tau + 0{\cdot}5367\tau^2.$$

(c) $\quad x_0^* = \dfrac{15(215-29\,e^2)}{4e} + \dfrac{5(8-e^2)}{e}\tau + \dfrac{35(21\,e^2-155)}{4e}\tau^2$

$$= 0{\cdot}9895 + 1{\cdot}1238\tau + 0{\cdot}5481\tau^2.$$

Note that the factor τ^2 has the effect of giving the errors at the ends of the interval $-1 \leqslant \tau \leqslant +1$ more significance (see the following table).

τ	-1	$-\tfrac{1}{2}$	0	$\tfrac{1}{2}$	1
e^τ	0·3679	0·6065	1·0000	1·6487	2·7183
$x_0(\tau)$	0·4294	0·5787	0·9963	1·6823	2·6366
$x_0^*(\tau)$	0·4138	0·5646	0·9895	1·6884	2·6614

2. A quantity ξ, which depends on the variable τ, is measured at the values $\tau = -2, -1, 0, 1, 2$ giving the values $\zeta = 4{\cdot}2, -1{\cdot}3, -0{\cdot}8, 8{\cdot}7, 9{\cdot}2$.

(a) Approximate to these experimental measurements by a polynomial of degree at most 3 using the method of Example 12.2;2.

(b) Use the orthogonal system of Exercise 12.1;3 to calculate the approximate polynomials of degrees at most 0, 1, 2, 3 and 4. Calculate the defect $\delta_r^2 = \sum\limits_{k=-2}^{2} [x_0^r(k) - \zeta_k]^2$ in each case. (x_0^r denotes the rth polynomial.)

Solution. $\quad x_0^0 = 4, \; x_0^1 = 4 + 2\tau, \; x_0^2 = 1 + 2\tau + 1{\cdot}5\tau^2,$

$$x_0^3 = 1 + 6{\cdot}25\tau + 1{\cdot}5\tau^2 - 1{\cdot}25\tau^3,$$

$$x_0^4 = -0{\cdot}8 + 6{\cdot}25\tau + 5{\cdot}375\tau^2 - 1{\cdot}25\tau^3 - 0{\cdot}875\tau^4.$$

The defects are 100·3, 60·3, 28·8, 6·3, 0.

3. Approximate to the solution of the system $\xi_1 = 5$, $\xi_2 = 7$, $\xi_1 + \xi_2 = 0$, $\xi_1 - \xi_2 = 1$ by the method of least squares.

Solution. $\quad \xi_1 = 2, \; \xi_2 = 2.$

Tchebychev's approximation (see the Exercise in 9.1) gave the solution $\xi_1 = 1$, $\xi_2 = 3$ for the same example. The absolute values of the residuals are

$|\xi_1-5|=4$, $|\xi_2-7|=4$, $|\xi_1+\xi_2|=4$, $|\xi_1-\xi_2-1|=3$ and the sum of the quadratic residuals is 57.

By the method of least squares the greatest residual is $|\xi_2-7|=5$, but, on the other hand, the sum of the quadratic residuals $5^2+4^2+3^2+1^2=51$ is smaller.

Problems

1. Use the Theorems in 7.1 to show that the system of equations (1) always has a solution.

2. Suppose $x(\tau) \in C$ (Example 2.1;7). Find the odd function $y(\tau) \in C$ for which $\int\limits_{-1}^{+1} [x(\tau)-y(\tau)]^2 d\tau$ is minimal (cf. Problem 12.1;5).

3. Suppose that the system of equations (4) is approximated by both Tchebychev's Method and the Method of Least Squares. Let the greatest absolute value of the residuals after the first approximation be ρ and after the second σ. Let δ^2 be the sum of the squares of the residuals from the Method of Least Squares. Show that $\delta^2 \leqslant m\rho^2 \leqslant m\sigma^2$.

4. Show that, if the method of least squares produces residuals which are all equal in absolute value, then both methods of approximation lead to the same result.

5. Show that the method of approximation in 12.2.1 is linear, i.e., if x_{01} and x_{02} are the best approximations to z_1, and $z_2 \in E$, then $\alpha_1 x_{01} + \alpha_2 x_{02}$ is the best approximation to $\alpha_1 z_1 + \alpha_2 z_2$.

12.3 Hilbert Spaces

If $\{e_1, e_2, e_3, \ldots\}$ is a complete orthonormal set in a Euclidean vector space E, then, for the partial sums $s_r = \sum\limits_{k=1}^{r} \lambda_k e_k$ of the Fourier series of a vector $z \in E$, we have

$$\lim_{r\to\infty} \|z-s_r\| = 0. \tag{1}$$

Now, if condition (1) is satisfied by an arbitrary sequence of vectors $s_1, s_2, s_3, \ldots \in E$, we say that the sequence is *convergent* and that z is its *limit vector*. (See Problem 2.) We then write

$$z = \lim_{r\to\infty} s_r. \tag{2}$$

If in particular

$$s_r = \sum_{k=1}^{r} \lambda_k e_k \text{ and } \lim_{r\to\infty} s_r = z,$$

then we also write

$$\sum_{k=1}^{\infty} \lambda_k e_k = z$$

and say that this Fourier series is *convergent* with the *sum z*. Thus an ortho-normal set in a Euclidean vector space E is complete if and only if every vector $z \in E$ is equal to the sum of its Fourier series.

Now suppose that (2) is satisfied by a sequence $s_1, s_2, s_3, \ldots \in E$ and let $\epsilon > 0$. By (1), there is a natural number n_0 such that $\|z - s_n\| \leqslant \frac{1}{2}\epsilon$ for all $n \geqslant n_0$. If $m, n \geqslant n_0$, then by Theorem 12.1;2

$$\|s_m - s_n\| = \|(z - s_n) - (z - s_m)\| \leqslant \|z - s_n\| + \|z - s_m\| \leqslant \epsilon.$$

A sequence of vectors $s_1, s_2, s_3, \ldots \in E$ with the property that, to each $\epsilon > 0$, there is an index n_0 such that $\|s_m - s_n\| \leqslant \epsilon$ for all $m, n \geqslant n_0$ is known as a *Cauchy sequence*. (In analogy with the real numbers.) Thus we have the following result.

Theorem 1. Every convergent sequence of vectors in a Euclidean vector space is a Cauchy sequence.

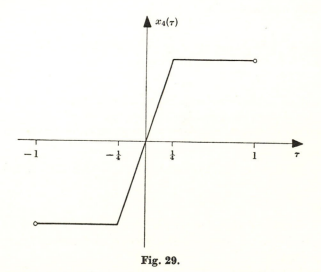

Fig. 29.

The following example will show that the converse of this theorem is not true in general. Let C be the Euclidean vector space of Example 12.1;3. For each natural number n, let $x_n \in C$ be the continuous function

$$x_n(\tau) = \begin{cases} -1 & \text{for } -1 \leqslant \tau \leqslant -\dfrac{1}{n} \\[2mm] \text{linear for } -\dfrac{1}{n} \leqslant \tau \leqslant +\dfrac{1}{n} \\[2mm] +1 & \text{for } \dfrac{1}{n} \leqslant \tau \leqslant 1 \end{cases}$$

e.g., for $n=4$, the graph of $x_4(\tau)$ is shown in Fig. 29.

Clearly $\|x_m - x_n\|^2 = \int\limits_{-1}^{+1} (x_m(\tau) - x_n(\tau))^2 \, d\tau$ becomes arbitrarily small when m and n are sufficiently large so that $x_1, x_2, x_3, \ldots$ is a Cauchy sequence in C. However this sequence is not convergent, because, if

$$\lim_{n \to \infty} \|z - x_n\|^2 = \lim_{n \to \infty} \int\limits_{-1}^{+1} (z(\tau) - x_n(\tau))^2 \, d\tau = 0,$$

then it is easy to verify that $z(\tau) = -1$ for $-1 \leqslant \tau < 0$ and $z(\tau) = +1$ for $0 < \tau \leqslant +1$. However there is no continuous function $z(\tau)$ with these properties.

The Euclidean vector spaces in which the converse of Theorem 1 is valid are of particular importance.

Definition 1. A Euclidean vector space in which every Cauchy sequence is convergent is known as a 'real Hilbert space'.

Theorem 2. An orthonormal set $\{e_1, e_2, e_3, \ldots\}$ in a Hilbert space E is complete if and only if, from $x \in E$ and $\lambda_k = (x, e_k) = 0$ for all $k = 1, 2, 3, \ldots$, it follows that $x = 0$ (i.e., if only the zero vector is orthogonal to every vector e_k).

Proof. 1. If the set is complete and $\lambda_k = (x, e_k) = 0$ for all $k = 1, 2, 3, \ldots$, then

$$x = \sum_{k=1}^{\infty} \lambda_k e_k = 0.$$

2. Suppose that, whenever $x \in E$ and $(x, e_k) = 0$ for all $k = 1, 2, 3, \ldots$, then $x = 0$. We have to show that every vector $z \in E$ is equal to the sum of its Fourier series.

2.1. Suppose that $z \in E$ and $\lambda_k = (z, e_k)$ $(k = 1, 2, 3, \ldots)$. We first show that the Fourier series of z is convergent, i.e., that there is a $z_0 \in E$ such that

$$z_0 = \sum_{k=1}^{\infty} \lambda_k e_k. \tag{3}$$

Suppose that s_m and s_n are partial sums of the series and that $m > n$, then

$$s_m - s_n = \sum_{k=n+1}^{m} \lambda_k e_k$$

and hence

$$\|s_m - s_n\|^2 = \sum_{k=n+1}^{m} \lambda_k^2.$$

Since the series which appears in Bessel's Inequality 12.2;(9) is convergent it follows that $s_1, s_2, s_3, \ldots$ is a Cauchy sequence. Hence the series (3) converges because E is a Hilbert space.

2.2. For the Fourier coefficients of z_0, we have

$$(z_0, e_k) = (s_r, e_k) + (z_0 - s_r, e_k) \qquad (r = 1, 2, 3, \ldots).$$

For $r \geqslant k$, $(s_r, e_k) = \lambda_k$. By Theorem 12.1;1, $|(z_0 - s_r, e_k)| \leqslant \|z_0 - s_r\|$. Hence, since $\lim_{r \to \infty} s_r = z_0$, it follows that $(z_0, e_k) = \lim_{r \to \infty} (s_r, e_k) = \lambda_k$, $(k = 1, 2, 3, \ldots)$. Thus z_0 has the same Fourier coefficients as z.

2.3. Hence $(z - z_0, e_k) = 0$ for all $k = 1, 2, 3, \ldots$ and it follows from the basic assumption that $z = z_0$, i.e., z is equal to the sum of its Fourier series.

We mention without proof the fact that every Hilbert space actually contains complete orthogonal sets.

We will also omit the proof of the next theorem since it would go beyond the scope of this book. However, the theorem is so important that it is well worth stating here.

Theorem 3. Every Euclidean vector space E can be embedded in a Hilbert space. More precisely: there is a Hilbert space H such that E is a subspace of H and the norm $\|x\|$ of any vector x in E is the same as its norm in H. The Hilbert space H can be chosen that, for every $z \in H$ and every $\epsilon > 0$, there is a vector $x \in E$ such that $\|x - z\| \leqslant \epsilon$.

The space H is known as the *completion* of E. It is uniquely determined by E except for isomorphic spaces.

A proof of Theorem 3 can be found in [7], p. 253.

If E is the space C of Example 12.1;3 then the completion H is the space L_2 of Lebesgue measurable functions $x(\tau)$ on the range $-1 \leqslant \tau \leqslant +1$ for which the integral $\int_{-1}^{+1} |x(\tau)|^2 \, d\tau$ exists. (See [7] p. 259.)

If we replace the boundaries -1, $+1$ by $-\pi$ and $+\pi$, then, by 3.1.2 (Example 2.1;8), the functions

$$e_0(\tau) = \frac{1}{\sqrt{(2\pi)}}$$

$$\left.\begin{array}{l} e_k(\tau) = \dfrac{1}{\sqrt{\pi}}\cos k\tau \\[2ex] f_k(\tau) = \dfrac{1}{\sqrt{\pi}}\sin k\tau \end{array}\right\} \quad k = 1,2,3,\ldots \tag{4}$$

form an orthonormal set in L_2. This is actually a complete set. The proof which uses Theorem 2 may be found in [18], p. 31.

Because (4) is complete, this means that every function $x \in L_2$ (and in particular every continuous function) can be arbitrarily closely approximated on the interval $-\pi \leqslant \tau \leqslant +\pi$ by the partial sums of its Fourier series

$$\frac{\lambda_0}{\sqrt{(2\pi)}} + \frac{1}{\sqrt{\pi}} \sum_{k=1}^{\infty} [\lambda_k \cos k\tau + \mu_k \sin k\tau]. \tag{5}$$

Here
$$\lambda_0 = \frac{1}{\sqrt{(2\pi)}} \int_{-\pi}^{\pi} x(\tau)\,d\tau$$

$$\left.\begin{array}{l} \lambda_k = \dfrac{1}{\sqrt{\pi}} \displaystyle\int_{-\pi}^{\pi} x(\tau)\cos k\tau\,d\tau \\[3ex] \mu_k = \dfrac{1}{\sqrt{\pi}} \displaystyle\int_{-\pi}^{\pi} x(\tau)\sin k\tau\,d\tau \end{array}\right\} \quad k = 1,2,3,\ldots$$

In analysis, it can be shown that there is a wide class of functions for which the sum of the series (5) is $x(\tau)$ in the usual sense (i.e., pointwise). For example, all differentiable functions on the interval $-\pi \leqslant \tau \leqslant +\tau$ have this property. (See [18], p. 55.)

Another complete orthonormal set in L_2 (where we again return to the interval $-1 \leqslant \tau \leqslant +1$) is given by the normalized Legendre polynomials $\overline{e}_k(k=0,1,2,\ldots)$ (see 12.1.2).

With the introduction of Hilbert spaces, Linear Algebra changes to Linear Analysis which represents its connection with Topology. However this new and rich theory cannot be followed any further here (see [21]).

Problems

1. Prove that every finite-dimensional Euclidean space is a Hilbert space.
2. Prove that a sequence of vectors in a Euclidean space has at most one limit vector.

3. Show that the scalar product in a Euclidean vector space is a continuous function, i.e.,

$$\text{if } \lim_{k \to \infty} x_k = x \quad \text{and} \quad \lim_{k \to \infty} y_k = y, \quad \text{then } (x,y) = \lim_{k \to \infty} (x_k, y_k).$$

4. Show that an orthonormal system in a Hilbert space is complete if and only if it is maximal, i.e., if it is not a proper subset of any orthonormal system.

5. Show that the subspace H of the space of sequences F (Example 2.1;3) which consists of all those sequences $x = (\xi_1, \xi_2, \ldots)$ for which $q(x) = \sum_{k=1}^{\infty} \xi_k^2$ converges is a Hilbert space with the fundamental form $q(x)$. (Cf. [7], p. 251.)

12.4 Unitary Vector Spaces

In this section we will describe the concepts and results in complex vector spaces which correspond to those in 12.1 and 12.2 for real vector spaces. We will again omit the proofs, because they are simple generalizations of those in the earlier sections.

Definition 1. A 'unitary vector space' is a complex vector space together with a positive definite Hermitian form $q(x)$ which is known as its fundamental form. (Cf. Definition 11.2;4.)

Unitary vector spaces are often also referred to as *Hermitian spaces*.

Example 1. The complex n-dimensional space of Example 2.1;12 is a unitary space with the fundamental form

$$q(x) = \sum_{k=1}^{n} \xi_k \bar{\xi}_k = \sum_{k=1}^{n} |\xi_k|^2.$$

Example 2. The complex vector space C of continuous complex-valued functions $x(\tau)$ on the interval $-1 \leqslant \tau \leqslant +1$ (Example 2.1;14) is a unitary space with

$$q(x) = \int_{-1}^{+1} x(\tau) \bar{x}(\tau) \, d\tau = \int_{-1}^{+1} |x(\tau)|^2 \, d\tau.$$

The length or norm of a vector x in a unitary space E is again defined to be the quantity $\|x\| = +\sqrt{\{q(x)\}}$. The zero vector is again the only vector which has norm 0.

The polar form of $q(x)$ is denoted by (x,y) and represents the scalar product of x and y. Thus this is a Hermitian quasi-bilinear form (Definition 11.2;2).

The Cauchy–Schwarz inequality and the triangle inequality are both still valid (Theorems 12.1;1 and 2).

Two vectors x, $y \in E$ are again said to be orthogonal if $(x,y)=0$. Since $(x,y)=0$ implies $(y,x)=\overline{(x,y)}=0$, orthogonality of vectors is a symmetric relation. The concepts of orthogonal and orthonormal sets (Definition 12.1;2) are unchanged and Theorems 12.1;5, 6, 7 are still valid with the appropriate meanings.

Example 3. In the unitary space C (Example 2), the set of functions

$$e_k(\tau) = \frac{1}{\sqrt{(2\pi)}} e^{ik\tau} \qquad (k = 0, \pm 1, \pm 2, \ldots)$$

is an orthonormal set (see Example 3.1;8) providing we replace the bounds ± 1 by $\pm \pi$.

The Schmidt Orthogonalization Method can be easily adapted to the situation in unitary spaces.

In place of orthogonal matrices, we have *unitary* matrices which are those matrices S such that

$$S^{-1} = \overline{S}'.$$

Thus a square matrix is unitary if its inverse is equal to the complex conjugate transposed matrix. Theorem 12.1;8 (change of basis) is also valid for unitary spaces if we replace 'orthogonal matrix' by 'unitary matrix'.

Suppose that an endomorphism $x \to f(x)$ of a finite-dimensional unitary vector space E is represented by $\eta = S\xi$ with respect to an orthonormal basis (cf. 5.3). Then

$$(x_1, x_2) = (f(x_1), f(x_2)) \text{ for all } x_1, x_2 \in E$$

(and hence $\|x\| = \|f(x)\|$ for all $x \in E$) if and only if S is unitary. An endomorphism with this property, i.e., which preserves scalar products and norms, is again referred to as a *rotation* of E. Since every unitary matrix is non-singular ($|\det S|^2 = \det(S\overline{S}') = 1$), it follows that a rotation is always an automorphism.

The set of all $n \times n$ unitary matrices is again a group with the usual matrix multiplication. Similarly the set of rotations of a unitary vector space of dimension n is a group and the two groups are isomorphic (cf. Theorems 12.1; 9 and 10).

The general approximation problem of 12.2.1 can be formulated in unitary spaces in the same way as in Euclidean spaces and the method of solution is also the same. In particular, 12.2;(1) is also valid in the unitary case where we must remember however that, for instance, (a_k, a_i) may not be replaced by (a_i, a_k) since these two numbers are complex conjugates. The same is true for the right-hand side (z, a_i).

The Fourier coefficients in the approximation with an orthonormal set are, as in 12.2;(7), $\lambda_i = (z, e_i)$ which may not be replaced by (e_i, z).

The defect of the best approximation with the first r vectors of an orthonormal set $\{e_1, e_2, e_3, \ldots\}$ in a unitary space is

$$\delta^2 = \|z - x_{0r}\|^2 = \left(z - \sum_i \lambda_i e_i, z - \sum_k \lambda_k e_k\right)$$

$$= (z, z) - \sum_i \lambda_i (e_i, z) - \sum_k \bar{\lambda}_k (z, e_k) + \sum_{ik} \lambda_i \bar{\lambda}_k (e_i, e_k)$$

$$= (z, z) - \sum_i \lambda_i \bar{\lambda}_i - \sum_k \lambda_k \bar{\lambda}_k + \sum_k \lambda_k \bar{\lambda}_k$$

$$= \|z\|^2 - \sum_{k=1}^{r} \lambda_k \bar{\lambda}_k$$

and Bessel's Inequality becomes

$$\sum_{k=1}^{\infty} \lambda_k \bar{\lambda}_k = \sum_{k=1}^{\infty} |\lambda_k|^2 \leqslant \|z\|^2.$$

Correspondingly, Parseval's equation (10) in the Definition 12.2;2 of a complete orthonormal set, now takes the form

$$\sum_{k=1}^{\infty} |\lambda_k|^2 = \|z\|^2.$$

We mention without proof that the orthonormal set of Example 3 is complete in the unitary space C.

Example 4. We will represent the continuous function $z(\tau) = |\tau|$ $(-\pi \leqslant \tau \leqslant +\pi)$ by its Fourier series with respect to the orthonormal set in Example 3. The Fourier coefficients are

$$\lambda_k = (z, e_k) = \frac{1}{\sqrt{(2\pi)}} \int_{-\pi}^{+\pi} |\tau| e^{-ik\tau} d\tau$$

and therefore

$$\lambda_0 = \frac{\pi^2}{\sqrt{(2\pi)}}, \qquad \lambda_k = -\frac{4}{k^2 \sqrt{(2\pi)}} \ (k \text{ odd}), \qquad \lambda_k = 0 \ (k \neq 0 \text{ and even}).$$

Hence the Fourier series is

$$|\tau| = \frac{\pi}{2} - \frac{4}{\pi} \sum_{k=1}^{\infty} \frac{1}{(2k-1)^2} \frac{(e^{i(2k-1)\tau} + e^{-i(2k-1)\tau})}{2}$$

$$= \frac{\pi}{2} - \frac{4}{\pi} \sum_{k=1}^{\infty} \frac{\cos(2k-1)\tau}{(2k-1)^2}$$

$$= \frac{\pi}{2} - \frac{4}{\pi} (\cos\tau + \tfrac{1}{9}\cos 3\tau + \tfrac{1}{25}\cos 5\tau + \ldots)$$

The function $z(\tau) = |\tau|$ can be approximated arbitrarily closely in the sense of square means on the interval $-\pi \leqslant \tau \leqslant +\pi$ by the partial sums of this series. Furthermore, it can also be shown that the series converges to $z(\tau)$ in the usual sense (and even uniformly).

Finally we point out that *complex Hilbert spaces*, which are a particular type of unitary space, can be defined in an analogous manner to Definition 12.3;1.

CHAPTER 13

EIGENVALUES AND EIGENVECTORS OF ENDOMORPHISMS OF A VECTOR SPACE

In this chapter, E will denote a real or complex vector space of dimension n, and f will denote an endomorphism of E, i.e., a linear mapping of E into itself. In connection with eigenvalues and eigenvectors, endomorphisms of E are also referred to as *linear transformations* or *operators* of E.

13.1 Eigenvectors and Eigenvalues

Definition 1. An 'eigenvector' of an endomorphism f of E is any non-zero vector $x \in E$ which is mapped onto a multiple of itself by f. If $f(x) = \lambda x$, then λ is referred to as the 'eigenvalue' of f corresponding to the eigenvector x.

If E is real, then every eigenvalue of f is a real number. If E is complex, then the eigenvalues are also complex. (We will consider the real numbers as being a special case of the complex numbers so that a number which is referred to as complex may also be real.)

If x is an eigenvector with the eigenvalue λ, then every scalar multiple of x is also an eigenvector with the eigenvalue λ. [$f(\alpha x) = \alpha f(x) = \lambda(\alpha x)$.]

The equation $f(x) = \lambda x$ can also be written in the form $(f - \lambda e)(x) = 0$ where e denotes the identity mapping on E. Thus λ is an eigenvalue of f if and only if $f - \lambda e$ is not 1-1, i.e., if it is not an automorphism. Now, if f is represented by the matrix A with respect to some basis of E, then $A - \lambda I$ is the matrix of $f - \lambda e$. Hence, by Theorem 5.2;13 we have the following result.

Theorem 1. If f is represented by the matrix A, then a scalar λ (which may be real or complex according as E is real or complex) is an eigenvalue of f if and only if $\det(A - \lambda I) = 0$.

Written out in full, the latter equation is

$$\begin{vmatrix} \alpha_{11} - \lambda & \alpha_{12} & \cdots & \alpha_{1n} \\ \alpha_{21} & \alpha_{22} - \lambda & \cdots & \alpha_{2n} \\ \cdots & \cdots & \cdots & \cdots \\ \alpha_{n1} & \alpha_{n2} & \cdots & \alpha_{nn} - \lambda \end{vmatrix} = 0$$

Clearly this is an algebraic equation of degree n for the eigenvalue λ and is referred to as the *characteristic equation* of f (or also of the matrix A). The left-hand side of the equation is known as the *characteristic polynomial* of f. However these terms will only be justified when we have shown that the characteristic polynomial of f is independent of the choice of basis. We can do this as follows. Under a change of basis the matrix A goes into a similar matrix of the form SAS^{-1} (see 5.4.;(10)). By Theorem 5.2;14 and 5.3;(4) the characteristic polynomial of this is $\det(SAS^{-1} - \lambda I) = \det(S(A - \lambda I)S^{-1}) = \det(A - \lambda I)$ and is therefore identical with that of A.

If λ is an eigenvalue of f, then the components $\xi_1, \ldots, \xi_n$ of an eigenvector x corresponding to λ are a solution of the homogeneous system of linear equations

$$(A - \lambda I)\,\boldsymbol{\xi} = 0. \tag{1}$$

Conversely every non-trivial solution of (1) represents an eigenvector corresponding to the eigenvalue λ.

If we consider a square matrix A independently of the endomorphism which it represents, then we refer to the complex roots of its characteristic equation as the *eigenvalues* of A and to every non-trivial complex solution $(\xi_1, \ldots, \xi_n) \in R_n$ of (1) as an *eigenvector* of A corresponding to the eigenvalue λ.

Theorem 2. *The set of all eigenvectors of f which correspond to the eigenvalue λ of f (together with the zero vector) is a subspace of E.*

Proof. If x_1 and x_2 are eigenvectors corresponding to the eigenvalue λ, then

$$f(\alpha_1 x_1 + \alpha_2 x_2) = \alpha_1 f(x_1) + \alpha_2 f(x_2) = \lambda(\alpha_1 x_1 + \alpha_2 x_2).$$

The assertion now follows by Theorem 2.4;1 since there is always at least one eigenvector corresponding to the eigenvalue λ.

The eigenvalue λ is said to be $(r-1)$ ply degenerate if $\dim L = r$, where L is the subspace of its eigenvectors.

Theorem 3. *If the eigenvectors $x_1, \ldots, x_r \in E$ of the endomorphism f correspond to distinct eigenvalues, then they are linearly independent.*

Proof. We will use induction on the number r. The theorem is obviously true for $r = 1$. Now suppose that it is true for r eigenvectors $x_1, \ldots, x_r$ and suppose x_{r+1} is a further eigenvector whose eigenvalue λ_{r+1} is distinct from $\lambda_1, \ldots, \lambda_r$. Now, if

$$\sum_{k=1}^{r} \alpha_k x_k + \alpha_{r+1} x_{r+1} = 0, \tag{2}$$

then, by applying f, we have

$$\sum_{k=1}^{r} \alpha_k \lambda_k x_k + \alpha_{r+1} \lambda_{r+1} x_{r+1} = 0.$$

Subtracting λ_{r+1} times equation (2) from this, we obtain

$$\sum_{k=1}^{r} \alpha_k (\lambda_{r+1} - \lambda_k) x_k = 0.$$

But, since the vectors $x_1, \ldots, x_r$ are linearly independent and $\lambda_{r+1} \neq \lambda_k$ for all $k = 1, \ldots, r$, it follows that $\alpha_1 = \ldots = \alpha_r = 0$ and hence $\alpha_{r+1} = 0$ (by (2)). Hence $x_1, \ldots, x_r, x_{r+1}$ are linearly independent.

From Theorem 3, it follows that E *has a basis consisting of eigenvectors* of f whenever the characteristic equation has n distinct real roots (when E is real) or n distinct complex roots (when E is complex).

Note however that these are not the only endomorphisms which have this property. For example, if $f = 0$, then $\lambda = 0$ is the only eigenvalue and every non-zero vector x is an eigenvector. On the other hand, not every endomorphism has a basis of eigenvectors as will be seen in Problem 9.

We will now investigate the matrix B which represents the endomorphism f with respect to a basis consisting of eigenvectors $\{g_1, \ldots, g_n\}$ (providing one exists). If $x = \sum_{k=1}^{n} \xi_k g_k$, then $f(x) = \sum_{k=1}^{n} (\lambda_k \xi_k) g_k$, where $\lambda_1, \ldots, \lambda_n$ are the eigenvalues. Thus the equations of the transformation are $\eta_k = \lambda_k \xi_k$ $(k = 1, \ldots, n)$, so that

$$B = \begin{pmatrix} \lambda_1 & 0 & \ldots & 0 \\ 0 & \lambda_2 & \ldots & 0 \\ 0 & 0 & \ldots & \lambda_n \end{pmatrix} \tag{3}$$

On the other hand we also know from 5.4;(10) that there is a matrix S such that $B = SAS^{-1}$, where A is the matrix of f with respect to the original basis. If, for a given matrix A, there is a matrix S such that $B = SAS^{-1}$ has the form (3), i.e., is diagonal, then we say that A is *reducible to diagonal form*.

Conversely, if the endomorphism f of E has a matrix B of the form (3) with respect to some basis, then this basis clearly consists of eigenvectors of f with the eigenvalues $\lambda_1, \ldots, \lambda_n$. Thus we have the following theorem.

Theorem 4. An endomorphism f has a basis of eigenvectors if and only if the matrix of f with respect to some basis is reducible to diagonal form. The eigenvalues of f will appear in the main diagonal of the reduced form of the matrix.

Theorem 5. If the endomorphism f has a basis of eigenvectors, then the eigenvalues of the basis vectors include all the eigenvalues of f.

Proof. Let $\{e_1,\ldots,e_n\}$ be a basis of eigenvectors and let λ_k be the eigenvalue which corresponds to e_k. If λ is any eigenvalue, x is a corresponding eigenvector and $\xi_1,\ldots,\xi_n$ are the components of x, then

$$\sum_{k=1}^n (\lambda\xi_k)\,e_k = \lambda x = f(x) = \sum_{k=1}^n \xi_k f(e_k) = \sum_{k=1}^n (\lambda_k\xi_k)\,e_k.$$

Therefore $(\lambda-\lambda_k)\,\xi_k=0$ for $k=1,\ldots,n$. Since $x\neq0$, it follows that λ must be equal to one of the eigenvalues $\lambda_1,\ldots,\lambda_n$.

Theorem 6. If the matrix A is reducible to diagonal form, then the main diagonal of the reduced form contains all the eigenvalues of A and in fact each appears as many times as its multiplicity as a root of the characteristic polynomial.

Proof. The fact that all the eigenvalues appear is a consequence of Theorems 4 and 5. The second part of the theorem follows from the fact that the characteristic polynomial has the factor $(\lambda_k-\lambda)$ for each diagonal element λ_k of the reduced matrix.

Theorem 7. If f is an automorphism of E and $\lambda_1,\ldots,\lambda_r$ are the eigenvalues of f, then $\lambda_1^{-1},\ldots,\lambda_r^{-1}$ are the eigenvalues of f^{-1}.

Proof. The characteristic equation $\det(A-\lambda I)=0$ is satisfied by each eigenvalue λ of f. Multiplying by $\lambda^{-n}\det A^{-1}$ ($\lambda\neq0$ because f is an automorphism), we have $\det(A^{-1}-\lambda^{-1}I)=0$, so that λ^{-1} is an eigenvalue of f^{-1}. To show that all the eigenvalues of f^{-1} are of this form, we merely have to reverse the roles of f and f^{-1} in this argument.

Problems

1. Show that, if $\lambda\neq0$ is an eigenvalue of the endomorphism f and x_0 is a corresponding eigenvector, then x_0 is also an eigenvector of f^{-1} corresponding to the eigenvalue λ^{-1} (providing f^{-1} exists).
2. Show that, if λ is an eigenvalue of f, then λ^2 is an eigenvalue of f^2. Is the converse also valid?
3. An endomorphism f for which $f^2=e$ is said to be an *involution*. What can be said about the eigenvalues of an involution?
4. What can be said about the eigenvalues of a projection? (Cf. Problem 5.3;6.)
5. Let ϕ be a permutation of $\{1,\ldots,n\}$ and define an endomorphism of the real vector space R_n by the rule $(\xi_1,\ldots,\xi_n)\to(\xi_{\phi(1)},\ldots,\xi_{\phi(n)})$. What are the eigenvalues of this endomorphism?
6. Find the eigenvalues and eigenvectors of a rotation of the Euclidean plane. (Cf. Problem 12.1;10.)

7. Show that every eigenvalue of a rotation of a Euclidean or unitary vector space has the absolute value 1.

8. Prove that a square matrix is non-singular if and only if it has no zero eigenvalues.

9. Show that $A = \begin{pmatrix} 1 & 0 \\ 1 & 1 \end{pmatrix}$ is an example of a matrix which represents an endomorphism for which there is no basis consisting of eigenvectors.

10. Suppose that f and g are endomorphisms of a finite-dimensional vector space. Show that fg and gf have the same characteristic equation. (See Problem 5.3;9.)

11. Suppose (E, F) is a dual pair of spaces, f is an endomorphism of E and g is the dual endomorphism. Show that f and g have the same eigenvalues. Show further that, if $E_\lambda \subseteq E$ and $F_\lambda \subseteq F$ are the subspaces which consist of the eigenvectors corresponding to the eigenvalue λ, then $\dim E_\lambda = \dim F_\lambda$.

13.2 Symmetric Endomorphisms of a Euclidean Vector Space

In this section we will consider a special type of endomorphism on a finite-dimensional Euclidean vector space E. In this case E forms a dual pair of spaces with itself with the scalar product $\langle x, y \rangle = (x, y)$ (see 6.2.3 and 12.1.1). Thus, to each endomorphism f of E, there corresponds the dual endomorphism g of E given by

$$(f(x), y) = (x, g(y)) \text{ for all } x, y \in E.$$

(see 6.2.5).

Definition 1. An endomorphism f of a Euclidean vector space E is said to be 'symmetric' if it is equal to its dual endomorphism, i.e., if $(f(x), y) = (x, f(y))$ for all $x, y \in E$.

Theorem 1. The endomorphism f is symmetric if and only if f is represented by a symmetric matrix with respect to some orthonormal basis of E. In this case, the matrix of f with respect to any orthonormal basis is symmetric.

Proof. An orthonormal basis of a Euclidean vector space is its own dual (see 6.2;(4)). Hence Theorem 1 follows immediately from Theorem 6.2;9.

Theorem 2. The roots of the characteristic equation of a symmetric endomorphism are all real. (It follows that every symmetric endomorphism has at least one eigenvector.)

Proof. Suppose that the symmetric endomorphism f is represented by the symmetric (and real) matrix A (Theorem 1). For each (real or complex) root λ of the characteristic equation, there is a matrix $\xi \neq 0$ such that

251

$A\xi = \lambda\xi$ (Theorem 7.1;1). It follows that $\bar{\xi}' A\xi = \lambda\bar{\xi}'\xi$, where $\bar{\xi}'$ is the conjugate complex matrix of ξ'. Transposing this and taking complex conjugates, we have $\bar{\xi}' A\xi = \bar{\lambda}\bar{\xi}'\xi$. Therefore $(\lambda - \bar{\lambda}) \sum_{k=1}^{n} |\xi_k|^2 = 0$ and hence $\lambda = \bar{\lambda}$.

Theorem 3. Eigenvectors which correspond to distinct eigenvalues of a symmetric endomorphism are orthogonal.

Proof. Suppose $f(x_1) = \lambda_1 x_1, f(x_2) = \lambda_2 x_2$, $\lambda_1 \neq \lambda_2$ and f is symmetric. Then $\lambda_1(x_1, x_2) = (f(x_1), x_2) = (x_1, f(x_2)) = \lambda_2(x_1, x_2)$ and hence $(x_1, x_2) = 0$.

Theorem 4. If x_1 is an eigenvector of the symmetric endomorphism f, then the orthogonal complement $S_1 = [L(x_1)]^\dagger$, which consists of all the vectors orthogonal to x_1, is mapped into itself by f.

Proof. Suppose $y \in S_1$, i.e., $(x_1, y) = 0$. Then $(x_1, f(y)) = (f(x_1), y) = \lambda_1(x_1, y) = 0$ and hence $f(y) \in S_1$.

Theorem 5. Given a symmetric endomorphism f of a finite-dimensional Euclidean vector space E, then there is an orthonormal basis of E consisting of eigenvectors of f.

Proof. By Theorem 2, f has an eigenvector x_1. The $(n-1)$-dimensional orthogonal complement S_1 of $L(x_1)$ is mapped into itself by f. So, if we restrict f to S_1, we obtain an endomorphism of S_1 which is again symmetric. Therefore, there is an eigenvector $x_2 \in S_1$ which is orthogonal to x_1.

Using this same argument for S_1 and x_2 (as for E and x_1) we can show that there is a further eigenvector x_3 which is orthogonal to both x_1 and x_2. We repeat this until $S_r = \{0\}$ for some r and then $x_1, \ldots, x_r$ is an orthonormal basis of E.

Theorem 6. Given a symmetric endomorphism f of E, then there is an orthonormal basis of E for which the matrix of f is diagonal.

Proof. Theorem 5.

Theorem 7. Given a real symmetric matrix A, then there is a real orthogonal matrix S such that $B = SAS^{-1}$ is diagonal.

Proof. We choose an orthonormal basis in a Euclidean vector space E and let A represent the endomorphism f with respect to this basis. Now, by Theorem 1, f is symmetric and therefore there is a further orthonormal basis of E for which the matrix of f is diagonal (Theorem 6). The matrix S, which corresponds to the change from one basis to the other, then satisfies the requirements of Theorem 7.

Problems

1. Prove that, if f is an endomorphism of a finite-dimensional Euclidean vector space which has an orthonormal basis of eigenvectors, then f is symmetric.

2. Show that, given any orthonormal basis $\{e_1, \ldots, e_n\}$ of the Euclidean vector space E and any real numbers $\lambda_1, \ldots, \lambda_n$, then there exists a symmetric endomorphism f of E for which $e_1, \ldots, e_n$ are eigenvectors corresponding to the eigenvalues $\lambda_1, \ldots, \lambda_n$.

3. Suppose that A is a symmetric and S is an orthogonal $n \times n$ matrix and further suppose that SAS^{-1} is diagonal. Show that the rows of S form an orthonormal basis consisting of eigenvectors of A.

4. Prove that the product of two symmetric endomorphisms f and g is symmetric if and only if $fg = gf$.

5. Prove that a projection f (cf. Problem 5.3;6) of a Euclidean vector space E is symmetric if and only if its kernel is the orthogonal complement of $f(E)$. (See Problem 6.2;14.)

6. Show that, if f is a symmetric endomorphism of a Euclidean vector space, then $g(x, y) = (f(x), y)$ is a symmetric bilinear form.

7. Show that, if f is a non-singular endomorphism of a Euclidean vector space and f^* is the dual endomorphism of f, then ff^* is symmetric and the quadratic form $q(x) = (ff^*(x), x)$ is positive definite.

8. Suppose that f is a symmetric endomorphism of a finite-dimensional vector space E and that $q(x) = (f(x), x)$ is positive definite. Show that there is a symmetric endomorphism g of E such that $f = g^2$.

9. Show that an endomorphism f of a Euclidean vector space is a rotation if and only if $ff^* = e$.

10. An endomorphism f of a Euclidean vector space E is said to be *normal* if it commutes with its dual endomorphism, $ff^* = f^*f$. Show that the symmetric endomorphisms and rotations of E are normal.

11. Show that every non-singular endomorphism f of a finite-dimensional Euclidean vector space can be represented in the form $f = gh$ where g is symmetric and h is a rotation. (Hint: By Problems 7 and 8, there is a symmetric endomorphism g such that $ff^* = g^2$. Then $h = g^{-1}f$ is a rotation.)

13.3 The Transformation of Quadratic Forms to Principal Axes

Theorem 1. Given a quadratic form $k(x)$ on a Euclidean vector space of dimension n, then there is an orthonormal basis with respect to which $k(x)$ is completely reduced. That is, $k(x)$ takes the form

$$k(x) = \sum_{k=1}^{n} \lambda_k \xi_k^2. \tag{1}$$

If $k(x) = \boldsymbol{\eta}' A \boldsymbol{\eta}$ *with respect to some orthonormal basis (with A symmetric), then the coefficients* λ_k *are the eigenvalues of A. Each distinct eigenvalue appears as many times as its multiplicity as a root of the characteristic polynomial.*

Proof. Let $k(x) = \boldsymbol{\eta}' A \boldsymbol{\eta}$ with respect to some orthonormal basis (A symmetric). By Theorem 13.2;7, there is an orthogonal matrix S such that $SAS^{-1} = SAS'$ is diagonal. If we change to a new orthonormal basis by the transformation $\boldsymbol{\xi} = S\boldsymbol{\eta}$, then $k(x)$ becomes $k(x) = \boldsymbol{\xi}'(SAS')\boldsymbol{\xi} = \sum_{k=1}^{n} \lambda_k \xi_k^2$ where the coefficients λ_k are the diagonal elements of SAS^{-1}, i.e., the eigenvalues of A (Theorem 13.1;6).

Note that in this way we have found a second proof for the reducibility of quadratic forms.

The process of changing to an orthonormal basis for which (1) is valid is known as the *transformation of $k(x)$ to principal axes*. In order to explain this terminology, it is helpful to consider *hyperquadric surfaces* in a vector space. If $k(x)$ is a quadratic form on a real vector space E, then the set of all vectors $x \in E$ for which

$$k(x) = \gamma = \text{constant} \tag{2}$$

is known as a hyperquadric surface in E. (For example, if E has dimension 2, then ellipses and hyperbolae are 'hyperquadric surfaces'.) Now, if E is also Euclidean, then the *semi-axes* of the surface (2) are those vectors in the surface for which the function $\|x\|^2$ is stationary as x varies, i.e., if the vector x is changed by an infinitely small amount dx, then the corresponding change in the square of the norm $d\|x\|^2 = 0$, whenever $dk(x) = 0$. (At this point, we have to extend some of the concepts from infinitesimal calculus to apply them to linear algebra but we will not go into the details of a rigorous justification.) If $d\xi_1, \ldots, d\xi_n$ are the components of dx, then we have

$$d\|x\|^2 = \sum_{k=1}^{n} \frac{\partial\|x\|^2}{\partial\xi_k} d\xi_k = 0 \quad \text{whenever} \quad dk(x) = \sum_{k=1}^{n} \frac{\partial k(x)}{\partial\xi_k} d\xi_k = 0. \tag{3}$$

Now suppose that $k(x)$ is given by (1) (with respect to an orthonormal basis). Then

$$\frac{\partial k(x)}{\partial\xi_k} = 2\lambda_k \xi_k \quad \text{and} \quad \frac{\partial\|x\|^2}{\partial\xi_k} = 2\xi_k \qquad (k = 1, \ldots, n).$$

With the help of Theorem 6.2;7, it follows from (3) that, for each semi-axis, there is a scalar θ such that

$$\xi_k = \theta\lambda_k \xi_k \qquad (k = 1, \ldots, n). \tag{4}$$

Conversely, every vector x in the surface, whose components satisfy (4) for some θ, is a semi-axis.

Hence, we find all the semi-axes of the hyperquadric given by (1) and (2) as follows.

1. For each basis vector e_k for which $\lambda_k \neq 0$ and $\dfrac{\gamma}{\lambda_k} > 0$, $\quad x = \pm \sqrt{\left(\dfrac{\gamma}{\lambda_k}\right)} [e_k]$

are semi-axes.

2. If there are several of the semi-axes in 1 which have the same norm, then any linear combination of these vectors which satisfies (2) is also a semi-axis.

We notice in particular that of two eigenvalues the one with the smaller absolute value corresponds to the larger semi-axis. Also we see that two semi-axes which have different norms are orthogonal.

It is this connection with the axes of the hyperquadric which is the reason for the terminology 'transformation of the quadratic form to principal axes'.

Theorem 2. A real symmetric matrix A is positive definite if and only if its eigenvalues are all strictly positive.

(A matrix A is positive definite if the corresponding quadratic form is positive definite.) The proof of the Theorem is immediate using (1).

Problems

1. Let A be a real symmetric $n \times n$ matrix. Show that $\boldsymbol{\xi}' A \boldsymbol{\xi}$ is positive definite if and only if there exists a non-singular real $n \times n$ matrix S such that $A = S'S$.

2. It is well-known that the extreme values of a differentiable function $f(\xi_1, \ldots, \xi_n)$ with the restrictions $g_k(\xi_1, \ldots, \xi_n) = 0$ $(k = 1, \ldots, r)$ can be found as follows. We introduce the function

$$h(\xi_1, \ldots, \xi_n) = f(\xi_1, \ldots, \xi_n) + \sum_{k=1}^{r} \lambda_k g_k(\xi_1, \ldots, \xi_n)$$

with the Lagrange multipliers $\lambda_1, \ldots, \lambda_r$ and put $\partial h/\partial \xi_i = 0$ $(i = 1, \ldots, n)$. Including the original restrictions, we then have $(n+r)$ equations for the $(n+r)$ unknowns ξ_i, λ_k. Justify this method on the basis of Theorem 6.2;7.

3. Let $p(x) = (f(x), x) = 1$ and $q(x) = (g(x), x) = 1$ be the equations of two hyperquadric surfaces F and G in a finite-dimensional Euclidean vector space E and let p, q be positive definite. Prove that F can be mapped onto G by a rotation of E if and only if the symmetric endomorphisms f and g have the same characteristic equation. (cf. 13.5; (1))

4. Prove the extremal property of the semi-axes of a hyperquadric surface without reference to a basis. (Hint: Let the equation of the surface be $k(x) = (f(x), x) = $ const., where f is symmetric. Then $\mathrm{d}k(x) = 2(f(x), \mathrm{d}x)$ and $\mathrm{d}\|x\|^2 = 2(x, \mathrm{d}x)$.)

5. Let x_0 be a vector in the surface $k(x) = (f(x), x) = 1$. Show that $f(x_0)$ is orthogonal to the surface at x_0, i.e., orthogonal to every infinitesimal vector $\mathrm{d}x$ for which $\mathrm{d}k(x_0) = 0$.

13.4 Self-Adjoint Endomorphisms of a Unitary Vector Space

Now, if we change from Euclidean to unitary spaces, then we must first notice that we cannot immediately construct the dual of an endomorphism of a unitary vector space, because the scalar product is not bilinear but only quasi-bilinear. We will therefore translate Definition 13.2;1 to this new context.

Definition 1. An endomorphism f of a unitary vector space E is said to be 'self-adjoint' if

$$(f(x), y) = (x, f(y)) \text{ for all } x, y \in E.$$

It is not difficult now to formulate the theorems proved in 13.2 in terms of self-adjoint endomorphisms.

Theorem 1. An endomorphism f of a finite-dimensional unitary vector space is self-adjoint if and only if it is represented by a Hermitian matrix with respect to some orthonormal basis. The same is then true for all orthonormal bases.

Proof. 1. Suppose f is self-adjoint and is given by $\boldsymbol{\xi}^* = A\boldsymbol{\xi}$ with respect to an orthonormal basis. Then

$$\boldsymbol{\xi}' A' \overline{\boldsymbol{\eta}} = \boldsymbol{\xi}^{*'} \overline{\boldsymbol{\eta}} = (f(x), y) = (x, f(y))$$
$$= \boldsymbol{\xi}' \overline{\boldsymbol{\eta}}^* = \boldsymbol{\xi}' \overline{A} \overline{\boldsymbol{\eta}}$$

for all $\boldsymbol{\xi}'$, $\boldsymbol{\eta}$. Therefore $A' = \overline{A}$.

2. Suppose A is Hermitian and f is the endomorphism which is given by $\boldsymbol{\xi}^* = A\boldsymbol{\xi}$ with respect to an orthonormal basis. Then

$$(f(x), y) = \boldsymbol{\xi}^{*'} \overline{\boldsymbol{\eta}} = \boldsymbol{\xi}' A' \overline{\boldsymbol{\eta}} = \boldsymbol{\xi}' \overline{A} \overline{\boldsymbol{\eta}} = \boldsymbol{\xi}' \overline{\boldsymbol{\eta}}^* = (x, f(y)).$$

Theorem 2. All the eigenvalues of a self-adjoint endomorphism f are real.

256

Proof. Let x be an eigenvector of f and let λ be the corresponding eigenvalue. Then

$$\lambda(x,x) = (f(x),x) = (x,f(x)) = \bar{\lambda}(x,x),$$

and therefore $\lambda = \bar{\lambda}$. Theorems 13.2;3, 4, 5 and their proofs are still valid for self-adjoint endomorphisms. In the following, they will be referred to as Theorems 13.4;3, 4, 5.

Thus in particular corresponding to each self-adjoint endomorphism there is an orthonormal basis consisting of eigenvectors.

Theorem 6. Given a self-adjoint endomorphism f, then there is an orthonormal basis such that the matrix of f with respect to this basis is a real diagonal matrix.

Proof. With respect to an orthonormal basis which consists of eigenvectors, the matrix of f will be a diagonal matrix. By Theorem 1, this matrix is Hermitian and therefore real.

Theorem 7. Given a Hermitian matrix A, then there is a unitary matrix S such that $B = SAS^{-1}$ is real and diagonal.

The proof is analogous to that of Theorem 13.2;7.
The complex version of Theorem 13.3;1 is as follows:

Theorem 8. Given a Hermitian form $k(x)$ on a unitary vector space of finite dimension, then there is an orthonormal basis such that $k(x)$ is completely reduced with respect to this basis, i.e., such that

$$k(x) = \sum_{k=1}^{n} \lambda_k |\xi_k|^2.$$

If $k(x) = \eta' A \bar{\eta}$ with respect to any orthonormal basis (A is Hermitian) then the coefficients λ_k are the eigenvalues of A.
The proof is analogous to that of Theorem 13.3;1.

Theorem 9. Let $q_1(x)$ and $q_2(x)$ be two quadratic (Hermitian) forms on a finite-dimensional real (complex) vector space E, and suppose that $q_1(x)$ is positive definite. Then there is a basis of E such that, with respect to this basis,

$$q_1(x) = \sum_{k=1}^{n} |\xi_k|^2 \quad \text{and} \quad q_2(x) = \sum_{k=1}^{n} \lambda_k |\xi_k|^2$$

for some coefficients λ_k.

Proof. We make E Euclidean (unitary) by using $q_1(x)$ as the fundamental form. Then there exists an orthonormal basis such that the second equation is satisfied (Theorem 13.3;1 and Theorem 8). At the same time the first equation is also satisfied because $q_1(x)$ is the fundamental form.

13.5 Extremal Properties of the Eigenvalues of Symmetric and Self-Adjoint Endomorphisms

Let E be a Euclidean (unitary) vector space of finite dimension and let f be a symmetric (self-adjoint) endomorphism of E. The (quasi-)bilinear form $g(x,y) = (f(x),y)$ is then symmetric (Hermitian), because

$$g(y,x) = (f(y),x) = (y,f(x)) = (\overline{f(x),y}) = \bar{g}(x,y).$$

Hence

$$k(x) = (f(x),x) \tag{1}$$

is a quadratic (Hermitian) form on E.

If $\{e_1,\ldots,e_n\}$ is an orthonormal basis of E in which all the vectors e_k are eigenvectors of f (Theorems 13.2;5 and 13.4;5), then f is given by

$$\xi_k \to \xi_k^* = \lambda_k \xi_k \qquad (k = 1,\ldots,n),$$

and

$$k(x) = (f(x),x) = \sum_{k=1}^n \lambda_k |\xi_k|^2 \tag{2}$$

which corresponds to 13.3;(1). Each eigenvalue λ_k appears in (2) as many times as its multiplicity as a root of the characteristic polynomial. In the following we will assume that $\lambda_1 \leqslant \lambda_2 \leqslant \ldots \leqslant \lambda_n$.

Now, if $x \in E$ is normalized, i.e., $\sum_{k=1}^n |\xi_k|^2 = 1$, then it follows from (2) that

$$\lambda_1 \leqslant (f(x),x) \leqslant \lambda_n. \tag{3}$$

When $x = e_1$, $(f(x),x) = \lambda_1$ and when $x = e_n$, $(f(x),x) = \lambda_n$. It follows that λ_1 and λ_n are the minimum and maximum values of $(f(x),x)$ as x runs through all the normalized vectors of E.

Theorem 1. If f is a symmetric (self-adjoint) endomorphism of a Euclidean (unitary) vector space of finite dimension, then the least eigenvalue λ_1 and largest eigenvalue λ_n of f are given by

$$\lambda_1 = \min_{\|x\|=1} (f(x),x), \qquad \lambda_n = \max_{\|x\|=1} (f(x),x). \tag{4}$$

The minimum (maximum) in (4) is attained by a normalized vector $x \in E$ if and only if x is an eigenvector corresponding to the eigenvalue $\lambda_1(\lambda_n)$.

Proof. Only the last part of the theorem is still to be proved.

1. Suppose x is a normalized eigenvector corresponding to the least eigenvalue λ_1. Then $(f(x), x) = (\lambda_1 x, x) = \lambda_1(x, x) = \lambda_1$.

2. Suppose $(f(x), x) = \lambda_1$ and x is normalized. Then by (2) it follows that all the components ξ_k of x for which $\lambda_k > \lambda_1$ are zero. Hence x is an eigenvector corresponding to the least eigenvalue λ_1.

3. The proof for the greatest eigenvalue λ_n is analogous to 1 and 2.

Now, in addition to the condition $\|x\| = 1$, we will also require x to be in the subspace $L(e_2, \ldots, e_n) = L_{n-1}$. Since this merely means that $\xi_1 = 0$, it follows by the same argument as above that the second eigenvalue λ_2 is given by

$$\lambda_2 = \min_{\substack{\|x\|=1 \\ x \in L_{n-1}}} (f(x), x).$$

Similarly

$$\lambda_{n-1} = \max_{\substack{\|x\|=1 \\ x \in L^*_{n-1}}} (f(x), x)$$

where $L^*_{n-1} = L(e_1, \ldots, e_{n-1})$.

Now in general let $L_{n-r+1} = L(e_r, \ldots, e_n)$, where the subscript $(n-r+1)$ denotes the dimension of this subspace. Since $x \in L_{n-r+1}$ if and only if $\xi_1 = \ldots = \xi_{r-1} = 0$, it again follows from (2) that

$$\lambda_r = \min_{\substack{\|x\|=1 \\ x \in L_{n-r+1}}} (f(x), x) \qquad (r = 1, \ldots, n)$$

and that this value is attained by a vector x satisfying the conditions, if and only if x is an eigenvector corresponding to the eigenvalue λ_r. If we also carry out the corresponding argument for the maximum, then we obtain the following result.

Theorem 2. Let f be a symmetric (self-adjoint) endomorphism and let $\{e_1, \ldots, e_n\}$ be an orthonormal basis of E which consists of eigenvectors of f and is so numbered, that the corresponding eigenvalues satisfy the condition $\lambda_1 \leqslant \ldots \leqslant \lambda_n$. If $L_{n-r+1} = L(e_r, \ldots, e_n)$ and $L^*_r = L(e_1, \ldots, e_r)$, then

$$\lambda_r = \min_{\substack{\|x\|=1 \\ x \in L_{n-r+1}}} (f(x), x) = \max_{\substack{\|x\|=1 \\ x \in L^*_r}} (f(x), x). \tag{5}$$

The minimum (maximum) is attained by a vector x satisfying the conditions if and only if x is an eigenvector corresponding to the eigenvalue λ_r.

Theorem 2 can be used to determine λ_r as the minimum when $e_1, \ldots, e_{r-1}$ are known, because L_{n-r+1} is then the orthogonal complement of $\{e_1, \ldots, e_{r-1}\}$ and the condition $x \in L_{n-r+1}$ can be replaced by 'x is orthogonal to

$e_1, \ldots, e_{r-1}$'. Similarly λ_r can be determined as the maximum when $e_{r+1}, \ldots, e_n$ are known (cf. 13.6.2).

Since the quotient $(f(\lambda x), \lambda x)(\lambda x, \lambda x)^{-1}$ is independent of the scalar λ, we can drop the condition $\|x\| = 1$ in the preceding discussion, by considering the extreme values of the quantity

$$\rho(x) = \frac{(f(x), x)}{(x, x)} = \frac{(f(x), x)}{\|x\|^2} , \quad (x \neq 0) \tag{6}$$

instead of those of $(f(x), x)$. This quantity is known as the *Rayleigh quotient* of the endomorphism f. In this way, for example, (4) is replaced by

$$\lambda_1 = \min_{x \neq 0} \rho(x), \qquad \lambda_n = \max_{x \neq 0} \rho(x). \tag{7}$$

If x is an eigenvector corresponding to the eigenvalue λ, then $\rho(x) = (\lambda x, x)(x, x)^{-1} = \lambda$.

Now we will construct the minimum in (5) for an arbitrary subspace M_{n-r+1} of dimension $(n-r+1)$ in place of the special subspace L_{n-r+1}. Let $\{e_1, \ldots, e_n\}$ again be a basis as in Theorem 2. If we put $N_r = L = L(e_1, \ldots, e_r)$, then by Theorem 3.2;8 there is a vector $x_0 \in M_{n-r+1} \cap N_r$ such that $\|x_0\| = 1$. Since $x_0 \in N_r$, the components $\xi_{r+1}, \ldots, \xi_n$ are all equal to zero for this vector. Hence

$$(f(x_0), x_0) = \sum_{k=1}^{r} \lambda_k |\xi_k|^2 \leqslant \lambda_r$$

and therefore, since $x_0 \in M_{n-r+1}$,

$$\min_{\substack{\|x\|=1 \\ x \in M_{n-r+1}}} (f(x), x) \leqslant \lambda_r. \tag{8}$$

If M_{n-r+1} is the special subspace L_{n-r+1}, then (8) is valid with the equality sign. Thus λ_r is the greatest value taken by the minimum in (8) as M_{n-r+1} runs through all the $(n-r+1)$-dimensional subspaces of E.

If we carry out the corresponding arguments for the maximum in (5), then we find the following theorem which is due to Fischer.

Theorem 3. For each $r = 1, \ldots, n$

$$\lambda_r = \max_{M_{n-r+1}} \min_{\substack{\|x\|=1 \\ x \in M_{n-r+1}}} (f(x), x) \tag{9}$$

and

$$\lambda_r = \min_{M_r} \max_{\substack{\|x\|=1 \\ x \in M_r}} (f(x), x) \tag{10}$$

where the maximum in (9) (the minimum in (10)) is taken over all subspaces M_{n-r+1} of E of dimension $(n-r+1)$ (M_r of E of dimension r). Each eigenvalue of f appears in the set $\{\lambda_1, \ldots, \lambda_n\}$ as many times as its multiplicity as a root of the characteristic polynomial.

In contrast to Theorem 2, Theorem 3 provides us with two representations of each eigenvalue which are independent of the eigenvectors corresponding to the other eigenvalues.

Since 13.3;(1) and 13.5;(2) are the same equation in the real case, the correspondence between the eigenvalues $\lambda_1, \ldots, \lambda_n$ of f and the semi-axes of the hyperquadric surface $k(x) = (f(x), x) = 1$, which was discussed in 13.3 also applies here. Consequently, Theorem 3 has the following geometrical interpretation.

We will assume for this example that $n = 3$ and that all the eigenvalues are strictly positive, which means that the surface is an ellipsoid. We obtain the middle eigenvalue λ_2 either from (9) or from (10) by putting $r = 2$. Then M_{n-r+1} in (9) and M_r in (10) both run through all 2-dimensional planes which pass through the centre of the ellipsoid. For each of these planes in the case of (9), we have to find the minimum of $(f(x), x)$, i.e., by 13.3 the greatest diameter (the major axis) of the cross-sectional ellipse. Then we look for the plane in which this greatest diameter is as small as possible. This is then equal to the middle axis of the ellipsoid and corresponds to the eigenvalue λ_2. On the other hand, in the case of (10) we first look for the smallest diameter (the minor axis) of the cross-sectional ellipse and then for the plane through the centre of the ellipsoid for which this is as large as possible. In this way we again find the middle axis of the ellipsoid.

Of the many important consequences of Fischer's Theorem, we will only mention the following result. (For further consequences see Problems 3 and 4.)

Theorem 4. Let A and B be symmetric or Hermitian matrices and suppose that the quadratic or Hermitian form $q(x) = \boldsymbol{\xi}' B \boldsymbol{\xi}$ corresponding to B takes no negative values. Then, if $\lambda_1 \leqslant \lambda_2 \leqslant \ldots \leqslant \lambda_n$ are the eigenvalues of A and $\mu_1 \leqslant \mu_2 \leqslant \ldots \leqslant \mu_n$ are those of $A + B$, then $\lambda_k \leqslant \mu_k \ (k = 1, \ldots, n)$.

Proof. Let f and g be the symmetric or self-adjoint endomorphisms represented by A and B with respect to an orthonormal basis of a Euclidean or unitary vector space. Each eigenvalue λ_r of A is then given by (10). Similarly each eigenvalue μ_r of $A + B$ is given by

$$\mu_r = \min_{M_r} \max_{\substack{\|x\|=1 \\ x \in M_r}} (f(x) + g(x), x). \tag{11}$$

It follows from the hypothesis about B that

$$(f(x) + g(x), x) = (f(x), x) + (g(x), x) \geqslant (f(x), x)$$

for all $x \in E$. From this it follows that

$$\max_{\substack{\|x\|=1 \\ x \in M_r}} (f(x) + g(x), x) \geqslant \max_{\substack{\|x\|=1 \\ x \in M_r}} (f(x), x) \quad \text{for all } M_r$$

261

and hence that the right-hand side of (11) is not less than the right-hand side of (10). This proves the theorem.

Problems

1. Let $\lambda_1(A)$ denote the least eigenvalue of the symmetric matrix A and let $\lambda_n(A)$ denote the greatest eigenvalue. Show that, if A and B are symmetric matrices and α, β are non-negative real numbers,

then
$$\lambda_1(\alpha A + \beta B) \geqslant \alpha\lambda_1(A) + \beta\lambda_1(B)$$
and
$$\lambda_n(\alpha A + \beta B) \leqslant \alpha\lambda_n(A) + \beta\lambda_n(B).$$

2. Suppose $q(x) = (f(x), x)$ (f symmetric) is positive definite. Prove that, if λ_1 is the least eigenvalue of f, then $\lambda_1^{-1} = \max\limits_{(f(x), x) = 1} \|x\|^2$. Find the analogous formula for the greatest eigenvalue. What is the connection between this result and section 13.3?

3. Let A, B be real symmetric $n \times n$ matrices, let $\lambda_1 \leqslant \ldots \leqslant \lambda_n$ be the eigenvalues of A, $\mu_1 \leqslant \ldots \leqslant \mu_n$ those of B and $\nu_1 \leqslant \ldots \leqslant \nu_n$ those of $A + B$. Prove that $\lambda_k + \mu_1 \leqslant \nu_k \leqslant \lambda_k + \mu_n$ ($k = 1, \ldots, n$).

4. Let A be a symmetric $n \times n$ matrix and let B be the $(n-1) \times (n-1)$ matrix which is found by leaving out the last row and column of A. Show that, if $\lambda_1 \leqslant \ldots \leqslant \lambda_n$ are the eigenvalues of A and $\mu_1 \leqslant \ldots \leqslant \mu_{n-1}$ are those of B, then $\lambda_k \leqslant \mu_k \leqslant \lambda_{k+1}$ ($k = 1, \ldots, n-1$).

13.6 Numerical Calculation of Eigenvalues and Eigenvectors

The calculation of eigenvalues and eigenvectors is a very exacting exercise in numerical analysis for which many methods have been developed. In the main these belong to two different groups. The methods in the first group first find the characteristic equation and then solve this to find the eigenvalues and finally the eigenvectors. In view of the large amount of arithmetic involved, there is always the risk in these methods that the results will be inaccurate due to rounding errors. The methods in the second group avoid the calculation of the characteristic polynomial and attempt to find the eigenvalues, and afterwards the eigenvectors, directly. Here we will only describe one simple method from each group.

13.6.1 Calculation of the Characteristic Polynomial and the Eigenvectors using the Gaussian Algorithm

Basically the calculation of the eigenvalues and eigenvectors of a matrix A is the solution of the homogeneous system of linear equations

$$(A - \lambda I)\boldsymbol{\xi} = 0 \tag{1}$$

and it is therefore natural to think of using the Gaussian Algorithm. However there is a difficulty in that λ is unknown. Nevertheless this does not stop us and we can still carry out the algorithm by allowing the elements of the exchange tableaux to be rational functions of the variable λ. As we will show, the algorithm can be so arranged that only whole rational functions of λ appear and it is easy to find the characteristic polynomial from these.

In the following tableaux we will use P_k to denote polynomials in λ with degree at most k. If the degree is exactly equal to k, then we will write P_k^+. If the symbol P_k or P_k^+ appears more than once in a tableau, then this will not necessarily mean that these polynomials are equal. The notation is only to give us information about the degrees. With these conventions, and if e.g. $n = 4$, the initial tableau for the system (1) may be written in the form

1st Tableau

	ξ_1	ξ_2	ξ_3	ξ_4
	P_1^+	P_0	P_0	P_0^*
	P_0	P_1^+	P_0	P_0
	P_0	P_0	P_1^+	P_0
	P_0	P_0	P_0	P_1^+
ξ_4	P_1^+	P_0	P_0	$*$

If all the elements off the main diagonal are zero, then the eigenvalues can be read off directly and the eigenvectors are easy to find. Otherwise suppose, for instance, that the element marked by * is not zero, so that we may use it as a pivot. The next tableau then has the following form.

2nd Tableau

	ξ_1	ξ_2	ξ_3
	P_1	P_1^+	P_0
	P_1	P_0	P_1^+
	P_2^+	P_1	P_1

(If we use another P_0 as pivot, we obtain a tableau of the same kind but with interchanged rows and columns.)

By adding suitable multiples of the first and second rows we can arrange that the last two elements of the third row are independent of λ. (This operation is equivalent to replacing the three linear equations represented by the tableau with three linearly independent combinations of them and has no effect on the solution.) Thus we will have a tableau of the form

2nd Tableau (modified)

	ξ_1	ξ_2	ξ_3
	P_1	P_1^+	P_0
	P_1	P_0	P_1^+
	P_2^+	P_0	P_0^*
ξ_3	P_2^+	P_0	*

If in this second tableau both of the P_0's in the third row are equal to zero, we will say that the method *degenerates*. However we will first assume that degeneration does not occur and choose one of these two elements, say the second, to be the next pivot. The next tableau has the form

3rd Tableau

	ξ_1	ξ_2
	P_2	P_1^+
	P_3^+	P_1

We can again add a multiple of the first row onto the second so that the P_1 becomes a P_0.

3rd Tableau (modified)

	ξ_1	ξ_2
	P_2	P_1^+
	P_3^+	P_0^*
ξ_2	P_3^+	*

We will again assume to start with that $P_0 \neq 0$, i.e., that degeneration does not occur, and we choose this element as the pivot.

4th Tableau

ξ_1
P_4^+

This corresponds to the equation $P_4^+(\lambda)\, \xi_1 = 0$. By 4.3.2, $P_4^+(\lambda)$ is a constant multiple of the characteristic polynomial and we find the eigenvalues by solving the algebraic equation $P_4^+(\lambda) = 0$. We then find the eigenvectors corresponding to an eigenvalue λ by giving ξ_1 an arbitrary non-zero value and calculating the other components from the cellar rows of the tableaux (putting λ equal to the eigenvalue in question). We see in particular that, if no degeneration occurs, then there is exactly one eigenvector (apart from scalar multiples) corresponding to each eigenvalue and hence the eigenvalues are not degenerate. The converse of this result is not true, as can be seen from Example 3.

This method can also be applied to the general case of dimension n providing no degeneration occurs (as we are initially assuming). From the first tableau (written out for dimension n), we obtain the second tableau in the following form by using the last element of the first row as the pivot.

ξ_1	ξ_2	ξ_3	$\cdots$	ξ_{n-1}
P_1	P_1^+	P_0	$\cdots$	P_0
P_1	P_0	P_1^+	$\cdots$	P_0
$\cdots$	$\cdots$	$\cdots$	$\cdots$	$\cdots$
P_1	P_0	P_0	$\cdots$	P_1^+
P_2^+	P_1	P_1	$\cdots$	P_1

By adding multiples of the first $n-1$ rows we can again arrange that in the last row there will be a P_2^+ in the first position and P_0's in the rest. If we now continue to choose for instance the element in the bottom right-hand corner as the pivot, then we will obtain the kth tableau in the form *kth Tableau* (modified)

$$
\begin{array}{|cccc|}
\hline
\xi_1 & \xi_2 & \cdots & \xi_{n-k+1} \\
\hline
P_{k-1} & P_1^+ & \cdots & P_0 \\
P_{k-1} & P_0 & \cdots & P_0 \\
\cdots & \cdots & \cdots & \cdots \\
P_{k-1} & P_0 & \cdots & P_1^+ \\
P_k^+ & P_0 & \cdots & P_0^* \\
\hline
\end{array}
\tag{2}
$$

and hence

$(k+1)st$ *Tableau*

$$
\begin{array}{|cccc|}
\hline
\xi_1 & \xi_2 & \cdots & \xi_{n-k} \\
\hline
P_k & P_1^+ & \cdots & P_0 \\
P_k & P_0 & \cdots & P_0 \\
\cdots & \cdots & \cdots & \cdots \\
P_{k+1}^+ & P_1 & \cdots & P_1 \\
\hline
\end{array}
$$

This can again be modified so that the last row contains all P_0's (except in the first position) and so on.

Example 1. To find the eigenvalues and eigenvectors of the matrix

$$
A = \begin{pmatrix} -5 & -8 & 6 \\ 4 & 13 & -12 \\ 5 & 10 & -8 \end{pmatrix}
$$

The calculation runs as follows.

| | ξ_1 | | | | ξ_2 | | ξ_3 | |
	1	λ	λ^2	λ^3	1	λ	1	λ
[1]	-5	-1	0	0	-8	0	6*	0
	4	0	0	0	13	-1	-12	0
	5	0	0	0	10	0	-8	-1
ξ_3	$\frac{5}{6}$	$\frac{1}{6}$	0	0	$\frac{4}{3}$	0		
[2]	-6	-2	0	0	-3	-1		
	10	13	1	0	4	8		
[3]	-6	-2	0	0	-3	-1		
	-38	-3	1	0	-20*	0		
ξ_2	$-\frac{38}{20}$	$-\frac{3}{20}$	$\frac{1}{20}$	0				
[4]	$-\frac{3}{10}$	$\frac{7}{20}$	0	$-\frac{1}{20}$				

Explanations

[1] is the first tableau ending with the cellar row which represents ξ_3.

[2] is the second tableau. Nothing needs to be said about the first row. The second row is constructed as follows. We multiply the polynomials $\frac{5}{6}+\frac{1}{6}\lambda$ and $\frac{4}{3}$ represented in the cellar row by $-8-\lambda$. Schematically this simply means that we add -8 times the cellar row and -1 times the cellar row shifted one place to the right onto the last row of [1]. The row has also been multiplied by 6 in order to obtain integer values.

[3] is the modified second tableau, whose second row is constructed by adding 8 times the first row onto the second row of [2].

[4] is the third tableau, which is constructed in a manner similar to that of the second row of [2].

From this, we obtain the characteristic equation

$$\lambda^3-7\lambda+6 = 0$$

where we have multiplied through by -20.

The solutions of this are $\lambda_1 = 1$, $\lambda_2 = 2$, $\lambda_3 = -3$ and the corresponding eigenvectors are $x_1 = (3, -6, -5)$, $x_2 = (2, -4, -3)$ and $x_3 = (1, -1, -1)$ (and scalar multiples of these). We find these by substituting arbitrary first components into the two cellar rows.

Example 2. To find the eigenvalues and eigenvectors of the matrix

$$A = \begin{pmatrix} 3 & 2 & 3 & 6 \\ 1 & 3 & 4 & 5 \\ 1 & 1 & 2 & 3 \\ -1 & -1 & -2 & -3 \end{pmatrix}$$

The calculating scheme is shown opposite in which the pivots are no longer chosen always to be in the same place as they were in the general description of the method. (The meaning of the last two columns will be explained later in 14.2.2.)

Thus the characteristic equation (multiplying by 4) is

$$\lambda^4 - 6\lambda^3 + 13\lambda^2 - 12\lambda + 4 = 0.$$

This has two double solutions $\lambda_1 = 1$ and $\lambda_2 = 2$. Thus the matrix A has two distinct eigenvalues. The eigenvectors are found, by choosing ξ_4 ($\neq 0$) to be arbitrary and substituting it in the cellar rows.

They are,   for $\lambda_1 = 1$, $x_1 = (1, 2, 0, -1)$

and   for $\lambda_2 = 2$, $x_2 = (3, 3, 1, -2)$.

After these examples we will now return to the case when the method degenerates. Suppose for instance that all the P_0's in the last row of tableau (2) are equal to zero. If all the other P_0's in (2) are also equal to zero, then the characteristic polynomial (apart from a scalar factor) is the product of P_k^+ and all the P_1^+'s so that the eigenvalues and then the eigenvectors can easily be found. Otherwise we choose a $P_0 \neq 0$ as the pivot. Since the tableau which is obtained from (2) by leaving out the last row and first column has the same form as the initial tableau, we can choose the later pivots in the same way as before providing no further degeneration occurs. In this latter case we finally obtain a tableau of the form

$$\begin{array}{c|cc} & \xi_1 & \xi_2 \\ \hline & & \\ & P_{n-2} & P_{n-k}^+ \\ & P_k^+ & 0 \end{array} \tag{3}$$

	ξ_1		ξ_2		ξ_3		ξ_4					$\lambda=1$	$\lambda=2$
	1	λ	1	λ	1	λ	1	λ	λ^2	λ^3	λ^4		
	3	-1	2		3		6					-1	-3
	1		3	-1	4		5					-2	-3
	1		1		2	-1	3					0	-1
	-1*		-1		-2		-2	-1				1	2
ξ_1			-1		-2		-2	-1				1	2
			-1	1	-3	2	0	-1	1			1	-1
			2	-1	2		3	-1				-1	-1
			0		0	-1	1	-1				1	1
			1		-1*		5	-4	1			2	0
			2	-1	2		3	-1				-1	-1
			0		0	-1	1	-1				1	1
ξ_3			1				5	-4	1			2	0
			4	-1			13	-9	2			3	-1
			0	-1			1	-6	4	-1		-1	1
			4	-1			13	-9	2			3	-1
			-4*				-12	3	2	-1		-4	2
ξ_2							-3	$\frac{3}{4}$	$\frac{1}{2}$	$-\frac{1}{4}$		-1	$\frac{1}{2}$
							1	-3	$\frac{13}{4}$	$-\frac{3}{2}$	$\frac{1}{4}$	0	0

(possibly with different variables in place of ξ_1, ξ_2). This corresponds to the system of equations

$$P_{n-2}(\lambda)\,\xi_1 + P_{n-k}^+(\lambda)\,\xi_2 = 0$$
$$P_k^+(\lambda)\,\xi_1 \qquad\qquad = 0 \tag{4}$$

The characteristic equation is $P_k^+(\lambda)P_{n-k}^+(\lambda)=0$. If we substitute an eigenvalue for λ in (4), then this system of equations has a one- or two-dimensional solution space depending on its rank. Corresponding to each solution $(\xi_1,\xi_2)\neq 0$ we can find an eigenvector by using the previous cellar rows.

Finally, if many degenerations occur in the method, then we proceed in a similar way. Instead of a 2×2 tableau, we will then arrive at a tableau which is at least 3×3 so that the characteristic polynomial will appear as the product of at least three factors.

Example 3. To find the eigenvalues and eigenvectors of the matrix

$$\begin{pmatrix} 4 & 0 & 0 & 4 \\ 4 & 8 & 4 & 4 \\ 8 & 3 & 12 & 4 \\ 15 & 0 & 0 & 8 \end{pmatrix}$$

We obtain the following scheme of calculations—the last column will be explained in 14.2.2.

	ξ_1			ξ_2			ξ_3		ξ_4		$\lambda=14$
	1	λ	λ^2	1	λ	λ^2	1	λ	1	λ	
	4	-1		0			0		4*		0
	4			8	-1		4		4		-2
	8			3			12	-1	4		-3
	15			0			0		8	-1	0
ξ_4	-1	$\frac{1}{4}$		0			0				0
	0	1		8	-1		4*				-2
	4	1		3			12	-1			-3
	7	3	$-\frac{1}{4}$	0			0				0
ξ_3	0	$-\frac{1}{4}$		-2	$\frac{1}{4}$						$\frac{1}{2}$
	4	-2	$\frac{1}{4}$	-21	5	$-\frac{1}{4}$					-4
	7	3	$-\frac{1}{4}$	0							0

The characteristic equation is $(\lambda^2-12\lambda-28)(\lambda^2-20\lambda+84)=0$. The solutions of this are $\lambda_1=-2$, $\lambda_2=6$ and the double solution $\lambda_3=14$. The eigenvectors are found as follows.

1. $\lambda_1=-2$. The last two rows of the calculation correspond to the equations

$$9\xi_1-32\xi_2 = 0$$
$$0\xi_1 \qquad = 0$$

Hence ξ_1 is a free variable and $\xi_2=\frac{9}{32}\xi_1$. By using the cellar row, we find the eigenvector $x_1=(64,18,-13,-96)$.

2. $\lambda_2=6$. It follows that

$$\xi_1+0\xi_2 = 0$$
$$16\xi_1 \qquad = 0$$

Hence $\xi_1=0$ and ξ_2 is a free variable and we find the eigenvector $x_2=(0,2,-1,0)$.

3. $\lambda_3=14$. It follows that

$$25\xi_1+0\xi_2 = 0$$
$$0\xi_1 \qquad = 0$$

Hence $\xi_1=0$ and ξ_2 is a free variable. We find the eigenvector $x_3=(0,2,3,0)$. Thus the matrix A has three non-degenerate eigenvalues.

Example 4. To find the eigenvalues and eigenvectors of the matrix

$$A = \begin{pmatrix} 3 & 6 & 6 & -2 \\ -2 & -5 & -6 & 2 \\ 2 & 4 & 5 & -2 \\ 4 & 6 & 6 & -3 \end{pmatrix}$$

We obtain the following scheme of calculations.

	ξ_1 1	λ	λ^2	ξ_2 1	λ	ξ_3 1	λ	λ^2	ξ_4 1	λ
	3	-1		6		6			-2*	
	-2			-5	-1	-6			2	
	2			4		5	-1		-2	
	4			6		6			-3	-1
ξ_4	$\frac{3}{2}$	$-\frac{1}{2}$		3		3				
	1	-1		1	-1	0				
	-1	1		-2		-1	-1			
	$-\frac{1}{2}$	0	$\frac{1}{2}$	-3	-3	-3	-3			
	1	-1		1	-1	0				
	-1	1		-2*		-1	-1			
	$-\frac{1}{2}$	0	$\frac{1}{2}$	0		0				
ξ_2	$-\frac{1}{2}$	$\frac{1}{2}$				$-\frac{1}{2}$	$-\frac{1}{2}$			
	$\frac{1}{2}$	0	$-\frac{1}{2}$			$-\frac{1}{2}$	0	$\frac{1}{2}$		
	$-\frac{1}{2}$	0	$\frac{1}{2}$			0				

The characteristic equation is $(\lambda^2-1)^2=0$. This has two double solutions $\lambda_1=-1$ and $\lambda_2=+1$. For each of these two eigenvalues, it follows from the last part of the scheme that

$$0\xi_1+0\xi_3 = 0$$
$$0\xi_1 \qquad = 0$$

so that for both eigenvalues we find the eigenvectors by choosing ξ_1 and ξ_3 to be arbitrary and calculating ξ_2 and ξ_4 from the cellar rows. To each eigenvalue there are two linearly independent eigenvectors given by,

for $\lambda_1 = -1,$ $x_{11} = (1,-1,0,-1),$ $x_{12} = (0,0,1,3)$
for $\lambda_2 = +1,$ $x_{21} = (1,0,0,1),$ $x_{22} = (0,-1,1,0).$

Thus the matrix A has two simply degenerate eigenvalues.

13.6.2   An Iterative Method

We will restrict the discussion to the real case and assume to start with that E is Euclidean. Further we will also assume that the endomorphism f has a basis $\{e_1,\ldots,e_n\}$ consisting of eigenvectors, which will be the case in particular when f is symmetric. In addition we will further suppose that the eigenvalue λ_n which corresponds to e_n has a greater absolute value than all the others, i.e.,

$$|\lambda_n| > |\lambda_k| \qquad (k = 1,\ldots,n-1). \tag{5}$$

Starting with any vector $x_0 \in E$ whose e_n-component ξ_n is not zero, we construct the following sequence of vectors.

$$x_1 = f(x_0), \qquad x_2 = f(x_1),\ldots, \quad x_k = f(x_{k-1}),\ldots$$

If $x_0 = \xi_1 e_1 + \xi_2 e_2 + \ldots + \xi_n e_n,$ then $x_1 = \lambda_1 \xi_1 e_1 + \lambda_2 \xi_2 e_2 + \ldots + \lambda_n \xi_n e_n$ and generally

$$\begin{aligned} x_k &= \lambda_1^k \xi_1 e_1 + \lambda_2^k \xi_2 e_2 + \ldots + \lambda_n^k \xi_n e_n \\ &= \lambda_n^k (\xi_n e_n + y_k) \end{aligned} \tag{6}$$

where
$$y_k = \left(\frac{\lambda_1}{\lambda_n}\right)^k \xi_1 e_1 + \ldots + \left(\frac{\lambda_{n-1}}{\lambda_n}\right)^k \xi_{n-1} e_{n-1}.$$

By Theorem 12.1;2, it follows that

$$|\lambda_n|^k \left(|\xi_n|\,\|e_n\| - \|y_k\|\right) \leqslant \|x_k\| \leqslant |\lambda_n|^k \left(|\xi_n|\,\|e_n\| + \|y_k\|\right). \tag{7}$$

In view of (5), the quotients

$$\left(\frac{\lambda_1}{\lambda_n}\right)^k, \ldots, \left(\frac{\lambda_{n-1}}{\lambda_n}\right)^k$$

tend to zero as k tends to infinity. From this it is easy to see that $\|y_k\|$ also tends to zero. Hence by (7)

$$\lim_{k\to\infty} \frac{\|x_k\|}{|\lambda_n|^k} = |\xi_n|\,\|e_n\|$$

and therefore

$$\lim_{k\to\infty} \frac{\|x_{k-1}\|}{|\lambda_n|^{k-1}} = |\xi_n|\,\|e_n\|.$$

273

Hence by division

$$\lim_{k \to \infty} \frac{\|x_k\|}{\|x_{k-1}\|} = |\lambda_n|. \tag{8}$$

From (6) it further follows that

$$\frac{x_k}{\lambda_n^k} - \xi_n e_n = y_k$$

and, since $\lim_{k \to \infty} \|y_k\| = 0$,

$$\lim_{k \to \infty} \frac{x_k}{\lambda_n^k} = \xi_n e_n \tag{9}$$

where convergence is defined in E as in 12.3;(2).

Thus the sequence of vectors

$$\frac{x_1}{\lambda_n}, \qquad \frac{x_2}{\lambda_n^2}, \qquad \frac{x_3}{\lambda_n^3}, \dots$$

converges to an eigenvector corresponding to the eigenvalue λ_n with the greatest absolute value.

Example 5. Suppose that the endomorphism f is represented by the matrix

$$A = \begin{pmatrix} 5 & -2 & -4 \\ -2 & 2 & 2 \\ -4 & 2 & 5 \end{pmatrix}$$

with respect to some basis. We choose x_0 to be the vector which has the components 1, 0, 0 with respect to this basis. Then the components of the vectors x_k are

$$\begin{aligned}
x_1 &= (5 &, -2 &, -4 &) \\
x_2 &= (45 &, -22 &, -44 &) \\
x_3 &= (445 &, -222 &, -444 &) \\
x_4 &= (4445, &-2222, &-4444)
\end{aligned}$$

etc. We see immediately that $\dfrac{\|x_k\|}{\|x_{k-1}\|}$ tends to 10, so that the eigenvalue with the greatest absolute value is $\lambda_3 = \pm 10$.

From (9), we also see that the corresponding eigenvector has the components 2, -1, -2 (apart from a scalar factor). Finally it is not possible that $\lambda_3 = -10$, because, in this case, for sufficiently large k, each component of x_k would change sign at each iteration. Hence $\lambda_3 = +10$.

The evaluation of the norm $\|x_k\|$ at each step involves a certain amount of calculation which is reduced if the given norm is replaced by a new one equal to the absolute value of the greatest component of x. We will denote this by

$[x]$. It is easy to verify that $\|x_k\|$, $\|e_n\|$ and $\|y_k\|$ may be replaced by $[x_k]$, $[e_n]$ and $[y_k]$ in the arguments (in particular (7)) which led to the conclusion (8), and hence we have

$$\lim_{k\to\infty} \frac{[x_k]}{[x_{k-1}]} = |\lambda_n|. \tag{10}$$

By introducing the new norm $[x_k]$, we have also made it possible to drop the condition that E should be Euclidean. (We remark however that E can always be made Euclidean by putting $\|x\|^2 = \sum_k \xi_k^2$ where ξ_k are the components of x with respect to some basis and f is represented by some matrix with respect to this basis.)

From the representation of y_k in (6), we see that the greater the quotient $\left|\dfrac{\lambda_n}{\lambda_{n-1}}\right|$ is, i.e., the more λ_n exceeds the other eigenvalues, then the more rapid is the convergence in (8), (9) and (10). From (9), we see that the components of x_k become very large or very small as k increases if $|\lambda_n| \neq 1$. In order to avoid this, we can divide x_k by $\|x_k\|$ or $[x_k]$ after

$$\frac{\|x_k\|}{\|x_{k-1}\|} \text{ or } \frac{[x_k]}{[x_{k-1}]}$$

has been calculated. In this way, we ensure that $\|x_k\| = 1$ or that $[x_k] = 1$. This has the added advantage of making it easier to recognize the eigenvector as a limit.

If E is Euclidean and f is symmetric then in general it is possible to find the eigenvalue λ_n to within a given degree of accuracy with less effort than is involved in (8) or (10) by using the Rayleigh quotient (see 13.5;(6)). If e_n is again an eigenvector corresponding to λ_n then

$$\lambda_n = \rho(e_n) = \frac{(e_n, f(e_n))}{(e_n, e_n)}.$$

Hence, the Rayleigh quotients

$$\rho(x_k) = \frac{(x_k, f(x_k))}{(x_k, x_k)} \qquad (k = 1, 2, 3, \ldots)$$

are approximations to λ_n. If we assume that $\|x_k\| = \|e_n\| = 1$ and $x_k = e_n + \mathrm{d}x_k$, then, after a short calculation, we obtain

$$\rho(x_k) - \lambda_n = (f(\mathrm{d}x_k) - \lambda_n \, \mathrm{d}x_k, \mathrm{d}x_k). \tag{11}$$

Since this is a quadratic form in $\mathrm{d}x_k$, $\rho(x_k)$ will approach λ_n particularly quickly when f is symmetric.

We will check this by looking again at Example 5. The matrix A is symmetric and hence it represents a symmetric endomorphism of R_n (where $(x, y) = \sum_k \xi_k \eta_k$). The calculation gives for instance

$$\rho(x_2) = \frac{\xi_2' A \xi_2}{\xi_2' \xi_2} = \frac{44445}{445} = 9\cdot998875 = \frac{[x_5]}{[x_4]}.$$

Finally we remark that the Rayleigh quotient also converges to the eigenvalue in the case of an arbitrary matrix A. However the extra rapidity of the convergence is lost when A is not symmetric.

In practical applications the main interest is often not in the eigenvalue with the greatest absolute value but in the one with the least absolute value. For example this is the case in oscillation problems, where the eigenvalues represent the eigenfrequencies. The iteration method can still be applied, by using Theorem 13.1;7. In view of this theorem, the eigenvalue of A with the least absolute value is equal to the reciprocal of the eigenvalue of A^{-1} with the greatest absolute value. There are methods of calculation which are based on this idea, but do not in fact require the inverse of the matrix to be completely calculated. (cf. [8], p. 286. For other numerical techniques for the calculation of eigenvalues see [27] chap. 10.)

Finally we will indicate how, after finding λ_n and a corresponding eigenvector e_n, it is possible to find the eigenvalue λ_{n-1}, assuming that f and A are symmetric. In this case, every eigenvector corresponding to λ_{n-1} is orthogonal to e_n and the orthogonal complement of $L(e_n)$ is mapped into itself by f (Theorem 13.2;4). If we assume that $|\lambda_{n-1}| > |\lambda_k|$ for $k = 1, \ldots, n-2$, then we can start with any vector x_0 such that $(x_0, e_n) = 0$ and again apply the iterative method already described for λ_n. In this way we will find λ_{n-1} and an eigenvector corresponding to this eigenvalue. However in actual numerical calculation there is always a risk that, during the method, the orthogonality to e_n will be lost due to rounding errors. (See Example 7.) In view of this possibility, we must check the orthogonality at each step and if necessary we must restore it by replacing x_k with $x_k - (x_k, e_n)e_n$ (assuming e_n is normalized).

The other eigenvalues $\lambda_{n-2}, \lambda_{n-3}, \ldots$ may also be found by the same method.

Example 6. We choose the same 3×3 matrix as in Example 1 and start with the arbitrarily chosen vector $x_0 = (1, 7, 3)$. Then $x_1 = (-43, 59, 51)$. In the following table, we list the results for several values of k writing the vectors x_k in normalized form so that $\|x_k\| = 1$. We will give the components to five decimal places although more places are effectively calculated.

k	x_k			$[x_k]/[x_{k-1}]^*$	$\rho(x_{k-1})$
5	−0·54815,	0·60559,	0·57687	−4·06489	−4·72404
10	0·58189,	−0·57278,	−0·57734	−2·77900	−2·83200
15	−0·57677,	0·57793,	0·57735	−3·01527	−3·02293
25	−0·57734,	0·57736,	0·57735	−3·00026	−3·00040
35	−0·57735,	0·57735,	0·57735	−3·00001	−3·00001

For convenience the method of evaluating $[x_k]/[x_{k-1}]^*$ has been adapted so that it is always the component of x_k which has the same index as the component of x_{k-1} with the greatest absolute value which is divided by the latter. This makes no difference to the final result.

From this table, we can see that $\rho(x_k)$ converges to the eigenvalue $\lambda_3 = -3$ and that x_k converges to the eigenvector $(-1,1,1)$. In this example $\rho(x_k)$ does not converge any more rapidly than $[x_k]/[x_{k-1}]^*$. Note that A is not symmetric.

Example 7.

$$A = \begin{pmatrix} 0·22 & 0·02 & 0·12 & 0·14 \\ 0·02 & 0·14 & 0·04 & -0·06 \\ 0·12 & 0·04 & 0·28 & 0·08 \\ 0·14 & -0·06 & 0·08 & 0·26 \end{pmatrix}$$

Note that A is symmetric. We start with $x_0 = (1,7,3,9)$

k	x_k	$[x_k]/[x_{k-1}]^*$	$\rho(x_{k-1})$
2	0·58045, 0·03537, 0·57421, 0·57629	0·43357	0·46796
5	0·57755, 0·00163, 0·57852, 0·57598	0·47939	0·47999
10	0·57735, 0·00004, 0·57739, 0·57731	0·47996	0·48000
15	0·57735, 0·00000, 0·57735, 0·57735	0·48000	0·48000

We see that 0·48 is an eigenvalue with the eigenvector $(1,0,1,1)$.

We will now repeat the method starting with $x_0 = (0,1,0,0)$, i.e., with a vector which is orthogonal to the eigenvector just found. Providing the rounding errors do not interfere, we should find the eigenvalue with the second greatest absolute value and a corresponding eigenvector.

k	x_k	$[x_k]/[x_{k-1}]^*$	$\rho(x_{k-1})$
5	0·01747, 0·59567, 0·55903, −0·57650	0·23231	0·23953
10	0·00056, 0·57791, 0·57679, −0·57735	0·23977	0·24000
11	0·00028, 0·57763, 0·57707, −0·57735	0·23988	0·24000
12	0·00014, 0·57749, 0·57721, −0·57735	0·23994	0·24000
30	−0·23477, 0·52746, 0·29269, −0·76223	0·26928	0·25132
40	−0·57735, 0·00127, −0·57608, −0·57862	0·47833	0·48000

As far as $k = 12$, the solution approaches the eigenvalue 0·24 and the eigenvector $(0, 1, 1, -1)$. However, the rounding errors then start to interfere
(nine significant places were used in the calculation) and we are led once
again to the greatest eigenvalue. Finally we note that, in this example,
the Rayleigh quotient always converges more rapidly than $[x_k]/[x_{k-1}]$.

Exercises

1. Use the Gaussian Algorithm to find the eigenvalues and eigenvectors of
the matrices

$$(a) \quad A = \begin{pmatrix} -8 & -27 & 9 \\ 9 & 28 & -9 \\ 9 & 27 & -8 \end{pmatrix} \qquad (b) \quad B = \begin{pmatrix} 5 & -4 & -2 & -1 \\ 6 & -6 & -1 & -2 \\ 2 & 0 & -1 & 2 \\ -4 & 4 & 2 & 2 \end{pmatrix}$$

Solutions. (a) $\lambda_1 = 10$ $x_1 = (1, -1, -1)$

$\lambda_2 = \lambda_3 = 1$ $x_2 = (1, 0, 1)$

$x_3 = (0, 1, 3)$

(b) $\lambda_1 = -2$ $x_1 = (1, 2, 0, -1)$

$\lambda_2 = -1$ $x_2 = (2, 3, 1, -2)$

$\lambda_3 = 1$ $x_3 = (3, 3, 1, -2)$

$\lambda_4 = 2$ $x_4 = (1, 1, 0, -1)$.

2. Use the iterative method to find the greatest eigenvalue and the corresponding eigenvector of the matrix A in Exercise 1.

Solution. Starting with $x_0 = (1,0,0)$, the method produces in sequence the vectors

$$x_1 = (-\ \ 8,\ \ 9,\ \ 9)$$
$$x_2 = (-\ 98,\ 99,\ 99)$$
$$x_3 = (-998, 999, 999)$$

From this, it is easy to see the eigenvalue 10 and the corresponding eigenvector $(-1,1,1)$.

CHAPTER 14

INVARIANT SUBSPACES,
CANONICAL FORMS OF MATRICES

If the finite-dimensional vector space E has a basis consisting of eigenvectors of the endomorphism f of E, then it is particularly easy to see how this endomorphism operates on E. Each vector in the basis is simply multiplied by the corresponding eigenvalue and then it is straightforward to find the image of an arbitrary vector $x \in E$ (see 13.6;(6)). By Theorem 13.1;4, the matrix which represents f with respect to some basis is reducible to diagonal form and hence is similar to a diagonal matrix. Apart from the order of the diagonal elements, this diagonal matrix is uniquely determined by f.

We will now investigate the situation in the case of an arbitrary endomorphism to see if there are any similar properties and also to see if every square matrix is similar to some matrix which has a simple canonical form.

14.1 Invariant Subspaces

14.1.1 Vector Spaces over an Arbitrary Field

So far we have considered *real* and *complex* vector spaces. The most important property of the real and complex numbers which we have used is the possibility of adding and multiplying them together (i.e., the existence of two binary operations on the set S of scalars) subject to the following axioms.

1. $(\alpha + \beta) + \gamma = \alpha + (\beta + \gamma)$.   Addition is associative.
2. $\alpha + \beta = \beta + \alpha$.   Addition is commutative.
3. There exists an element $0 \in S$ such that

$$0 + \alpha = \alpha \text{ for all } \alpha \in S.$$

4. Corresponding to each element $\alpha \in S$, there is an element $-\alpha \in S$ such that $\alpha + (-\alpha) = 0$.
5. $(\alpha\beta)\gamma = \alpha(\beta\gamma)$.   Multiplication is associative.
6. $\alpha\beta = \beta\alpha$.   Multiplication is commutative.
7. There exists an element $1 \in S$ such that

$$1\alpha = \alpha \text{ for all } \alpha \in S.$$

8. Corresponding to each element $\alpha(\neq 0) \in S$, there is an element α^{-1} such that $\alpha\alpha^{-1} = 1$.

9. $\alpha(\beta+\gamma) = \alpha\beta + \alpha\gamma.$ Distributive.

10. $1 \neq 0$.

An algebraic structure which has two binary operations satisfying axioms 1, ..., 10 is known as a *field*. The sets of real and of complex numbers with their usual addition and multiplication are two examples of fields, but of course there are infinitely many other fields of which one well-known example is the field of rational numbers.

We can now redefine the concept of a vector space, replacing the fields of real or of complex numbers by an arbitrary field F. We will then speak of a vector space over the field F, and refer to F as the ground field of the vector space. The elements of the field F are the scalars for all vector spaces over F.

A great many of the previous results are also valid for vector spaces over an arbitrary field F. In this last chapter, we will now deal basically with vector spaces over arbitrary fields because it is only in this case that the theory attains its true value while the main results take on a very special form when applied to real or complex vector spaces.

14.1.2 Polynomials

A *polynomial over a field F* is an expression of the form

$$\mathbf{f} = \alpha_0 + \alpha_1\xi + \ldots + \alpha_n\xi^n = \sum_{k=0}^{n} \alpha_k\xi^k \tag{1}$$

where $\alpha_k \in F$ $(k=0,\ldots,n)$. $\alpha_1, \ldots, \alpha_n$ are referred to as the *coefficients* of the polynomial $\mathbf{f}$. If $\alpha_n \neq 0$, then n is referred to as the *degree* of $\mathbf{f}$. The polynomial $\mathbf{f}$ is said to be *monic* if $\alpha_n = 1$. The symbol ξ is only introduced for convenience in the definition of the multiplication of polynomials and its actual nature is not relevant to the polynomial $\mathbf{f}$.

If $\mathbf{g} = \sum_{l=0}^{m} \beta_l\xi^l$ is a second polynomial over F (and $n \geqslant m$ say), then the sum of $\mathbf{f}$ and $\mathbf{g}$ is defined by

$$\mathbf{f}+\mathbf{g} = \sum_{k=0}^{n} (\alpha_k+\beta_k)\xi^k \tag{2}$$

where $\beta_{m+1} = \ldots = \beta_n = 0$.

The product of $\mathbf{f}$ and $\mathbf{g}$ is defined by

$$\mathbf{f}\mathbf{g} = \sum_{k=0}^{n}\sum_{l=0}^{m} \alpha_k\beta_l\xi^{k+l}. \tag{3}$$

281

If **f** and **ġ** have degrees n and m then **fġ** has degree $(m+n)$.

It is easy to verify that these operations of addition and multiplication of the polynomials over F satisfy the axioms 1, ..., 6 and 9 in the definition of a field (cf. 14.1.1). Thus the polynomials over a field F form an algebraic structure which is known as a *commutative ring*. (Fields are a special type of commutative ring.) We refer to this ring as the ring of polynomials over the field F.

The elements $\alpha \in F$ can be considered as polynomials over F, in that they are the polynomials of degree 0 (except $\alpha = 0$, which has no degree). If we apply Definition (3) to this special case, then we obtain the following rule for the multiplication of the polynomial (1) by an element $\alpha \in F$.

$$\alpha \mathbf{f} = \sum_{k=0}^{n} (\alpha \alpha_k) \, \xi^k \tag{4}$$

Division in the ring of polynomials over the field F has many properties which are similar to the properties of division in the ring of integers. We will now set out the most important of these, but we will not include their proofs which may be found in [24] chapter 3.

The polynomial **f** is said to *divide* the polynomial **ġ**, in symbols **f|ġ**, if there is a polynomial **h** such that **ġ = fh**. If **f|ġ** and **f** is neither equal to α**ġ** ($\alpha \in F$) nor of degree 0, then **f** is said to be a proper divisor of **ġ**. Every polynomial of degree 0 (i.e., every $\alpha \in F$, $\alpha \neq 0$) is a divisor of every polynomial. If **f|ġ**, then **ġ** will also be referred to as a *multiple* of **f**.

A *greatest common divisor* (g.c.d.) of two polynomials **f** and **ġ** is any polynomial **h** which has the following two properties.

1. **h|f** and **h|ġ**.
2. Every common divisor of **f** and **ġ** is also a divisor of **h**.

Apart from a factor $\alpha \in F$, the g.c.d. **h** of **f** and **ġ** is uniquely determined by **f** and **ġ** and there exist polynomials **f₁** and **ġ₁**, such that

$$\mathbf{h} = \mathbf{f}_1 \mathbf{f} + \mathbf{g}_1 \mathbf{g}. \tag{5}$$

If 1 is a g.c.d. of **f** and **ġ**, then **f** and **ġ** are said to be *relatively prime*.

A *least common multiple* (l.c.m.) of two polynomials **f** and **ġ** is any polynomial **h** such that

1. **f|h** and **ġ|h**.
2. Every common multiple of **f** and **ġ** is also a multiple of **h**.

Apart from a factor $\alpha \in F$, the l.c.m. **h** of **f** and **ġ** is also uniquely determined by **f** and **ġ**.

The polynomials which correspond to the prime integers are the *irreducible* polynomials. A polynomial is said to be irreducible if it has no proper divisors and is not of degree 0. If **f** is irreducible, then so is α**f** for all $\alpha \in F$, $\alpha \neq 0$.

Every polynomial has a representation as a product of irreducible polynomials which is unique in the following sense. If

$$\mathbf{f} = \mathbf{p}_1 \mathbf{p}_2 \cdots \mathbf{p}_r = \mathbf{q}_1 \mathbf{q}_2 \cdots \mathbf{q}_s$$

are two such representations, then $r = s$ and, with a suitable renumbering, $\mathbf{p}_k$ is identical with $\mathbf{q}_k$ except for a factor $\alpha_k \in F$ $(k = 1, \ldots, r)$. We see that the polynomials of degree 0 play the same part in the ring of polynomials as the two numbers ± 1 in the ring of integers. They are known as the *units* of the ring. If $\mathbf{f}$ is monic, and the irreducible factors $\mathbf{p}_1, \ldots, \mathbf{p}_r$ are also monic, then these are uniquely determined by $\mathbf{f}$.

An *ideal* in the ring P of polynomials over the field F is a subset J of P with the following properties.

1. With respect to addition, J is a subgroup of P
(i.e., $J \neq \varnothing$ and, if $\mathbf{f}, \mathbf{g} \in J$, then $\mathbf{f} - \mathbf{g} \in J$ see [24] p. 372).

2. If $\mathbf{f} \in J$ and $\mathbf{g} \in P$, then $\mathbf{f}\mathbf{g} \in J$.

If an ideal J of P does not consist only of the zero polynomial (zero ideal), then it contains a polynomial $\mathbf{f}_0 \neq 0$ of minimum degree. Apart from a factor $\alpha \in F$, this is uniquely determined by J. Because, if $\mathbf{g}_0 \neq 0$ is a second polynomial of minimum degree in J, then, by (5), the g.c.d. $\mathbf{h}$ of $\mathbf{f}_0$ and $\mathbf{g}_0$ is also in J and therefore has the same degree as $\mathbf{f}_0$ and $\mathbf{g}_0$. It follows that $\mathbf{f}_0$ and $\mathbf{g}_0$ only differ by a factor $\alpha \in F$. Now, if $\mathbf{f} \in J$, then again by (5), the g.c.d. of $\mathbf{f}$ and $\mathbf{f}_0$ is also in J and hence $\mathbf{f}_0 | \mathbf{f}$. Thus every polynomial $\mathbf{f} \in J$ is a multiple of $\mathbf{f}_0$, i.e., $\mathbf{f}_0$ generates the ideal J. An ideal which is generated by a single element is generally referred to as a *principal ideal*. Thus every ideal in the ring of polynomials over a field F is principal.

14.1.3 Minimal Polynomials and Annihilators

Now let E be a vector space of finite dimension n over a field F, let f be an endomorphism of E and let $\mathbf{f}$ be the polynomial (1) over F. By 5.3.1

$$\mathbf{f}_f = \sum_{k=0}^{n} \alpha_k f^k \tag{6}$$

is again an endomorphism of E (f^k means the product of k factors each equal to f and $\alpha_0 f^0$ is the identity endomorphism multiplied by α_0). The endomorphism $\mathbf{f}_f$ is constructed by substituting f for ξ in the polynomial $\mathbf{f}$. If $\mathbf{g}$ is a second polynomial over F, then we can also form $\mathbf{g}_f$ and we have

$$(\mathbf{f} + \mathbf{g})_f = \mathbf{f}_f + \mathbf{g}_f \tag{7}$$

$$(\mathbf{f}\mathbf{g})_f = \mathbf{f}_f \mathbf{g}_f.$$

The polynomial $\mathbf{f}=1$ corresponds to the identity endomorphism and the polynomial $\mathbf{f}=0$ corresponds to the zero-endomorphism. Now (7) means that the mapping $\mathbf{f} \rightarrow \mathbf{f}_f$ is a homomorphic mapping of the ring of polynomials into the ring of endomorphisms of E (see 5.3.1).

Definition 1. A subspace H of E is said to be 'f-invariant' if $f(H)$ is contained in H, i.e., if H is mapped into itself by the endomorphism f.

If H is f-invariant, then it is easy to see that $\mathbf{f}_f(H) \subseteq H$ for all polynomials $\mathbf{f}$ over F.

Now, if H is an f-invariant subspace of E and L is an arbitrary subspace of E, then we consider the set $J \subseteq P$ of all polynomials $\mathbf{f}$ over F such that $\mathbf{f}_f(L) \subseteq H$. Then this set J is an ideal. Because, if $\mathbf{f} \in J$ and $\mathbf{g} \in J$, then $(\mathbf{f}-\mathbf{g})_f(L)=\mathbf{f}_f(L)-\mathbf{g}_f(L) \subseteq H$. Further, if $\mathbf{f} \in J$ and $\mathbf{g} \in P$, then $(\mathbf{gf})_f(L)=\mathbf{g}_f\,\mathbf{f}_f(L) \subseteq \mathbf{g}_f(H) \subseteq H$. Finally $0 \in J$ and therefore $J \neq \varnothing$.

This ideal J is not the zero-ideal. Because, by Theorem 5.2;11, the n^2+1 endomorphisms $f^0=1,f,f^2, \ldots ,f^{n^2}$ are linearly dependent which means that there is a polynomial $\mathbf{f} \neq 0$ such that $\mathbf{f}_f=0$. Hence $\mathbf{f}_f(L) \subseteq H$ and therefore $\mathbf{f} \in J$. The polynomial $\mathbf{f}_0 \neq 0$ which generates J is the one which is uniquely determined (apart from a factor $\alpha \in F$) as the polynomial with the least degree such that $\mathbf{f}_{0f}(L) \subseteq H$.

We now consider the following two special cases.

1. $L=E$, $H=\{0\}$. Then $\mathbf{f}_0$ is the unique (apart from a factor $\alpha \in F$) polynomial of the least degree such that $\mathbf{f}_{0f}=0$ (the zero-endomorphism).

2. Let $x \in E$. If we put $L=L(x)$ and $H=\{0\}$, then $\mathbf{f}_0$ is the unique (apart from a factor $\alpha \in F$) polynomial of the least degree such that $\mathbf{f}_{0f}(x)=0$.

Definition 2. The 'minimal polynomial' of the endomorphism f is the unique monic polynomial $\mathbf{m}$ of the least degree such that $\mathbf{m}_f=0$. The 'f-annihilator' of a vector $x \in E$ is the unique monic polynomial $\mathbf{h}$ of the least degree such that $\mathbf{h}_f(x)=0$.

The f-annihilator of $x=0$ is the polynomial 1.

Theorem 1. If s is the degree of the f-annihilator $\mathbf{h}$ of $x \in E$ and $t \geqslant s$, then the vectors $x, f(x), \ldots, f^{s-1}(x)$, are linearly independent, and the vectors $x, f(x), \ldots, f^{s-1}(x), f^t(x)$ are linearly dependent.

Proof. 1. If the first set of vectors was linearly dependent, then there would be a polynomial $\mathbf{k}$ of degree $<s$ such that $\mathbf{k}_f(x)=0$ and this contradicts the definition of $\mathbf{h}$.

2. If $t=s$, the second part of the theorem follows from the fact that $\mathbf{h}_f(x)$ is a linear combination of $x, \ldots, f^s(x)$. Now suppose that this part of the

theorem is also true for some $t \geqslant s$. Then $f^t(x) \in L(x, \ldots, f^{s-1}(x))$. Therefore $f^{t+1}(x) \in L(f(x), \ldots, f^s(x)) \subseteq L(x, \ldots, f^{s-1}(x))$ so that the assertion is also true for $t+1$ and hence by induction for all t.

14.1.4 Invariant Subspaces

*Theorem 2. If **f** is a polynomial over the field F and f is an endomorphism of E, then* $\ker \mathbf{f}_f$ *is an f-invariant subspace of E.*
(ker $\mathbf{f}_f$ denotes the kernel of the endomorphism $\mathbf{f}_f$ (see Definition 5.1;2).)

Proof. If $x \in \ker \mathbf{f}_f$, then $\mathbf{f}_f(x) = 0$ and hence

$$\mathbf{f}_f(f(x)) = f(\mathbf{f}_f(x)) = f(0) = 0, \text{ i.e., } f(x) \in \ker \mathbf{f}_f.$$

We note that in general not all of the f-invariant subspaces can be found in this way, for example, when $f = e$ (the identity endomorphism), $\ker \mathbf{f}_e$ is either $\{0\}$ or E depending on the polynomial **f**, whereas in fact every subspace of E is e-invariant.

Theorem 3. If $\mathbf{f} | \mathbf{g}$, *then* $\ker \mathbf{f}_f \subseteq \ker \mathbf{g}_f$.

Proof. If $x \in \ker \mathbf{f}_f$, then $\mathbf{f}_f(x) = 0$ and hence, if $\mathbf{g} = \mathbf{hf}$, $\mathbf{g}_f(x) = \mathbf{h}_f \mathbf{f}_f(x) = 0$, i.e., $x \in \ker \mathbf{g}_f$.

Theorem 4. If **f** *is a proper divisor of* **g** *and* **g** *is a divisor of the minimal polynomial* **m** *of f, then* $\ker \mathbf{f}_f$ *is a proper subspace of $\ker \mathbf{g}_f$.*

Proof. We put $\mathbf{m} = \mathbf{hg}$ and $\mathbf{f}_1 = \mathbf{hf}$. Then $\mathbf{f}_1$ is a proper divisor of **m** and therefore $\deg \mathbf{f}_1 < \deg \mathbf{m}$. Hence there is an $x_0 \in E$ such that $\mathbf{f}_{1f}(x_0) \neq 0$. If we put $y_0 = \mathbf{h}_f(x_0)$, then $\mathbf{f}_f(y_0) = (\mathbf{hf})_f(x_0) = \mathbf{f}_{1f}(x_0) \neq 0$ and hence $y_0 \notin \ker \mathbf{f}_f$. On the other hand $\mathbf{g}_f(y_0) = (\mathbf{hg})_f(x_0) = \mathbf{m}_f(x_0) = 0$ and hence $y_0 \in \ker \mathbf{g}_f$. The assertion follows from this and Theorem 3.

Theorem 5. If **h** *is a g.c.d. of* **f** *and* **g**, *then*

$$\ker \mathbf{h}_f = \ker \mathbf{f}_f \cap \ker \mathbf{g}_f.$$

Proof. 1. From Theorem 3, it follows that

$$\ker \mathbf{h}_g \subseteq \ker \mathbf{f}_f \cap \ker \mathbf{g}_f.$$

2. Suppose $x \in \ker \mathbf{f}_f \cap \ker \mathbf{g}_f$, i.e., $\mathbf{f}_f(x) = \mathbf{g}_f(x) = 0$. Using (5), it follows that

$$\mathbf{h}_f(x) = \mathbf{f}_{1f} \mathbf{f}_f(x) + \mathbf{g}_{1f} \mathbf{g}_f(x) = 0$$

and therefore that $x \in \ker \mathbf{h}_f$.

285

Theorem 6. If **h** *is an l.c.m. of* **f** *and* **g** *then*

$$\ker \mathbf{h}_f = \ker \mathbf{f}_f + \ker \mathbf{g}_f.$$

Proof. 1. From Theorem 3, it follows that $\ker \mathbf{f}_f \subseteq \ker \mathbf{h}_f$ and $\ker \mathbf{g}_f \subseteq \ker \mathbf{h}_f$ and therefore $\ker \mathbf{f}_f + \ker \mathbf{g}_f \subseteq \ker \mathbf{h}_f$.

2. Suppose $x \in \ker \mathbf{h}_f$. If we put $\mathbf{h} = \mathbf{f}_1 \mathbf{f} = \mathbf{g}_1 \mathbf{g}$ then 1 is a g.c.d. of $\mathbf{f}_1$ and $\mathbf{g}_1$. Therefore by (5) there exist $\mathbf{f}_2, \mathbf{g}_2$ such that $\mathbf{f}_2 \mathbf{f}_1 + \mathbf{g}_2 \mathbf{g}_1 = 1$. Putting $x_1 = \mathbf{f}_{2f} \mathbf{f}_{1f}(x)$, $\mathbf{f}_f(x_1) = (\mathbf{f}_2 \mathbf{h})_f (x) = 0$ and therefore $x_1 \in \ker \mathbf{f}_f$. Similarly $x_2 = \mathbf{g}_{2f} \mathbf{g}_{1f}(x) \in \ker \mathbf{g}_f$. Further $x_1 + x_2 = e(x) = x$. Hence $x \in \ker \mathbf{f}_f + \ker \mathbf{g}_f$.

Theorem 7. If **f** *and* **g** *are relatively prime, then*

$$\ker (\mathbf{f}\mathbf{g})_f = \ker \mathbf{f}_f \oplus \ker \mathbf{g}_f.$$

Proof. This follows from Theorems 5 and 6, remembering that 1 is a g.c.d. and **fg** is an l.c.m. of **f** and **g**.

Definition 3. A subspace L of E is said to be 'f-irreducible' if L is f-invariant and there is no decomposition $L = L_1 \oplus L_2$ *of L into a direct sum of non-zero f-invariant subspaces.*

Theorem 8. If f is an endomorphism of the vector space E of finite dimension n, then E is equal to the direct sum of f-irreducible subspaces.

Proof. 1. For $n = 1$, the theorem is obvious because E is itself f-irreducible.

2. We therefore use induction on the dimension n. Suppose that the theorem is true for all spaces of dimension less than n, and suppose that $\dim E = n$. There is nothing to prove if E is already f-irreducible. Suppose therefore that there is a decomposition $E = L_1 \oplus L_2$, in which each L_i is f-invariant, each $L_i \neq \{0\}$ and therefore $\dim L_i \leqslant n - 1$ for each $i = 1, 2$. From the induction hypothesis, the theorem is now true for L_1 and L_2 and hence this gives a decomposition of E.

Definition 4. A subspace L of E is said to be 'f-cyclic' if there is an element $a \in L$ *such that* $L = L(a, f(a), f^2(a), \ldots)$. *We refer to a as an 'f-generator' of L.*

Clearly an f-cyclic subspace is also f-invariant. Each vector $a \in E$ is an f-generator of an f-cyclic subspace, viz., $L = L(a, f(a), f^2(a), \ldots)$. If $a \in E$ is an f-generator of $L \subseteq E$, then, corresponding to each $x \in L$, there is a polynomial **k** such that $x = \mathbf{k}_f(a)$.

The next theorem follows immediately from Theorem 1.

Theorem 9. If a is an f-generator of the f-cyclic subspace L of E, then the dimension of L is equal to the degree of the f-annihilator of a.

Finally it is clear that the f-annihilator of a is the minimal polynomial of the restriction of the endomorphism f to the subspace L $(a, f(a), ...)$.

Theorem 10. Suppose that the endomorphism f of the vector space E of dimension n has the minimal polynomial $\mathbf{m} = \mathbf{p}^s$ where $\mathbf{p}$ is an irreducible polynomial. Then E is the direct sum of f-cyclic subspaces.

Proof. 1. If $\dim E = 1$, then E itself is f-cyclic and the theorem is true. We therefore use induction on the dimension n. Thus we assume that the theorem is true for all vector spaces E with $\dim E \leqslant n - 1$.

2. The f-annihilator $\mathbf{h}$ of a vector $a \in E$ is a divisor of the minimal polynomial $\mathbf{m}$ of f. Because, from $\mathbf{m}_f(a) = 0$, it follows that $\mathbf{m}$ is in the principal ideal generated by $\mathbf{h}$. Hence $\mathbf{h}$ has the form $\mathbf{h} = \mathbf{p}^k$ where $k \leqslant s$. We now choose $a_0 \in E$, so that the corresponding exponent k_0 is as large as possible. Then $0 < k_0 = s$.

Let $L_0 \subseteq E$ be the f-cyclic subspace generated by a_0. By Theorem 9, $\dim L_0 = k_0 \deg \mathbf{p} \geqslant 1$. Thus, if $\bar{E}$ is the quotient space E/L_0, we have $\dim \bar{E} < \dim E = n$.

3. An endomorphism $\bar{f}$ of $\bar{E}$ is defined by the mapping $\bar{x} \rightarrow \bar{f}(\bar{x}) = \overline{f(x)}$. ($\bar{x} \in \bar{E}$ is the canonical image of $x \in E$. Cf. Example 5.1;1.) Because, if $\bar{x} = \bar{y}$, then $x - y \in L_0$ and $f(x) - f(y) \in L_0$ which means that $\overline{f(x)} = \overline{f(y)}$.

Now, for any polynomial $\mathbf{g}$, $\mathbf{g}_f(\bar{x}) = \overline{\mathbf{g}_f(x)}$ and therefore in particular $\mathbf{m}_f(\bar{x}) = \overline{\mathbf{m}_f(x)} = 0$ for all $\bar{x} \in \bar{E}$. Hence the minimal polynomial $\overline{\mathbf{m}}$ of $\bar{f}$ is a divisor of $\mathbf{m}$ and therefore $\overline{\mathbf{m}} = \mathbf{p}^{\bar{s}}$ for some $\bar{s} \leqslant s$.

4. By 2 and 3 and the induction hypothesis, Theorem 10 is true for $\bar{E}$ with the endomorphism $\bar{f}$. Thus there is a decomposition

$$\bar{E} = \bar{L}_1 \oplus \dots \oplus \bar{L}_r \tag{8}$$

of $\bar{E}$ into a direct sum of $\bar{f}$-cyclic subspaces $\bar{L}_1, ..., \bar{L}_r$.

Let $\bar{a}_1$ be an $\bar{f}$-generator of $\bar{L}_1$ and let $a_1' \in \bar{a}_1 \subseteq E$. By 2, the f-annihilator of a_1' has the form $\mathbf{p}^{s_1'}$ where $s_1' \leqslant k_0 = s$. Since it follows from $\mathbf{p}_f^{s_1'}(a_1') = 0$ that $\mathbf{p}_{\bar{f}}^{s_1'}(\bar{a}_1) = \overline{\mathbf{p}_f^{s_1'}(a_1')} = 0$, the $\bar{f}$-annihilator of $\bar{a}_1$ is a divisor of $\mathbf{p}^{s_1'}$ and therefore has the form $\mathbf{p}^{\bar{s}_1}$ where $\bar{s}_1 \leqslant s_1' \leqslant k_0 = s$.

5. From $\mathbf{p}_f^{s_1'}(a_1') = 0$, it follows that

$$\mathbf{p}_f^{k_0}(a_1') = \mathbf{p}_f^{k_0 - \bar{s}_1} \mathbf{p}_f^{\bar{s}_1}(a_1') = 0. \tag{9}$$

287

Since $\mathbf{p}_f^{\delta_1}(\bar{a}_1) = 0$, $\mathbf{p}_f^{\delta_1}(a_1') \in L_0$ and therefore there exists a polynomial $\mathbf{k}$ such that $\mathbf{p}_f^{\delta_1}(a_1') = \mathbf{k}_f(a_0)$. Therefore $\mathbf{p}^{k_0} | \mathbf{p}^{k_0 - \delta_1} \mathbf{k}$, because $\mathbf{p}^{k_0}$ is the f-annihilator of a_0. Hence there exists a polynomial $\mathbf{k}^*$ such that $\mathbf{k} = \mathbf{p}^{\delta_1} \mathbf{k}^*$ and it follows that

$$\mathbf{p}_f^{\delta_1}(a_1') = \mathbf{p}_f^{\delta_1} \mathbf{k}_f^*(a_0),$$

i.e., there exists an $a_0' = \mathbf{k}_f^*(a_0) \in L_0$ such that

$$\mathbf{p}_f^{\delta_1}(a_1') = \mathbf{p}_f^{\delta_1}(a_0'). \tag{10}$$

6. We now put $a_1 = a_1' - a_0' \in \bar{a}_1 \subseteq E$ and obtain $\mathbf{p}_f^{\delta_1}(a_1) = 0$ so that the f-annihilator of a_1 is a divisor of $\mathbf{p}^{\delta_1}$. By 4 however (putting a_1 in the place of a_1') it is also a multiple of $\mathbf{p}^{\delta_1}$ and therefore it is equal to $\mathbf{p}^{\delta_1}$.

7. Now suppose that $L_1 \subseteq E$ is the f-cyclic subspace generated by a_1. If we denote the degree of $\mathbf{p}^{\delta_1}$ by t_1, then, by Theorem 1, $\{a_1, f(a_1), \ldots, f^{t_1-1}(a_1)\}$ is a basis of L_1 and $\{\bar{a}_1, \bar{f}(\bar{a}_1), \ldots, \bar{f}^{t_1-1}(\bar{a}_1)\}$ is a basis of $\bar{L}_1$. The canonical mapping $x \in L_1 \to \bar{x} \in \bar{L}_1$ is therefore an isomorphism of L_1 onto $\bar{L}_1$.

8. If we carry out the same arguments for $\bar{L}_2, \ldots, \bar{L}_r$ as for $\bar{L}_1$, then we will obtain, in addition to L_1, the f-cyclic subspaces $L_2, \ldots, L_r \subseteq E$ and we assert that

$$E = L_0 \oplus L_1 \oplus \ldots \oplus L_r. \tag{11}$$

This will complete the proof of Theorem 10.

8.1. Suppose $x \in E$. Then there is a decomposition of $\bar{x} \in \bar{E}$ into the form $\bar{x} = \bar{x}_1 + \ldots + \bar{x}_r$ where $\bar{x}_k \in \bar{L}_k$ for $k = 1, \ldots, r$. If $x_k \in L_k$ is the inverse image of $\bar{x}_k$ (which according to 7 is uniquely determined by $\bar{x}_k$) for $k = 1, \ldots, r$, then there is an $x_0 \in L_0$ such that $x = x_0 + x_1 + \ldots + x_r$. Therefore $E = L_0 + L_1 + \ldots + L_r$.

8.2. Finally to prove that this sum is direct, we suppose that there is a relation

$$x_0 + x_1 + \ldots + x_r = 0, \qquad \text{where } x_k \in L_k \quad (k = 0, \ldots, r)$$

(see Problem 2.4;7). It follows that $\bar{x}_1 + \ldots + \bar{x}_r = 0$ where $\bar{x}_k \in L_k (k = 1, \ldots, r)$. In view of (8), it follows that $\bar{x}_1 = \ldots = \bar{x}_r = 0$ and therefore by 7 that $x_1 = \ldots = x_r = 0$ and finally that $x_0 = 0$.

Theorem 11. Let E be a finite-dimensional vector space and let f be an endomorphism of E. Then E is the direct sum of non-zero f-cyclic subspaces which are f-irreducible, i.e.,

$$E = L_1 \oplus L_2 \oplus \ldots \oplus L_r, \text{ where } L_1, \ldots, L_r \neq \{0\}. \tag{12}$$

If $\mathbf{m}$ is the minimal polynomial of f, then the minimal polynomial of the restriction of f to L_k is of the form $\mathbf{m}_k = \mathbf{p}_k^{\sigma_k}$ where $\mathbf{p}_k$ is irreducible, $\sigma_k \geqslant 1$ and $\mathbf{p}_k^{\sigma_k} | \mathbf{m}$ $(k = 1, \ldots, r)$.

Proof. 1. By Theorem 8, there is a decomposition of the form (12) in which $L_1, \ldots, L_r$ are f-irreducible. The minimal polynomial $\mathbf{m}_k$ of the restriction of f to L_k is obviously a divisor of $\mathbf{m}$. If $\mathbf{m}_k$ was not of the form $\mathbf{p}_k^{q_k}$, then $\mathbf{m}_k$ would be equal to $\mathbf{g}\mathbf{f}$ for some relatively prime factors $\mathbf{g}$ and $\mathbf{f}$. By Theorem 7, applied to L_k (instead of E), then $L_k = \ker \mathbf{g}_f + \ker \mathbf{f}_f$ and, by Theorem 4, L_k would be f-reducible. Finally $\mathbf{m}_k$ cannot be equal to 1, because $L_k \neq \{0\}$.

2. By Theorem 10, L_k is a direct sum of f-cyclic subspaces. But because L_k itself is f-irreducible, this means that L_k must be f-cyclic $(k = 1, \ldots, r)$.

Definition 5. *Two f-invariant subspaces L_1, $L_2 \subseteq E$ are said to be 'f-equivalent', if there is an isomorphism h of L_1 onto L_2 which commutes with f, i.e., for which $hf(x) = fh(x)$ for all $x \in L_1$.*

Since $h^{-1}f(y) = h^{-1}fhh^{-1}(y) = fh^{-1}(y)$ $(y \in L_2)$, f-equivalence is a symmetric relationship.

Theorem 12. *Two f-cyclic subspaces L_1 and L_2 of E are f-equivalent if and only if the restrictions of f to L_1 and L_2 have the same minimal polynomial.*

Proof. 1. Suppose that L_1 and L_2 are f-equivalent and that h is an isomorphism of L_1 onto L_2 which commutes with f. Then, for every polynomial $\mathbf{g}$, $\mathbf{g}_f$ also commutes with h. Let $\mathbf{m}_1$ and $\mathbf{m}_2$ be the minimal polynomials of the restrictions of f to L_1 and L_2. Then $h\mathbf{m}_{1f}(x) = \mathbf{m}_{1f}h(x) = 0$ for all $x \in L_1$, i.e., $\mathbf{m}_{1f}(y) = 0$ for all $y \in L_2$. Therefore $\mathbf{m}_2|\mathbf{m}_1$. Similarly $\mathbf{m}_1|\mathbf{m}_2$ and hence $\mathbf{m}_1 = \mathbf{m}_2$.

2. Suppose that the restrictions of f to L_1 and L_2 have the same minimal polynomial $\mathbf{m}_0$ and let a_1, a_2 be f-generators of L_1, L_2. Every vector $x \in L_1$ can then be written in the form $x = \mathbf{f}_f(a_1)$ for some suitable polynomial $\mathbf{f}$ and $\mathbf{f}_f(a_1) = \mathbf{g}_f(a_1)$ if and only if $\mathbf{m}_0|(\mathbf{f} - \mathbf{g})$. The same is true for L_2 and a_2. Hence an isomorphism h of L_1 onto L_2 is defined by $h\mathbf{f}_f(a_1) = \mathbf{f}_f(a_2) = \mathbf{f}_f h(a_1)$ (for all polynomials $\mathbf{f}$) and this has the property that $hf(x) = fh(x)$ for all $x \in L_1$, i.e., h commutes with f.

The decomposition (12) of the vector space E into the f-irreducible subspaces $L_1, \ldots, L_r$ is not unique in general. For example, in the case of the identity endomorphism of E, every 1-dimensional subspace is invariant. However, the decomposition is unique in a weaker sense which will be explained in the following theorem.

Theorem 13. *With the same conditions as in Theorem 11, suppose that, in addition to (12),*

$$E = L_1^* \oplus \ldots \oplus L_s^*$$

289

is also a decomposition of the same kind. Then $r = s$ and with a suitable re-numbering L_k and L_k^* are f-equivalent $(k = 1, \ldots, r)$.

Proof. 1. Let $\mathbf{m} = \mathbf{p}_1^{s_1} \mathbf{p}_2^{s_2} \ldots \mathbf{p}_l^{s_l}$ be the minimal polynomial of f, where $\mathbf{p}_1, \mathbf{p}_2, \ldots, \mathbf{p}_l$ are distinct monic irreducible polynomials. By Theorem 11, the minimal polynomial $\mathbf{m}_k$ of the restriction of f to the subspace L_k in (12) has the form $\mathbf{m}_k = \mathbf{p}_\lambda^\sigma$ where $1 \leqslant \lambda \leqslant l$ and $1 \leqslant \sigma \leqslant s_\lambda$. By Theorem 12, Theorem 13 will be proved if we can show that, for each pair (λ, σ), the number $n(\lambda, \sigma)$ of L_k with $\mathbf{m}_k = \mathbf{p}_\lambda^\sigma$ is uniquely determined by f.

2. We will now calculate the rank of the endomorphism $\mathbf{p}_{\lambda f}^\sigma$ for $\lambda = 1, \ldots, l$ and arbitrary $\sigma \geqslant 1$. Since $L_1, \ldots, L_r$ are f-invariant, we have

$$\operatorname{rank} \mathbf{p}_{\lambda f}^\sigma = \dim \mathbf{p}_{\lambda f}^\sigma(E) = \sum_{k=1}^r \dim \mathbf{p}_{\lambda f}^\sigma(L_k). \tag{13}$$

In order to evaluate the individual terms in (13), we will distinguish two cases.

1st Case. Suppose $\mathbf{m}_k = \mathbf{p}_{\lambda_k}^{\sigma_k}$ where $\lambda_k \neq \lambda$. Then the g.c.d. of $\mathbf{m}_k$ and $\mathbf{p}_\lambda^\sigma$ is 1 and hence, by Theorem 5, $\ker \mathbf{m}_{kf} \cap \ker \mathbf{p}_{\lambda f}^\sigma = \ker 1_f = \{0\}$. ($1_f$ is the identity endomorphism.) Since $L_k \subseteq \ker \mathbf{m}_{kf}$, it follows in particular that $L_k \cap \ker \mathbf{p}_{\lambda f}^\sigma = \{0\}$, i.e., L_k is mapped onto itself by $\mathbf{p}_{\lambda f}^\sigma$ (Theorem 5.1;8). Hence $\dim \mathbf{p}_{\lambda f}^\sigma(L_k) = \dim L_k$.

2nd Case. Suppose $\mathbf{m}_k = \mathbf{p}_{\lambda_k}^{\sigma_k}$ and $\lambda_k = \lambda$.

2.1. If $\sigma > \sigma_k$, then $\mathbf{p}_{\lambda f}^\sigma(L_k) = \mathbf{p}_{\lambda f}^{\sigma - \sigma_k} \mathbf{p}_{\lambda f}^{\sigma_k}(L_k) = \{0\}$ and therefore $\dim \mathbf{p}_{\lambda f}^\sigma(L_k) = 0$.

2.2. If $\sigma \leqslant \sigma_k$, then $\mathbf{p}_{\lambda f}^\sigma(L_k) = \mathbf{p}_{\lambda f}^{\sigma_k - \sigma} \mathbf{p}_{\lambda f}^\sigma(L_k) = \{0\}$. Therefore the minimal polynomial of the restriction of f to $\mathbf{p}_{\lambda f}^\sigma(L_k)$ is a divisor of $\mathbf{p}_\lambda^{\sigma_k - \sigma}$. However it cannot be a proper divisor, because then $\mathbf{p}_{\lambda f}^\mu(L_k) = \{0\}$ for some $\mu < \sigma_k$. Therefore

$$\dim \mathbf{p}_{\lambda f}^\sigma(L_k) = \deg \mathbf{p}_\lambda^{\sigma_k - \sigma} = (\sigma_k - \sigma) \deg \mathbf{p}_\lambda = \dim L_k - \sigma \deg \mathbf{p}_\lambda.$$

(Cf. Theorem 9.)

Collecting all this together, we have, for each of the terms in (13),

$$\dim \mathbf{p}_{\lambda f}^\sigma(L_k) = \dim L_k - \delta_{\lambda, \lambda_k} \min(\sigma, \sigma_k) \deg \mathbf{p}_\lambda$$

and by (13)

$$\operatorname{rank} \mathbf{p}_{\lambda f}^\sigma = \dim E - \deg \mathbf{p}_\lambda \sum_{k=1}^r \delta_{\lambda, \lambda_k} \min(\sigma, \sigma_k), \tag{14}$$

($\delta_{\lambda, \lambda_k}$ is the Kronecker delta).

Similarly

$$\operatorname{rank} \mathbf{p}_{\lambda f}^{\sigma-1} = \dim E - \deg \mathbf{p}_\lambda \sum_{k=1}^{r} \delta_{\lambda, \lambda_k} \min(\sigma-1, \sigma_k).$$

Subtracting, we have

$$\operatorname{rank} \mathbf{p}_{\lambda f}^{\sigma-1} - \operatorname{rank} \mathbf{p}_{\lambda f}^{\sigma} = \deg \mathbf{p}_\lambda \sum_{k=1}^{r} \delta_{\lambda, \lambda_k} \epsilon(\sigma \leqslant \sigma_k)$$

$$= \deg \mathbf{p}_\lambda[n(\lambda, \sigma) + n(\lambda, \sigma+1) + \ldots + n(\lambda, s_\lambda)]$$

$$(1 \leqslant \lambda \leqslant l, 1 \leqslant \sigma \leqslant s_\lambda), \tag{15}$$

where $\epsilon(\sigma \leqslant \sigma_k) = 1$ if $\sigma \leqslant \sigma_k$ and $=0$ if $\sigma > \sigma_k$.

From (15) it further follows that

$$n(\lambda, \sigma) = \frac{1}{\deg \mathbf{p}_\lambda}[\operatorname{rank} \mathbf{p}_{\lambda f}^{\sigma-1} - 2\operatorname{rank} \mathbf{p}_{\lambda f}^{\sigma} + \operatorname{rank} \mathbf{p}_{\lambda f}^{\sigma+1}]. \tag{16}$$

Hence the numbers $n(\lambda, \sigma)$ $(1 \leqslant \lambda \leqslant l, 1 \leqslant \sigma \leqslant s_\lambda)$ are uniquely determined by the irreducible factors $\mathbf{p}_\lambda$ of the minimal polynomial $\mathbf{m}$ of f and therefore by f itself. This completes the proof of the theorem.

From this proof, it is also easy to see that, corresponding to each λ $(1 \leqslant \lambda \leqslant l)$ there is an L_k such that $\mathbf{m}_k = \mathbf{p}_\lambda^{s_\lambda}$. Otherwise the minimal polynomial $\mathbf{m}$ of f would only have a lower power of $\mathbf{p}_\lambda$ as a factor. Using (14), we now have the following characterization of the exponents s_λ.

Theorem 14. The exponent s_λ $(1 \leqslant \lambda \leqslant l)$ is the least natural number μ such that $\operatorname{rank} \mathbf{p}_{\lambda f}^{\mu} = \operatorname{rank} \mathbf{p}_{\lambda f}^{\mu+1}$.

Problems

1. Let E be a vector space and let x_0 be an eigenvector of the endomorphism f corresponding to the eigenvalue λ. Further let $\mathbf{f}$ be a polynomial. Show that $\mathbf{f}(\lambda)$ is an eigenvalue of $\mathbf{f}_f$.

2. Let f be an endomorphism of an n-dimensional vector space E. Show that, if there is an integer k such that $f^k = 0$, then $f^n = 0$.

3. Show that, if A is a complex $n \times n$ matrix and $A^k = I$ for some integer k, then A is reducible to diagonal form.

4. Let f be a proper rotation of a Euclidean plane through an angle ω. Show that the minimal polynomial of f is $\xi^2 - 2\xi \cos \omega + 1$ providing $\omega \neq 0, \pi$. What is the minimal polynomial if $\omega = 0$ or $\omega = \pi$?

5. Prove that an endomorphism of a finite-dimensional vector space is non-singular if and only if the constant term in its minimal polynomial is not zero.

291

6. Let f be an endomorphism of a finite-dimensional vector space E and let $\mathbf{m}$ be the minimal polynomial of f. Prove that E is f-cyclic if and only if the degree of $\mathbf{m}$ is equal to $\dim E$. (Hint. In the decomposition (12) of E, let x_k be an f-generator of L_k $(k=1,\ldots,r)$. Then $z=x_1+\ldots+x_r$ is an f-generator of E if $\deg \mathbf{m} = \dim E$.)

14.2 Canonical Forms of Matrices

14.2.1 The General Canonical Form

Let $A = (\alpha_{ik})$ be an $n \times n$ matrix whose elements are from a field F. Let E be an n-dimensional vector space over the same field F (for example E could be the obvious generalization of the space of n-tuples R_n). If $\{e_1,\ldots,e_n\}$ is a basis of E, then

$$f(e_i) = \sum_{k=1}^{n} \alpha_{ik} e_k \tag{1}$$

defines an endomorphism f of E (Theorem 5.1;3). In the decomposition 14.1;(12) of E, let a_k be an f-generator for L_k $(k=1,\ldots,r)$. If m_k is the dimension of L_k (i.e., the degree of the minimal polynomial $\mathbf{m}_k$), then by Theorem 14.1;1 the vectors

$$a_k, f(a_k), \ldots, f^{m_k-1}(a_k) \tag{2}$$

are a basis of L_k, because $\mathbf{m}_k$ is also the f-annihilator of a_k. By combining these bases of $L_1, \ldots, L_r$, we obtain a basis of E. We now look for the images of these basis vectors under the endomorphism f. Since $f(f^i(a_k)) = f^{i+1}(a_k)$, each basis vector goes into the next for $i = 0, \ldots, m_k - 2$.
Further $f(f^{m_k-1}(a_k)) = f^{m_k}(a_k)$ and this is a linear combination of the vectors in (2). Suppose

$$f^{m_k}(a_k) = -\gamma_{k,\,m_k-1} f^{m_k-1}(a_k) - \ldots - \gamma_{k,\,1} f(a_k) - \gamma_{k,\,0} a_k.$$

It follows that

$$\sum_{i=0}^{m_k} \gamma_{ki} f^i(a_k) = 0,$$

where $\gamma_{k,\,m_k} = 1$ and hence the polynomial

$$\sum_{i=0}^{m_k} \gamma_{ki} \xi^i \tag{3}$$

is a multiple of the minimal polynomial $\mathbf{m}_k$. However, since it has the same degree as $\mathbf{m}_k$, (3) must actually be the minimal polynomial and in particular a power of an irreducible polynomial.

Thus the part of the matrix of the endomorphism f which corresponds to the basis vectors (2) of L_k has the form

$$C_k = \begin{pmatrix} 0 & 1 & 0 & \dots & 0 \\ 0 & 0 & 1 & \dots & 0 \\ \dots & \dots & \dots & \dots & \dots \\ 0 & 0 & 0 & \dots & 1 \\ -\gamma_{k0} & -\gamma_{k1} & -\gamma_{k2} & \dots & -\gamma_{k, m_k-1} \end{pmatrix} \qquad (4)$$

This contains the coefficients which express the images of the basis vectors (2) as linear combinations of themselves, where the elements $-\gamma_{ki}$ in the last row are such that the polynomial (3) formed from them (with $\gamma_{k, m_k} = 1$) is a power of an irreducible polynomial.

It follows that, for the full vector space E, the matrix of f has the form

$$C = \begin{bmatrix} C_1 & 0 & \dots & 0 \\ 0 & C_2 & \dots & 0 \\ \dots & \dots & \dots & \dots \\ 0 & 0 & \dots & C_r \end{bmatrix} \qquad (5)$$

where each of the blocks C_k has the form (4). By Theorem 14.1;13, the blocks $C_1, \dots, C_r$ are uniquely determined by A, apart from their order. Hence we have the following result.

Theorem 1. Given a square matrix A with elements from a field F, then there is a similar matrix $C = SAS^{-1}$ which has the normal form (5). Except for their order, the blocks $C_1, \dots, C_r$ are uniquely determined by A.

Theorem 2. The minimal polynomial of an endomorphism divides the characteristic polynomial.

Proof. 1. We first calculate the characteristic polynomial of a block C_k. This is

$$D_m(\gamma_0, \dots, \gamma_{m-1}; \mu) = \begin{vmatrix} -\mu & 1 & 0 & \dots & 0 \\ 0 & -\mu & 1 & \dots & 0 \\ \dots & \dots & \dots & \dots & \dots \\ 0 & 0 & 0 & \dots & 1 \\ -\gamma_0 & -\gamma_1 & -\gamma_2 & \dots & -\gamma_{m-1}-\mu \end{vmatrix}$$

Expanding about the first column (Theorem 4.3;1) we have

$$D_m(\gamma_0, \ldots, \gamma_{m-1}; \mu) = -\mu D_{m-1}(\gamma_1, \ldots, \gamma_{m-1}; \mu) - (-1)^{m-1} \gamma_0.$$

From this recurrence relation, it follows that

$$D_m(\gamma_0, \ldots, \gamma_{m-1}; \mu) = (-1)^m (\mu^m + \gamma_{m-1} \mu^{m-1} + \ldots + \gamma_1 \mu + \gamma_0),$$

because the polynomial on the right-hand side satisfies the same recurrence relation and, for $m=1$, the characteristic polynomial is $-(\mu + \gamma_0)$.

Therefore by (3) the characteristic polynomial of C_k is equal to the minimal polynomial $\mathbf{m}_k = \mathbf{p}_{\lambda_k}^{q_k}$ of the restriction of f to L_k.

2. The characteristic polynomial of C is now equal to the product of those for each of the individual blocks C_k. Since, corresponding to each λ $(0 \leqslant \lambda \leqslant l)$, there is a subspace L_k such that $\mathbf{m}_k = \mathbf{p}_\lambda^{s_\lambda}$, the theorem follows immediately from $\mathbf{m} = \mathbf{p}_1^{s_1} \ldots \mathbf{p}_r^{s_r}$. (cf. the remark before Theorem 14.1;14.)

Theorem 3. (Cayley–Hamilton). If $\mathbf{h}$ is the characteristic polynomial of the endomorphism f of a finite-dimensional vector space, then $\mathbf{h}_f = 0$, i.e., every endomorphism satisfies its own characteristic equation. The same is also true for every square matrix.

Proof. The fact that $\mathbf{h}_f = 0$ is a direct consequence of Theorem 2 and $\mathbf{m}_f = 0$. The same assertion for square matrices follows by Theorem 5.2;11, and the fact that, if A is the matrix of f, then A^k is the matrix of f^k.

14.2.2   The Jordan Canonical Form

Now suppose that the irreducible factors $\mathbf{p}_\lambda$ $(0 \leqslant \lambda \leqslant l)$ of the minimal polynomial of f are linear, i.e., they are of degree 1.

$$\mathbf{p}_{\lambda f} = f - \mu_\lambda e \ (0 \leqslant \lambda \leqslant l); \quad \mathbf{m}_f = (f - \mu_1 e)^{s_1} \ldots (f - \mu_l e)^{s_l}. \qquad (6)$$

This will always be the case for any polynomial $\mathbf{m}$, if the field F is algebraically closed ([19], p. 200) and in particular therefore when E is a complex vector space.

The minimal polynomial $\mathbf{m}_k$ of the restriction of f to the subspace L_k then has the form $\mathbf{m}_k = (f - \mu_k e)^{\sigma_k}$ (For the sake of simplicity we are writing μ_k in place of μ_{λ_k}.) Hence dim $L_k = \sigma_k$. If a_k is an f-generator of L_k, then the vectors

$$a_k, (f - \mu_k e)(a_k), \ldots, (f - \mu_k e)^{\sigma_k - 1}(a_k) \qquad (7)$$

are linearly independent (otherwise the degree of the f-annihilator of a_k would be less than σ_k) and therefore form a basis of L_k. For $0 \leqslant \sigma \leqslant \sigma_k - 1$, $(f - \mu_k e)^\sigma (a_k)$ is mapped by f onto $(f - \mu_k e)^{\sigma+1}(a_k) + \mu_k (f - \mu_k e)^\sigma (a_k)$. If $\sigma = \sigma_k - 1$, then the first term is zero. Hence, corresponding to the subspace L_k, there is a matrix of the form

$$
J_k = \begin{pmatrix}
\mu_k & 1 & 0 & \ldots & 0 & 0 \\
0 & \mu_k & 1 & \ldots & 0 & 0 \\
\ldots & \ldots & \ldots & \ldots & \ldots & \ldots \\
0 & 0 & 0 & \ldots & \mu_k & 1 \\
0 & 0 & 0 & \ldots & 0 & \mu_k
\end{pmatrix} \tag{8}
$$

If we combine the bases (7) of the subspaces L_k into a basis of E, then the matrix of the endomorphism f has the form

$$
J = \begin{bmatrix}
\begin{array}{c|c|c|c}
J_1 & 0 & \ldots & 0 \\
\hline
0 & J_2 & \ldots & 0 \\
\hline
\ldots & \ldots & \ldots & \ldots \\
\hline
0 & 0 & \ldots & J_r
\end{array}
\end{bmatrix} \tag{9}
$$

in which the blocks $J_1, \ldots, J_r$ each have the form (8). This matrix J is known as the Jordan canonical form. The characteristic polynomial of the block J_k in (8) is clearly $(\xi - \mu_k)^{\sigma_k}$ and that of (9) is therefore $\mathbf{h} = (\xi - \mu_1)^{\sigma_1} \ldots (\xi - \mu_r)^{\sigma_r}$. The diagonal elements $\mu_1, \ldots, \mu_r$ are therefore the eigenvalues of f and in fact each eigenvalue appears in J just as often as its multiplicity as a root of the characteristic polynomial. Since the subspaces L_k which correspond to the blocks J_k are uniquely determined apart from f-equivalence (Theorem 14.1;13), and since the restrictions of f to two f-equivalent subspaces have the same characteristic equation, then J is uniquely determined by A except for the order of the blocks J_k. This proves the following result.

Theorem 4. Every square matrix A with elements from an algebraically closed field F is similar to a matrix $J = SAS^{-1}$ in the Jordan canonical form. The diagonal elements of the Jordan canonical form are the eigenvalues of A and each appears as often as its multiplicity as a root of the characteristic polynomial. The Jordan canonical form is uniquely determined by A except for the order of the blocks (8).

Consider now the special case when there is a basis consisting of eigenvectors of f, i.e., when E is a direct sum of 1-dimensional f-invariant subspaces. This is clearly the case if and only if all the blocks (8) in (9) consist of only one element μ_k, which means that the minimal polynomial of f factorizes into distinct linear factors. Thus we have

Theorem 5. Given an endomorphism f, then there is a basis consisting of eigenvectors of f if and only if the minimal polynomial of f factorizes into a product of distinct linear factors.

By Theorem 13.4;6, we know that the situation of Theorem 5 occurs in particular when E is a unitary complex vector space and f is self-adjoint. We will now also consider the case when f is a rotation of a unitary vector space (see 12.4). The last of the basis vectors which correspond to a Jordan block J_k (8) is an eigenvector corresponding to the eigenvalue μ_k and therefore spans a 1-dimensional f-invariant subspace L. The orthogonal complement $L^\dagger$ is also f-invariant, because orthogonal vectors stay orthogonal under a rotation. The restriction of f to $L^\dagger$ is again a rotation so that $L^\dagger$ again contains an eigenvector. If we continue this process, then we find that, corresponding to a rotation of a finite-dimensional unitary vector space, there is an orthonormal basis with respect to which the rotation is represented by a diagonal matrix J (the Jordan canonical form). Since J must be unitary, it follows that the diagonal elements must have the absolute value 1, i.e., they are of the form $\mu_k = e^{i\phi_k}$.

The method of calculating eigenvalues and eigenvectors described in 13.6.1 can also be used to find the Jordan canonical form and the invariant subspaces of a matrix A or of the endomorphism f of R_n represented by A.

We will first do this for the Example 13.6;2. Since the eigenvalues $\lambda_1 = 1$ and $\lambda_2 = 2$ are double solutions of the characteristic equation and are non-degenerate, the Jordan canonical form must be

$$J = \begin{pmatrix} 1 & 1 & 0 & 0 \\ 0 & 1 & 0 & 0 \\ 0 & 0 & 2 & 1 \\ 0 & 0 & 0 & 2 \end{pmatrix}$$

The two Jordan blocks of J correspond to two f-irreducible subspaces L_1 and L_2. Every vector $y_1 \in R_4$ for which

$$(f - \lambda_1 e)(y_1) = x_1 \qquad (x_1 \text{ an eigenvector of } \lambda_1)$$

is an f-generator of L_1 and similarly for L_2. Thus to find an f-generator of L_1, we can simply solve the system of equations

$$(A - \lambda_1 I)\eta_1 = \xi_1 \tag{10}$$

for η_1. In practice, we do this by adjoining an extra column to the scheme of calculations which has already been done in Example 13.6;2 (the second last column). At the top are the components of the vector x_1 (multiplied by -1). The last number in this column is 0, so that ξ_4 can be chosen arbitrarily. With $\xi_4 = 0$, we find the f-generator $y_1 = (0, -1, 1, 0)$ for L_1 by using the cellar rows. Hence L_1 is the linear hull of $x_1 = (1, 2, 0, -1)$ and $y_1 = (0, -1, 1, 0)$.

The corresponding calculations for L_2 are carried out in the last column and we find $y_2 = (\frac{1}{2}, \frac{1}{2}, \frac{1}{2}, 0)$.

It is also straightforward to find a matrix T such that $T^{-1}AT$ is the Jordan canonical form. Because it is easy to see that

$$
T = (\xi_1, \eta_1, \xi_2, \eta_2) = \begin{pmatrix} 1 & 0 & 3 & \frac{1}{2} \\ 2 & -1 & 3 & \frac{1}{2} \\ 0 & 1 & 1 & \frac{1}{2} \\ -1 & 0 & -2 & 0 \end{pmatrix}
$$

has this property and we can also verify this by calculation.

In Example 13.6;3, the eigenvalue $\lambda_3 = 14$ is a double solution of the characteristic equation and is not degenerate so that there must be a corresponding 2-dimensional f-irreducible subspace L_3. Hence the Jordan canonical form is

$$
J = \begin{pmatrix} -2 & 0 & 0 & 0 \\ 0 & 6 & 0 & 0 \\ 0 & 0 & 14 & 1 \\ 0 & 0 & 0 & 14 \end{pmatrix}
$$

The calculation of an f-generator of L_3 is already prepared in the last column in the scheme of the calculations and we find $y_3 = (-8, 0, 3, -20)$.

If there are f-irreducible subspaces which have dimension greater than 2, then we can also find a basis for each of these. From the general theory, after the solution of (10), we would solve the system

$$
(A - \lambda_1 I)\zeta_1 = \eta_1
$$

to find a vector z_1 which, in the case when $\dim L_1 = 3$, is an f-generator and which forms a basis with x_1 and y_1. If the dimension is greater than 3, then we proceed similarly. Actually we do not need to know the dimension in this, because, as soon as we have found an f-generator, the next system of equations is no longer soluble. This will be the case when the bottom element of the column in question is not zero.

14.2.3 Endomorphisms of a Real Vector Space, Real Matrices

If F is the field of real numbers, then every irreducible polynomial with coefficients from F is either of degree 1 or of degree 2. Because every real polynomial f of degree at least 2 either has a real root α or a pair of conjugate complex roots α, $\bar{\alpha}$. In the first case, if the degree of f is at least 2, then

$(\xi - \alpha)$ is a proper divisor of f. In the second case, if the degree is at least 3, then $(\xi - \alpha)(\xi - \bar{\alpha})$, i.e., a polynomial with real coefficients, is a proper divisor of f.

Now, if L_k is an f-irreducible subspace of E which corresponds to a linear factor of the minimal polynomial $\mathbf{m}$, i.e., if $\mathbf{m}_k = (\xi - \alpha)^\sigma$, then, by 14.2.2, there is a basis of L_k such that the matrix of f is in the Jordan canonical form. On the other hand, if $\mathbf{m}_k = (\alpha_{k0} + \alpha_{k1}\xi + \xi^2)^\sigma$ where $\mathbf{p} = \alpha_{k0} + \alpha_{k1}\xi + \xi^2$ is irreducible, then $\dim L_k = 2\sigma$. If a_k is an f-generator of L_k, then the 2σ vectors

$$a_k, \, \mathbf{p}_f(a_k), \, \mathbf{p}_f^2(a_k), \, \ldots, \, \mathbf{p}_f^{\sigma-1}(a_k)$$
$$f(a_k), \, \mathbf{p}_f f(a_k), \, \mathbf{p}_f^2 f(a_k), \ldots, \, \mathbf{p}_f^{\sigma-1} f(a_k) \tag{11}$$

form a basis of L_k. Now, for $0 \leqslant i \leqslant \sigma - 1$,

$$\begin{aligned}
f\mathbf{p}_f^i f(a_k) &= f^2 \mathbf{p}_f^i(a_k) \\
&= (\mathbf{p}_f - \alpha_{k1}f - \alpha_{k0})\,\mathbf{p}_f^i(a_k) \\
&= \mathbf{p}_f^{i+1}(a_k) - \alpha_{k1}\,\mathbf{p}_f^i f(a_k) - \alpha_{k0}\,\mathbf{p}_f^i(a_k)
\end{aligned}$$

Also, if $i = \sigma - 1$, then $\mathbf{p}_f^{i+1}(a_k) = \mathbf{m}_{kf}(a_k) = 0$.

Thus we find the images of the basis vectors (11), by using the matrix

$$R_k = \begin{bmatrix}
0 & 1 & 0 & 0 & \ldots & 0 & 0 & 0 & 0 \\
-\alpha_{k0} & -\alpha_{k1} & 1 & 0 & \ldots & 0 & 0 & 0 & 0 \\
0 & 0 & 0 & 1 & \ldots & 0 & 0 & 0 & 0 \\
0 & 0 & -\alpha_{k0} & -\alpha_{k1} & \ldots & 0 & 0 & 0 & 0 \\
\ldots & \ldots & \ldots & \ldots & \ldots & \ldots & \ldots & \ldots & \ldots \\
0 & 0 & 0 & 0 & \ldots & 0 & 1 & 0 & 0 \\
0 & 0 & 0 & 0 & \ldots & -\alpha_{k0} & -\alpha_{k1} & 1 & 0 \\
0 & 0 & 0 & 0 & \ldots & 0 & 0 & 0 & 1 \\
0 & 0 & 0 & 0 & \ldots & 0 & 0 & -\alpha_{k0} & -\alpha_{k1}
\end{bmatrix} \tag{12}$$

If we combine these bases of the subspaces L_k into a basis of E, then, with respect to this basis, the matrix of f has the form

$$R = \begin{bmatrix}
R_1 & 0 & \ldots & 0 \\
0 & R_2 & \ldots & 0 \\
\ldots & \ldots & \ldots & \ldots \\
0 & 0 & \ldots & R_r
\end{bmatrix} \tag{13}$$

where the blocks R_r have either the Jordan canonical form or the form (12).

Hence in particular every real matrix A is similar to a matrix R of the form (12), i.e., there is a real non-singular matrix S such that $SAS^{-1} = R$.

If A is symmetric, then only 1-dimensional Jordan blocks occur because, in this case, there is a basis consisting of eigenvectors. (Theorem 13.2;7.)

Finally we will consider the case of a rotation of a finite-dimensional Euclidean vector space or, equivalently, the case of an orthogonal matrix A. If the subspace L_k corresponds to a block in the Jordan canonical form (8), then the last of the corresponding basis vectors (7) is again an eigenvector which spans a 1-dimensional f-invariant subspace $L_k^* \subseteq L_k$. On the other hand, if L_k corresponds to a block of the form (12), then the last two of the corresponding basis vectors (11) span a 2-dimensional f-invariant subspace $L_k^* \subseteq L_k$. In both cases L_k must be equal to L_k^*, because the orthogonal complement of L_k^* in L_k is also f-invariant (orthogonal vectors are mapped onto orthogonal vectors by a rotation). Further, since L_k is the direct sum of L_k^* and its orthogonal complement (Theorem 12.1;3), and L_k is f-irreducible, it follows that $L_k = L_k^*$. Hence it follows that all f-irreducible subspaces L_k have the dimension 1 or 2. Furthermore, they can be chosen to be mutually orthogonal, because the orthogonal complement of an f-invariant subspace is also f-invariant. If we choose them in this way and if we further choose an orthonormal basis of each L_k then the matrix of f will be orthogonal and therefore will have the form

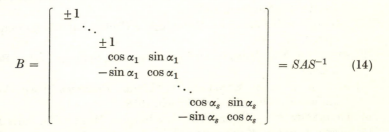

$$B = \begin{bmatrix} \pm 1 & & & & & & \\ & \ddots & & & & & \\ & & \pm 1 & & & & \\ & & & \cos\alpha_1 & \sin\alpha_1 & & \\ & & & -\sin\alpha_1 & \cos\alpha_1 & & \\ & & & & & \ddots & \\ & & & & & & \cos\alpha_s & \sin\alpha_s \\ & & & & & & -\sin\alpha_s & \cos\alpha_s \end{bmatrix} = SAS^{-1} \quad (14)$$

where all the elements except those shown are equal to zero. (See Problems 12.1;10 and 13.1;6). In this, S is real and the canonical form (14) is uniquely determined by f or A apart from the order of the 1- and 2-dimensional blocks.

BIBLIOGRAPHY

A Selection of Books on Linear Algebra

[1] GELFAND I. M. *Lectures on Linear Algebra*, Interscience Publishers, New York, 1961.

[2] GRÄUB W. *Lineare Algebra*, Springer Verlag, Berlin, 1958.

[3] HADLEY G. *Linear Algebra*, Addison Wesley Publ., London, 1961.

[4] HALMOS P. R. *Finite-dimensional Vector Spaces*, D. van Nostrand Comp., Princeton, 1958.

[5] HOFFMANN K. AND KUNZE R. *Linear Algebra*, Prentice-Hall Inc. 1961.

[6] KOWALSKY J. *Linear Algebra*, Walter de Gruyter Co., Berlin, 1963.

[7] SHILOV G. E. *An Introduction to the Theory of Linear Spaces*, Prentice-Hall Inc. 1961.

[8] ZURMÜHL R. *Matrizen*, Springer Verlag, Berlin, 1958.

Other General Works

[9] BOURBAKI N. *Eléments de mathématique*, Paris, Hermann Cie.

[10] DANTZIG G. B. *Linear Programming and Extensions*, Princeton University Press, Princeton, 1963.

[11] HALMOS P. R. *Introduction to Hilbert Space*, Chelsea Publishing Company, New York, 1951.

[12] HERMES H. *Einführung in die Verbandstheorie*, Springer Verlag, Berlin, 1955.

[13] KRELLE W. UND KÜNZI H. P. *Linear Programmierung*, Verlag Industrielle Organisation, Zürich, 1958.

[14] LICHNEROWICZ A. *Algebre et analyse lineares*, Masson Cie, Paris 1947.

[15] ROTHE R. *Höhere Mathematik für Mathematiker, Physiker, Ingenieure*, B. G. Teubner, Stuttgart, 1947.

[16] STIEFEL E. *Einführung in die numerische Mathematik*, B. G. Teubner, Stuttgart, 1961.

[17] SPEISER A. *Theorie der Gruppen von endlicher Ordnung*, Springer Verlag, Berlin, 1927.

[18] TRICOMI F. G. *Vorlesungen über Orthogonalreihen*, Springer Verlag, Berlin, 1955.

[19] VAN DER WAERDEN B. L. *Moderne Algebra Vol. 1*, Berlin, 1959.

[20] VAN DER WAERDEN B. L. *Mathematische Statistik*, Springer Verlag, Berlin, 1957.

[21] DIEUDONNÉ J. *Foundations of Modern Analysis*, Academic Press, New York, 1960.

[22] KUHN H. W. AND TUCKER A. W. 'Linear Inequalities and Related Systems', *Annals of Mathematics Studies*, **38**, Princeton, 1956.

[23] ABIAN A. *The Theory of Sets and Transfinite Arithmetic*, W. B. Saunders Co., Philadelphia, 1965.

[24] BIRKHOFF G. AND MACLANE S. *A Survey of Modern Algebra*, Macmillan Co., New York, 1953.

[25] LEDERMANN W. *Introduction to the Theory of Finite Groups*, Oliver and Boyd, Edinburgh, 1961.

[26] MCKINSEY J. *Introduction to the Theory of Games*, McGraw-Hill Book Co. Inc., New York, 1952.

[27] RALSTON A. *A First Course in Numerical Analysis*, McGraw-Hill Book Co. Inc., New York, 1965.

[28] SZEGŐ G. *Orthogonal Polynomials*, American Mathematical Society, New York, 1959.

[29] FELLER W. *Probability Theory and its Applications*, Vol. 1., Wiley, New York, 1950.

INDEX

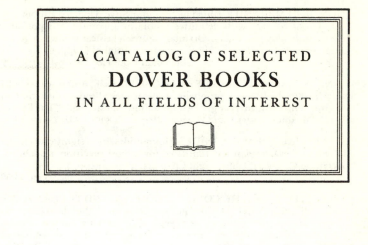

A CATALOG OF SELECTED

DOVER BOOKS

IN ALL FIELDS OF INTEREST

A CATALOG OF SELECTED DOVER
BOOKS IN ALL FIELDS OF INTEREST

DRAWINGS OF REMBRANDT, edited by Seymour Slive. Updated Lippmann, Hofstede de Groot edition, with definitive scholarly apparatus. All portraits, biblical sketches, landscapes, nudes. Oriental figures, classical studies, together with selection of work by followers. 550 illustrations. Total of 630pp. 9⅛ × 12¼.
21485-0, 21486-9 Pa., Two-vol. set $25.00

GHOST AND HORROR STORIES OF AMBROSE BIERCE, Ambrose Bierce. 24 tales vividly imagined, strangely prophetic, and decades ahead of their time in technical skill: "The Damned Thing," "An Inhabitant of Carcosa," "The Eyes of the Panther," "Moxon's Master," and 20 more. 199pp. 5⅜ × 8½. 20767-6 Pa. $3.95

ETHICAL WRITINGS OF MAIMONIDES, Maimonides. Most significant ethical works of great medieval sage, newly translated for utmost precision, readability. Laws Concerning Character Traits, Eight Chapters, more. 192pp. 5⅜ × 8½.
24522-5 Pa. $4.50

THE EXPLORATION OF THE COLORADO RIVER AND ITS CANYONS, J. W. Powell. Full text of Powell's 1,000-mile expedition down the fabled Colorado in 1869. Superb account of terrain, geology, vegetation, Indians, famine, mutiny, treacherous rapids, mighty canyons, during exploration of last unknown part of continental U.S. 400pp. 5⅜ × 8½. 20094-9 Pa. $6.95

HISTORY OF PHILOSOPHY, Julián Marías. Clearest one-volume history on the market. Every major philosopher and dozens of others, to Existentialism and later. 505pp. 5⅜ × 8½. 21739-6 Pa. $8.50

ALL ABOUT LIGHTNING, Martin A. Uman. Highly readable non-technical survey of nature and causes of lightning, thunderstorms, ball lightning, St. Elmo's Fire, much more. Illustrated. 192pp. 5⅜ × 8½. 25237-X Pa. $5.95

SAILING ALONE AROUND THE WORLD, Captain Joshua Slocum. First man to sail around the world, alone, in small boat. One of great feats of seamanship told in delightful manner. 67 illustrations. 294pp. 5⅜ × 8½. 20326-3 Pa. $4.95

LETTERS AND NOTES ON THE MANNERS, CUSTOMS AND CONDI-TIONS OF THE NORTH AMERICAN INDIANS, George Catlin. Classic account of life among Plains Indians: ceremonies, hunt, warfare, etc. 312 plates. 572pp. of text. 6⅛ × 9¼. 22118-0, 22119-9 Pa. Two-vol. set $15.90

ALASKA: The Harriman Expedition, 1899, John Burroughs, John Muir, et al. Informative, engrossing accounts of two-month, 9,000-mile expedition. Native peoples, wildlife, forests, geography, salmon industry, glaciers, more. Profusely illustrated. 240 black-and-white line drawings. 124 black-and-white photographs. 3 maps. Index. 576pp. 5⅜ × 8½. 25109-8 Pa. $11.95

THE BOOK OF BEASTS: Being a Translation from a Latin Bestiary of the Twelfth Century, T. H. White. Wonderful catalog real and fanciful beasts: manticore, griffin, phoenix, amphivius, jaculus, many more. White's witty erudite commentary on scientific, historical aspects. Fascinating glimpse of medieval mind. Illustrated. 296pp. 5⅜ × 8¼. (Available in U.S. only) 24609-4 Pa. $5.95

FRANK LLOYD WRIGHT: ARCHITECTURE AND NATURE With 160 Illustrations, Donald Hoffmann. Profusely illustrated study of influence of nature—especially prairie—on Wright's designs for Fallingwater, Robie House, Guggenheim Museum, other masterpieces. 96pp. 9¼ × 10¾. 25098-9 Pa. $7.95

FRANK LLOYD WRIGHT'S FALLINGWATER, Donald Hoffmann. Wright's famous waterfall house: planning and construction of organic idea. History of site, owners, Wright's personal involvement. Photographs of various stages of building. Preface by Edgar Kaufmann, Jr. 100 illustrations. 112pp. 9¼ × 10.
23671-4 Pa. $7.95

YEARS WITH FRANK LLOYD WRIGHT: Apprentice to Genius, Edgar Tafel. Insightful memoir by a former apprentice presents a revealing portrait of Wright the man, the inspired teacher, the greatest American architect. 372 black-and-white illustrations. Preface. Index. vi + 228pp. 8¼ × 11. 24801-1 Pa. $9.95

THE STORY OF KING ARTHUR AND HIS KNIGHTS, Howard Pyle. Enchanting version of King Arthur fable has delighted generations with imaginative narratives of exciting adventures and unforgettable illustrations by the author. 41 illustrations. xviii + 313pp. 6⅛ × 9¼. 21445-1 Pa. $5.95

THE GODS OF THE EGYPTIANS, E. A. Wallis Budge. Thorough coverage of numerous gods of ancient Egypt by foremost Egyptologist. Information on evolution of cults, rites and gods; the cult of Osiris; the Book of the Dead and its rites; the sacred animals and birds; Heaven and Hell; and more. 956pp. 6⅛ × 9¼.
22055-9, 22056-7 Pa., Two-vol. set $21.90

A THEOLOGICO-POLITICAL TREATISE, Benedict Spinoza. Also contains unfinished *Political Treatise*. Great classic on religious liberty, theory of government on common consent. R. Elwes translation. Total of 421pp. 5⅜ × 8½.
20249-6 Pa. $6.95

INCIDENTS OF TRAVEL IN CENTRAL AMERICA, CHIAPAS, AND YUCATAN, John L. Stephens. Almost single-handed discovery of Maya culture; exploration of ruined cities, monuments, temples; customs of Indians. 115 drawings. 892pp. 5⅜ × 8½. 22404-X, 22405-8 Pa., Two-vol. set $15.90

LOS CAPRICHOS, Francisco Goya. 80 plates of wild, grotesque monsters and caricatures. Prado manuscript included. 183pp. 6⅜ × 9⅜. 22384-1 Pa. $4.95

AUTOBIOGRAPHY: The Story of My Experiments with Truth, Mohandas K. Gandhi. Not hagiography, but Gandhi in his own words. Boyhood, legal studies, purification, the growth of the Satyagraha (nonviolent protest) movement. Critical, inspiring work of the man who freed India. 480pp. 5⅜ × 8½. (Available in U.S. only)
24593-4 Pa. $6.95

ILLUSTRATED DICTIONARY OF HISTORIC ARCHITECTURE, edited by Cyril M. Harris. Extraordinary compendium of clear, concise definitions for over 5,000 important architectural terms complemented by over 2,000 line drawings. Covers full spectrum of architecture from ancient ruins to 20th-century Modernism. Preface. 592pp. 7½ × 9⅜. 24444-X Pa. $14.95

THE NIGHT BEFORE CHRISTMAS, Clement Moore. Full text, and woodcuts from original 1848 book. Also critical, historical material. 19 illustrations. 40pp. 4⅝ × 6. 22797-9 Pa. $2.50

THE LESSON OF JAPANESE ARCHITECTURE: 165 Photographs, Jiro Harada. Memorable gallery of 165 photographs taken in the 1930's of exquisite Japanese homes of the well-to-do and historic buildings. 13 line diagrams. 192pp. 8⅜ × 11¼. 24778-3 Pa. $8.95

THE AUTOBIOGRAPHY OF CHARLES DARWIN AND SELECTED LETTERS, edited by Francis Darwin. The fascinating life of eccentric genius composed of an intimate memoir by Darwin (intended for his children); commentary by his son, Francis; hundreds of fragments from notebooks, journals, papers; and letters to and from Lyell, Hooker, Huxley, Wallace and Henslow. xi + 365pp. 5⅜ × 8.
20479-0 Pa. $5.95

WONDERS OF THE SKY: Observing Rainbows, Comets, Eclipses, the Stars and Other Phenomena, Fred Schaaf. Charming, easy-to-read poetic guide to all manner of celestial events visible to the naked eye. Mock suns, glories, Belt of Venus, more. Illustrated. 299pp. 5¼ × 8¼. 24402-4 Pa. $7.95

BURNHAM'S CELESTIAL HANDBOOK, Robert Burnham, Jr. Thorough guide to the stars beyond our solar system. Exhaustive treatment. Alphabetical by constellation: Andromeda to Cetus in Vol. 1; Chamaeleon to Orion in Vol. 2; and Pavo to Vulpecula in Vol. 3. Hundreds of illustrations. Index in Vol. 3. 2,000pp. 6⅛ × 9¼. 23567-X, 23568-8, 23673-0 Pa., Three-vol. set $37.85

STAR NAMES: Their Lore and Meaning, Richard Hinckley Allen. Fascinating history of names various cultures have given to constellations and literary and folkloristic uses that have been made of stars. Indexes to subjects. Arabic and Greek names. Biblical references. Bibliography. 563pp. 5⅜ × 8½. 21079-0 Pa. $7.95

THIRTY YEARS THAT SHOOK PHYSICS: The Story of Quantum Theory, George Gamow. Lucid, accessible introduction to influential theory of energy and matter. Careful explanations of Dirac's anti-particles, Bohr's model of the atom, much more. 12 plates. Numerous drawings. 240pp. 5⅜ × 8½. 24895-X Pa. $4.95

CHINESE DOMESTIC FURNITURE IN PHOTOGRAPHS AND MEASURED DRAWINGS, Gustav Ecke. A rare volume, now affordably priced for antique collectors, furniture buffs and art historians. Detailed review of styles ranging from early Shang to late Ming. Unabridged republication. 161 black-and-white drawings, photos. Total of 224pp. 8⅜ × 11¼. (Available in U.S. only) 25171-3 Pa. $12.95

VINCENT VAN GOGH: A Biography, Julius Meier-Graefe. Dynamic, penetrating study of artist's life, relationship with brother, Theo, painting techniques, travels, more. Readable, engrossing. 160pp. 5⅜ × 8½. (Available in U.S. only)
25253-1 Pa. $3.95

HOW TO WRITE, Gertrude Stein. Gertrude Stein claimed anyone could understand her unconventional writing—here are clues to help. Fascinating improvisations, language experiments, explanations illuminate Stein's craft and the art of writing. Total of 414pp. 4⅜ × 6⅜. 23144-5 Pa. $5.95

ADVENTURES AT SEA IN THE GREAT AGE OF SAIL: Five Firsthand Narratives, edited by Elliot Snow. Rare true accounts of exploration, whaling, shipwreck, fierce natives, trade, shipboard life, more. 33 illustrations. Introduction. 353pp. 5⅜ × 8½. 25177-2 Pa. $7.95

THE HERBAL OR GENERAL HISTORY OF PLANTS, John Gerard. Classic descriptions of about 2,850 plants—with over 2,700 illustrations—includes Latin and English names, physical descriptions, varieties, time and place of growth, more. 2,706 illustrations. xlv + 1,678pp. 8½ × 12¼. 23147-X Cloth. $75.00

DOROTHY AND THE WIZARD IN OZ, L. Frank Baum. Dorothy and the Wizard visit the center of the Earth, where people are vegetables, glass houses grow and Oz characters reappear. Classic sequel to *Wizard of Oz*. 256pp. 5⅜ × 8. 24714-7 Pa. $4.95

SONGS OF EXPERIENCE: Facsimile Reproduction with 26 Plates in Full Color, William Blake. This facsimile of Blake's original "Illuminated Book" reproduces 26 full-color plates from a rare 1826 edition. Includes "The Tyger," "London," "Holy Thursday," and other immortal poems. 26 color plates. Printed text of poems. 48pp. 5¼ × 7. 24636-1 Pa. $3.50

SONGS OF INNOCENCE, William Blake. The first and most popular of Blake's famous "Illuminated Books," in a facsimile edition reproducing all 31 brightly colored plates. Additional printed text of each poem. 64pp. 5¼ × 7. 22764-2 Pa. $3.50

PRECIOUS STONES, Max Bauer. Classic, thorough study of diamonds, rubies, emeralds, garnets, etc.: physical character, occurrence, properties, use, similar topics. 20 plates, 8 in color. 94 figures. 659pp. 6⅛ × 9¼. 21910-0, 21911-9 Pa., Two-vol. set $15.90

ENCYCLOPEDIA OF VICTORIAN NEEDLEWORK, S. F. A. Caulfeild and Blanche Saward. Full, precise descriptions of stitches, techniques for dozens of needlecrafts—most exhaustive reference of its kind. Over 800 figures. Total of 679pp. 8⅛ × 11. Two volumes. Vol. 1 22800-2 Pa. $11.95
Vol. 2 22801-0 Pa. $11.95

THE MARVELOUS LAND OF OZ, L. Frank Baum. Second Oz book, the Scarecrow and Tin Woodman are back with hero named Tip, Oz magic. 136 illustrations. 287pp. 5⅜ × 8½. 20692-0 Pa. $5.95

WILD FOWL DECOYS, Joel Barber. Basic book on the subject, by foremost authority and collector. Reveals history of decoy making and rigging, place in American culture, different kinds of decoys, how to make them, and how to use them. 140 plates. 156pp. 7⅞ × 10¾. 20011-6 Pa. $8.95

HISTORY OF LACE, Mrs. Bury Palliser. Definitive, profusely illustrated chronicle of lace from earliest times to late 19th century. Laces of Italy, Greece, England, France, Belgium, etc. Landmark of needlework scholarship. 266 illustrations. 672pp. 6⅛ × 9¼. 24742-2 Pa. $14.95

ILLUSTRATED GUIDE TO SHAKER FURNITURE, Robert Meader. All furniture and appurtenances, with much on unknown local styles. 235 photos. 146pp. 9 × 12. 22819-3 Pa. $7.95

WHALE SHIPS AND WHALING: A Pictorial Survey, George Francis Dow. Over 200 vintage engravings, drawings, photographs of barks, brigs, cutters, other vessels. Also harpoons, lances, whaling guns, many other artifacts. Comprehensive text by foremost authority. 207 black-and-white illustrations. 288pp. 6 × 9. 24808-9 Pa. $8.95

THE BERTRAMS, Anthony Trollope. Powerful portrayal of blind self-will and thwarted ambition includes one of Trollope's most heartrending love stories. 497pp. 5⅜ × 8½. 25119-5 Pa. $8.95

ADVENTURES WITH A HAND LENS, Richard Headstrom. Clearly written guide to observing and studying flowers and grasses, fish scales, moth and insect wings, egg cases, buds, feathers, seeds, leaf scars, moss, molds, ferns, common crystals, etc.—all with an ordinary, inexpensive magnifying glass. 209 exact line drawings aid in your discoveries. 220pp. 5⅜ × 8½. 23330-8 Pa. $4.50

RODIN ON ART AND ARTISTS, Auguste Rodin. Great sculptor's candid, wide-ranging comments on meaning of art; great artists; relation of sculpture to poetry, painting, music; philosophy of life, more. 76 superb black-and-white illustrations of Rodin's sculpture, drawings and prints. 119pp. 8⅜ × 11¼. 24487-3 Pa. $6.95

FIFTY CLASSIC FRENCH FILMS, 1912–1982: A Pictorial Record, Anthony Slide. Memorable stills from Grand Illusion, Beauty and the Beast, Hiroshima, Mon Amour, many more. Credits, plot synopses, reviews, etc. 160pp. 8¼ × 11. 25256-6 Pa. $11.95

THE PRINCIPLES OF PSYCHOLOGY, William James. Famous long course complete, unabridged. Stream of thought, time perception, memory, experimental methods; great work decades ahead of its time. 94 figures. 1,391pp. 5⅜ × 8½. 20381-6, 20382-4 Pa., Two-vol. set $19.90

BODIES IN A BOOKSHOP, R. T. Campbell. Challenging mystery of blackmail and murder with ingenious plot and superbly drawn characters. In the best tradition of British suspense fiction. 192pp. 5⅜ × 8½. 24720-1 Pa. $3.95

CALLAS: PORTRAIT OF A PRIMA DONNA, George Jellinek. Renowned commentator on the musical scene chronicles incredible career and life of the most controversial, fascinating, influential operatic personality of our time. 64 black-and-white photographs. 416pp. 5⅜ × 8¼. 25047-4 Pa. $7.95

GEOMETRY, RELATIVITY AND THE FOURTH DIMENSION, Rudolph Rucker. Exposition of fourth dimension, concepts of relativity as Flatland characters continue adventures. Popular, easily followed yet accurate, profound. 141 illustrations. 133pp. 5⅜ × 8½. 23400-2 Pa. $3.50

HOUSEHOLD STORIES BY THE BROTHERS GRIMM, with pictures by Walter Crane. 53 classic stories—Rumpelstiltskin, Rapunzel, Hansel and Gretel, the Fisherman and his Wife, Snow White, Tom Thumb, Sleeping Beauty, Cinderella, and so much more—lavishly illustrated with original 19th century drawings. 114 illustrations. x + 269pp. 5⅜ × 8½. 21080-4 Pa. $4.50

SUNDIALS, Albert Waugh. Far and away the best, most thorough coverage of ideas, mathematics concerned, types, construction, adjusting anywhere. Over 100 illustrations. 230pp. 5⅜ × 8½. 22947-5 Pa. $4.50

PICTURE HISTORY OF THE NORMANDIE: With 190 Illustrations, Frank O. Braynard. Full story of legendary French ocean liner: Art Deco interiors, design innovations, furnishings, celebrities, maiden voyage, tragic fire, much more. Extensive text. 144pp. 8⅜ × 11¾. 25257-4 Pa. $9.95

THE FIRST AMERICAN COOKBOOK: A Facsimile of "American Cookery," 1796, Amelia Simmons. Facsimile of the first American-written cookbook published in the United States contains authentic recipes for colonial favorites—pumpkin pudding, winter squash pudding, spruce beer, Indian slapjacks, and more. Introductory Essay and Glossary of colonial cooking terms. 80pp. 5⅜ × 8½. 24710-4 Pa. $3.50

101 PUZZLES IN THOUGHT AND LOGIC, C. R. Wylie, Jr. Solve murders and robberies, find out which fishermen are liars, how a blind man could possibly identify a color—purely by your own reasoning! 107pp. 5⅜ × 8½. 20367-0 Pa. $2.50

THE BOOK OF WORLD-FAMOUS MUSIC—CLASSICAL, POPULAR AND FOLK, James J. Fuld. Revised and enlarged republication of landmark work in musico-bibliography. Full information about nearly 1,000 songs and compositions including first lines of music and lyrics. New supplement. Index. 800pp. 5⅜ × 8¼. 24857-7 Pa. $14.95

ANTHROPOLOGY AND MODERN LIFE, Franz Boas. Great anthropologist's classic treatise on race and culture. Introduction by Ruth Bunzel. Only inexpensive paperback edition. 255pp. 5⅜ × 8½. 25245-0 Pa. $5.95

THE TALE OF PETER RABBIT, Beatrix Potter. The inimitable Peter's terrifying adventure in Mr. McGregor's garden, with all 27 wonderful, full-color Potter illustrations. 55pp. 4¼ × 5½. (Available in U.S. only) 22827-4 Pa. $1.75

THREE PROPHETIC SCIENCE FICTION NOVELS, H. G. Wells. *When the Sleeper Wakes, A Story of the Days to Come* and *The Time Machine* (full version). 335pp. 5⅜ × 8½. (Available in U.S. only) 20605-X Pa. $5.95

APICIUS COOKERY AND DINING IN IMPERIAL ROME, edited and translated by Joseph Dommers Vehling. Oldest known cookbook in existence offers readers a clear picture of what foods Romans ate, how they prepared them, etc. 49 illustrations. 301pp. 6⅛ × 9¼. 23563-7 Pa. $6.50

SHAKESPEARE LEXICON AND QUOTATION DICTIONARY, Alexander Schmidt. Full definitions, locations, shades of meaning of every word in plays and poems. More than 50,000 exact quotations. 1,485pp. 6½ × 9¼. 22726-X, 22727-8 Pa., Two-vol. set $27.90

THE WORLD'S GREAT SPEECHES, edited by Lewis Copeland and Lawrence W. Lamm. Vast collection of 278 speeches from Greeks to 1970. Powerful and effective models; unique look at history. 842pp. 5⅜ × 8½. 20468-5 Pa. $11.95

THE BLUE FAIRY BOOK, Andrew Lang. The first, most famous collection, with many familiar tales: Little Red Riding Hood, Aladdin and the Wonderful Lamp, Puss in Boots, Sleeping Beauty, Hansel and Gretel, Rumpelstiltskin; 37 in all. 138 illustrations. 390pp. 5⅜ × 8½. 21437-0 Pa. $5.95

THE STORY OF THE CHAMPIONS OF THE ROUND TABLE, Howard Pyle. Sir Launcelot, Sir Tristram and Sir Percival in spirited adventures of love and triumph retold in Pyle's inimitable style. 50 drawings, 31 full-page. xviii + 329pp. 6½ × 9¼. 21883-X Pa. $6.95

AUDUBON AND HIS JOURNALS, Maria Audubon. Unmatched two-volume portrait of the great artist, naturalist and author contains his journals, an excellent biography by his granddaughter, expert annotations by the noted ornithologist, Dr. Elliott Coues, and 37 superb illustrations. Total of 1,200pp. 5⅜ × 8.
Vol. I 25143-8 Pa. $8.95
Vol. II 25144-6 Pa. $8.95

GREAT DINOSAUR HUNTERS AND THEIR DISCOVERIES, Edwin H. Colbert. Fascinating, lavishly illustrated chronicle of dinosaur research, 1820's to 1960. Achievements of Cope, Marsh, Brown, Buckland, Mantell, Huxley, many others. 384pp. 5¼ × 8¼. 24701-5 Pa. $6.95

THE TASTEMAKERS, Russell Lynes. Informal, illustrated social history of American taste 1850's–1950's. First popularized categories Highbrow, Lowbrow, Middlebrow. 129 illustrations. New (1979) afterword. 384pp. 6 × 9.
23993-4 Pa. $6.95

DOUBLE CROSS PURPOSES, Ronald A. Knox. A treasure hunt in the Scottish Highlands, an old map, unidentified corpse, surprise discoveries keep reader guessing in this cleverly intricate tale of financial skullduggery. 2 black-and-white maps. 320pp. 5⅜ × 8½. (Available in U.S. only) 25032-6 Pa. $5.95

AUTHENTIC VICTORIAN DECORATION AND ORNAMENTATION IN FULL COLOR: 46 Plates from "Studies in Design," Christopher Dresser. Superb full-color lithographs reproduced from rare original portfolio of a major Victorian designer. 48pp. 9¼ × 12¼. 25083-0 Pa. $7.95

PRIMITIVE ART, Franz Boas. Remains the best text ever prepared on subject, thoroughly discussing Indian, African, Asian, Australian, and, especially, Northern American primitive art. Over 950 illustrations show ceramics, masks, totem poles, weapons, textiles, paintings, much more. 376pp. 5⅜ × 8. 20025-6 Pa. $6.95

SIDELIGHTS ON RELATIVITY, Albert Einstein. Unabridged republication of two lectures delivered by the great physicist in 1920–21. *Ether and Relativity* and *Geometry and Experience.* Elegant ideas in non-mathematical form, accessible to intelligent layman. vi + 56pp. 5⅜ × 8½. 24511-X Pa. $2.95

THE WIT AND HUMOR OF OSCAR WILDE, edited by Alvin Redman. More than 1,000 ripostes, paradoxes, wisecracks: Work is the curse of the drinking classes, I can resist everything except temptation, etc. 258pp. 5⅜ × 8½. 20602-5 Pa. $4.50

ADVENTURES WITH A MICROSCOPE, Richard Headstrom. 59 adventures with clothing fibers, protozoa, ferns and lichens, roots and leaves, much more. 142 illustrations. 232pp. 5⅜ × 8½. 23471-1 Pa. $3.95

PLANTS OF THE BIBLE, Harold N. Moldenke and Alma L. Moldenke. Standard reference to all 230 plants mentioned in Scriptures. Latin name, biblical reference, uses, modern identity, much more. Unsurpassed encyclopedic resource for scholars, botanists, nature lovers, students of Bible. Bibliography. Indexes. 123 black-and-white illustrations. 384pp. 6 × 9. 25069-5 Pa. $8.95

FAMOUS AMERICAN WOMEN: A Biographical Dictionary from Colonial Times to the Present, Robert McHenry, ed. From Pocahontas to Rosa Parks, 1,035 distinguished American women documented in separate biographical entries. Accurate, up-to-date data, numerous categories, spans 400 years. Indices. 493pp. 6½ × 9¼. 24523-3 Pa. $9.95

THE FABULOUS INTERIORS OF THE GREAT OCEAN LINERS IN HISTORIC PHOTOGRAPHS, William H. Miller, Jr. Some 200 superb photographs capture exquisite interiors of world's great "floating palaces"—1890's to 1980's: *Titanic, Ile de France, Queen Elizabeth, United States, Europa,* more. Approx. 200 black-and-white photographs. Captions. Text. Introduction. 160pp. 8⅜ × 11¾. 24756-2 Pa. $9.95

THE GREAT LUXURY LINERS, 1927-1954: A Photographic Record, William H. Miller, Jr. Nostalgic tribute to heyday of ocean liners. 186 photos of Ile de France, Normandie, Leviathan, Queen Elizabeth, United States, many others. Interior and exterior views. Introduction. Captions. 160pp. 9 × 12. 24056-8 Pa. $9.95

A NATURAL HISTORY OF THE DUCKS, John Charles Phillips. Great landmark of ornithology offers complete detailed coverage of nearly 200 species and subspecies of ducks: gadwall, sheldrake, merganser, pintail, many more. 74 full-color plates, 102 black-and-white. Bibliography. Total of 1,920pp. 8⅜ × 11¼. 25141-1, 25142-X Cloth. Two-vol. set $100.00

THE SEAWEED HANDBOOK: An Illustrated Guide to Seaweeds from North Carolina to Canada, Thomas F. Lee. Concise reference covers 78 species. Scientific and common names, habitat, distribution, more. Finding keys for easy identification. 224pp. 5⅜ × 8½. 25215-9 Pa. $5.95

THE TEN BOOKS OF ARCHITECTURE: The 1755 Leoni Edition, Leon Battista Alberti. Rare classic helped introduce the glories of ancient architecture to the Renaissance. 68 black-and-white plates. 336pp. 8⅜ × 11¼. 25239-6 Pa. $14.95

MISS MACKENZIE, Anthony Trollope. Minor masterpieces by Victorian master unmasks many truths about life in 19th-century England. First inexpensive edition in years. 392pp. 5⅜ × 8½. 25201-9 Pa. $7.95

THE RIME OF THE ANCIENT MARINER, Gustave Doré, Samuel Taylor Coleridge. Dramatic engravings considered by many to be his greatest work. The terrifying space of the open sea, the storms and whirlpools of an unknown ocean, the ice of Antarctica, more—all rendered in a powerful, chilling manner. Full text. 38 plates. 77pp. 9¼ × 12. 22305-1 Pa. $4.95

THE EXPEDITIONS OF ZEBULON MONTGOMERY PIKE, Zebulon Montgomery Pike. Fascinating first-hand accounts (1805-6) of exploration of Mississippi River, Indian wars, capture by Spanish dragoons, much more. 1,088pp. 5⅜ × 8½. 25254-X, 25255-8 Pa. Two-vol. set $23.90

A CONCISE HISTORY OF PHOTOGRAPHY: Third Revised Edition, Helmut Gernsheim. Best one-volume history—camera obscura, photochemistry, daguerreotypes, evolution of cameras, film, more. Also artistic aspects—landscape, portraits, fine art, etc. 281 black-and-white photographs. 26 in color. 176pp. 8⅜ × 11¼. 25128-4 Pa. $12.95

THE DORÉ BIBLE ILLUSTRATIONS, Gustave Doré. 241 detailed plates from the Bible: the Creation scenes, Adam and Eve, Flood, Babylon, battle sequences, life of Jesus, etc. Each plate is accompanied by the verses from the King James version of the Bible. 241pp. 9 × 12. 23004-X Pa. $8.95

HUGGER-MUGGER IN THE LOUVRE, Elliot Paul. Second Homer Evans mystery-comedy. Theft at the Louvre involves sleuth in hilarious, madcap caper. "A knockout."—Books. 336pp. 5⅜ × 8½. 25185-3 Pa. $5.95

FLATLAND, E. A. Abbott. Intriguing and enormously popular science-fiction classic explores the complexities of trying to survive as a two-dimensional being in a three-dimensional world. Amusingly illustrated by the author. 16 illustrations. 103pp. 5⅜ × 8½. 20001-9 Pa. $2.25

THE HISTORY OF THE LEWIS AND CLARK EXPEDITION, Meriwether Lewis and William Clark, edited by Elliott Coues. Classic edition of Lewis and Clark's day-by-day journals that later became the basis for U.S. claims to Oregon and the West. Accurate and invaluable geographical, botanical, biological, meteorological and anthropological material. Total of 1,508pp. 5⅜ × 8½.
21268-8, 21269-6, 21270-X Pa. Three-vol. set $25.50

LANGUAGE, TRUTH AND LOGIC, Alfred J. Ayer. Famous, clear introduction to Vienna, Cambridge schools of Logical Positivism. Role of philosophy, elimination of metaphysics, nature of analysis, etc. 160pp. 5⅜ × 8½. (Available in U.S. and Canada only) 20010-8 Pa. $2.95

MATHEMATICS FOR THE NONMATHEMATICIAN, Morris Kline. Detailed, college-level treatment of mathematics in cultural and historical context, with numerous exercises. For liberal arts students. Preface. Recommended Reading Lists. Tables. Index. Numerous black-and-white figures. xvi + 641pp. 5⅜ × 8½.
24823-2 Pa. $11.95

28 SCIENCE FICTION STORIES, H. G. Wells. Novels, *Star Begotten* and *Men Like Gods,* plus 26 short stories: "Empire of the Ants," "A Story of the Stone Age," "The Stolen Bacillus," "In the Abyss," etc. 915pp. 5⅜ × 8½. (Available in U.S. only)
20265-8 Cloth. $10.95

HANDBOOK OF PICTORIAL SYMBOLS, Rudolph Modley. 3,250 signs and symbols, many systems in full; official or heavy commercial use. Arranged by subject. Most in Pictorial Archive series. 143pp. 8⅛ × 11. 23357-X Pa. $5.95

INCIDENTS OF TRAVEL IN YUCATAN, John L. Stephens. Classic (1843) exploration of jungles of Yucatan, looking for evidences of Maya civilization. Travel adventures, Mexican and Indian culture, etc. Total of 669pp. 5⅜ × 8½.
20926-1, 20927-X Pa., Two-vol. set $9.90

DEGAS: An Intimate Portrait, Ambroise Vollard. Charming, anecdotal memoir by famous art dealer of one of the greatest 19th-century French painters. 14 black-and-white illustrations. Introduction by Harold L. Van Doren. 96pp. 5⅜ × 8½.
25131-4 Pa. $3.95

PERSONAL NARRATIVE OF A PILGRIMAGE TO ALMANDINAH AND MECCAH, Richard Burton. Great travel classic by remarkably colorful personality. Burton, disguised as a Moroccan, visited sacred shrines of Islam, narrowly escaping death. 47 illustrations. 959pp. 5⅜ × 8½. 21217-3, 21218-1 Pa., Two-vol. set $17.90

PHRASE AND WORD ORIGINS, A. H. Holt. Entertaining, reliable, modern study of more than 1,200 colorful words, phrases, origins and histories. Much unexpected information. 254pp. 5⅜ × 8½. 20758-7 Pa. $5.95

THE RED THUMB MARK, R. Austin Freeman. In this first Dr. Thorndyke case, the great scientific detective draws fascinating conclusions from the nature of a single fingerprint. Exciting story, authentic science. 320pp. 5⅜ × 8½. (Available in U.S. only) 25210-8 Pa. $5.95

AN EGYPTIAN HIEROGLYPHIC DICTIONARY, E. A. Wallis Budge. Monumental work containing about 25,000 words or terms that occur in texts ranging from 3000 B.C. to 600 A.D. Each entry consists of a transliteration of the word, the word in hieroglyphs, and the meaning in English. 1,314pp. 6⅜ × 10.
23615-3, 23616-1 Pa., Two-vol. set $27.90

THE COMPLEAT STRATEGYST: Being a Primer on the Theory of Games of Strategy, J. D. Williams. Highly entertaining classic describes, with many illustrated examples, how to select best strategies in conflict situations. Prefaces. Appendices. xvi + 268pp. 5⅜ × 8½. 25101-2 Pa. $5.95

THE ROAD TO OZ, L. Frank Baum. Dorothy meets the Shaggy Man, little Button-Bright and the Rainbow's beautiful daughter in this delightful trip to the magical Land of Oz. 272pp. 5⅜ × 8. 25208-6 Pa. $4.95

POINT AND LINE TO PLANE, Wassily Kandinsky. Seminal exposition of role of point, line, other elements in non-objective painting. Essential to understanding 20th-century art. 127 illustrations. 192pp. 6½ × 9¼. 23808-3 Pa. $4.50

LADY ANNA, Anthony Trollope. Moving chronicle of Countess Lovel's bitter struggle to win for herself and daughter Anna their rightful rank and fortune—perhaps at cost of sanity itself. 384pp. 5⅜ × 8½. 24669-8 Pa. $6.95

EGYPTIAN MAGIC, E. A. Wallis Budge. Sums up all that is known about magic in Ancient Egypt: the role of magic in controlling the gods, powerful amulets that warded off evil spirits, scarabs of immortality, use of wax images, formulas and spells, the secret name, much more. 253pp. 5⅜ × 8½. 22681-6 Pa. $4.50

THE DANCE OF SIVA, Ananda Coomaraswamy. Preeminent authority unfolds the vast metaphysic of India: the revelation of her art, conception of the universe, social organization, etc. 27 reproductions of art masterpieces. 192pp. 5⅜ × 8½.
24817-8 Pa. $5.95

CHRISTMAS CUSTOMS AND TRADITIONS, Clement A. Miles. Origin, evolution, significance of religious, secular practices. Caroling, gifts, yule logs, much more. Full, scholarly yet fascinating; non-sectarian. 400pp. 5⅜ × 8½.
23354-5 Pa. $6.50

THE HUMAN FIGURE IN MOTION, Eadweard Muybridge. More than 4,500 stopped-action photos, in action series, showing undraped men, women, children jumping, lying down, throwing, sitting, wrestling, carrying, etc. 390pp. 7⅞ × 10⅝.
20204-6 Cloth. $19.95

THE MAN WHO WAS THURSDAY, Gilbert Keith Chesterton. Witty, fast-paced novel about a club of anarchists in turn-of-the-century London. Brilliant social, religious, philosophical speculations. 128pp. 5⅜ × 8½.
25121-7 Pa. $3.95

A CEZANNE SKETCHBOOK: Figures, Portraits, Landscapes and Still Lifes, Paul Cezanne. Great artist experiments with tonal effects, light, mass, other qualities in over 100 drawings. A revealing view of developing master painter, precursor of Cubism. 102 black-and-white illustrations. 144pp. 8¾ × 6⅝.
24790-2 Pa. $5.95

AN ENCYCLOPEDIA OF BATTLES: Accounts of Over 1,560 Battles from 1479 B.C. to the Present, David Eggenberger. Presents essential details of every major battle in recorded history, from the first battle of Megiddo in 1479 B.C. to Grenada in 1984. List of Battle Maps. New Appendix covering the years 1967–1984. Index. 99 illustrations. 544pp. 6½ × 9¼.
24913-1 Pa. $14.95

AN ETYMOLOGICAL DICTIONARY OF MODERN ENGLISH, Ernest Weekley. Richest, fullest work, by foremost British lexicographer. Detailed word histories. Inexhaustible. Total of 856pp. 6½ × 9¼.
21873-2, 21874-0 Pa., Two-vol. set $17.00

WEBSTER'S AMERICAN MILITARY BIOGRAPHIES, edited by Robert McHenry. Over 1,000 figures who shaped 3 centuries of American military history. Detailed biographies of Nathan Hale, Douglas MacArthur, Mary Hallaren, others. Chronologies of engagements, more. Introduction. Addenda. 1,033 entries in alphabetical order. xi + 548pp. 6½ × 9¼. (Available in U.S. only)
24758-9 Pa. $11.95

LIFE IN ANCIENT EGYPT, Adolf Erman. Detailed older account, with much not in more recent books: domestic life, religion, magic, medicine, commerce, and whatever else needed for complete picture. Many illustrations. 597pp. 5⅜ × 8½.
22632-8 Pa. $8.95

HISTORIC COSTUME IN PICTURES, Braun & Schneider. Over 1,450 costumed figures shown, covering a wide variety of peoples: kings, emperors, nobles, priests, servants, soldiers, scholars, townsfolk, peasants, merchants, courtiers, cavaliers, and more. 256pp. 8⅜ × 11¼.
23150-X Pa. $7.95

THE NOTEBOOKS OF LEONARDO DA VINCI, edited by J. P. Richter. Extracts from manuscripts reveal great genius; on painting, sculpture, anatomy, sciences, geography, etc. Both Italian and English. 186 ms. pages reproduced, plus 500 additional drawings, including studies for *Last Supper*, *Sforza* monument, etc. 860pp. 7⅞ × 10¾. (Available in U.S. only) 22572-0, 22573-9 Pa., Two-vol. set $25.90

THE ART NOUVEAU STYLE BOOK OF ALPHONSE MUCHA: All 72 Plates from "Documents Decoratifs" in Original Color, Alphonse Mucha. Rare copyright-free design portfolio by high priest of Art Nouveau. Jewelry, wallpaper, stained glass, furniture, figure studies, plant and animal motifs, etc. Only complete one-volume edition. 80pp. 9⅜ × 12¼.   24044-4 Pa. $8.95

ANIMALS: 1,419 COPYRIGHT-FREE ILLUSTRATIONS OF MAMMALS, BIRDS, FISH, INSECTS, ETC., edited by Jim Harter. Clear wood engravings present, in extremely lifelike poses, over 1,000 species of animals. One of the most extensive pictorial sourcebooks of its kind. Captions. Index. 284pp. 9 × 12.
23766-4 Pa. $9.95

OBELISTS FLY HIGH, C. Daly King. Masterpiece of American detective fiction, long out of print, involves murder on a 1935 transcontinental flight—"a very thrilling story"—NY Times. Unabridged and unaltered republication of the edition published by William Collins Sons & Co. Ltd., London, 1935. 288pp. 5⅜ × 8½. (Available in U.S. only)   25036-9 Pa. $4.95

VICTORIAN AND EDWARDIAN FASHION: A Photographic Survey, Alison Gernsheim. First fashion history completely illustrated by contemporary photographs. Full text plus 235 photos, 1840–1914, in which many celebrities appear. 240pp. 6½ × 9¼.   24205-6 Pa. $6.00

THE ART OF THE FRENCH ILLUSTRATED BOOK, 1700–1914, Gordon N. Ray. Over 630 superb book illustrations by Fragonard, Delacroix, Daumier, Doré, Grandville, Manet, Mucha, Steinlen, Toulouse-Lautrec and many others. Preface. Introduction. 633 halftones. Indices of artists, authors & titles, binders and provenances. Appendices. Bibliography. 608pp. 8⅜ × 11¼.   25086-5 Pa. $24.95

THE WONDERFUL WIZARD OF OZ, L. Frank Baum. Facsimile in full color of America's finest children's classic. 143 illustrations by W. W. Denslow. 267pp. 5⅜ × 8½.   20691-2 Pa. $5.95

FRONTIERS OF MODERN PHYSICS: New Perspectives on Cosmology, Relativity, Black Holes and Extraterrestrial Intelligence, Tony Rothman, et al. For the intelligent layman. Subjects include: cosmological models of the universe; black holes; the neutrino; the search for extraterrestrial intelligence. Introduction. 46 black-and-white illustrations. 192pp. 5⅜ × 8½.   24587-X Pa. $6.95

THE FRIENDLY STARS, Martha Evans Martin & Donald Howard Menzel. Classic text marshalls the stars together in an engaging, non-technical survey, presenting them as sources of beauty in night sky. 23 illustrations. Foreword. 2 star charts. Index. 147pp. 5⅜ × 8½.   21099-5 Pa. $3.50

FADS AND FALLACIES IN THE NAME OF SCIENCE, Martin Gardner. Fair, witty appraisal of cranks, quacks, and quackeries of science and pseudoscience: hollow earth, Velikovsky, orgone energy, Dianetics, flying saucers, Bridey Murphy, food and medical fads, etc. Revised, expanded In the Name of Science. "A very able and even-tempered presentation."—The New Yorker. 363pp. 5⅜ × 8.
20394-8 Pa. $6.50

ANCIENT EGYPT: ITS CULTURE AND HISTORY, J. E Manchip White. From pre-dynastics through Ptolemies: society, history, political structure, religion, daily life, literature, cultural heritage. 48 plates. 217pp. 5⅜ × 8½.   22548-8 Pa. $4.95

SIR HARRY HOTSPUR OF HUMBLETHWAITE, Anthony Trollope. Incisive, unconventional psychological study of a conflict between a wealthy baronet, his idealistic daughter, and their scapegrace cousin. The 1870 novel in its first inexpensive edition in years. 250pp. 5⅜ × 8½. 24953-0 Pa. $5.95

LASERS AND HOLOGRAPHY, Winston E. Kock. Sound introduction to burgeoning field, expanded (1981) for second edition. Wave patterns, coherence, lasers, diffraction, zone plates, properties of holograms, recent advances. 84 illustrations. 160pp. 5⅜ × 8¼. (Except in United Kingdom) 24041-X Pa. $3.50

INTRODUCTION TO ARTIFICIAL INTELLIGENCE: SECOND, ENLARGED EDITION, Philip C. Jackson, Jr. Comprehensive survey of artificial intelligence—the study of how machines (computers) can be made to act intelligently. Includes introductory and advanced material. Extensive notes updating the main text. 132 black-and-white illustrations. 512pp. 5⅜ × 8½. 24864-X Pa. $8.95

HISTORY OF INDIAN AND INDONESIAN ART, Ananda K. Coomaraswamy. Over 400 illustrations illuminate classic study of Indian art from earliest Harappa finds to early 20th century. Provides philosophical, religious and social insights. 304pp. 6⅜ × 9⅜. 25005-9 Pa. $8.95

THE GOLEM, Gustav Meyrink. Most famous supernatural novel in modern European literature, set in Ghetto of Old Prague around 1890. Compelling story of mystical experiences, strange transformations, profound terror. 13 black-and-white illustrations. 224pp. 5⅜ × 8½. (Available in U.S. only) 25025-3 Pa. $5.95

ARMADALE, Wilkie Collins. Third great mystery novel by the author of *The Woman in White* and *The Moonstone*. Original magazine version with 40 illustrations. 597pp. 5⅜ × 8½. 23429-0 Pa. $9.95

PICTORIAL ENCYCLOPEDIA OF HISTORIC ARCHITECTURAL PLANS, DETAILS AND ELEMENTS: With 1,880 Line Drawings of Arches, Domes, Doorways, Facades, Gables, Windows, etc., John Theodore Haneman. Sourcebook of inspiration for architects, designers, others. Bibliography. Captions. 141pp. 9 × 12. 24605-1 Pa. $6.95

BENCHLEY LOST AND FOUND, Robert Benchley. Finest humor from early 30's, about pet peeves, child psychologists, post office and others. Mostly unavailable elsewhere. 73 illustrations by Peter Arno and others. 183pp. 5⅜ × 8½. 22410-4 Pa. $3.95

ERTÉ GRAPHICS, Erté. Collection of striking color graphics: *Seasons, Alphabet, Numerals, Aces* and *Precious Stones*. 50 plates, including 4 on covers. 48pp. 9⅜ × 12¼. 23580-7 Pa. $6.95

THE JOURNAL OF HENRY D. THOREAU, edited by Bradford Torrey, F. H. Allen. Complete reprinting of 14 volumes, 1837–61, over two million words; the sourcebooks for *Walden*, etc. Definitive. All original sketches, plus 75 photographs. 1,804pp. 8½ × 12¼. 20312-3, 20313-1 Cloth., Two-vol. set $80.00

CASTLES: THEIR CONSTRUCTION AND HISTORY, Sidney Toy. Traces castle development from ancient roots. Nearly 200 photographs and drawings illustrate moats, keeps, baileys, many other features. Caernarvon, Dover Castles, Hadrian's Wall, Tower of London, dozens more. 256pp. 5⅜ × 8¼. 24898-4 Pa. $5.95

AMERICAN CLIPPER SHIPS: 1833–1858, Octavius T. Howe & Frederick C. Matthews. Fully-illustrated, encyclopedic review of 352 clipper ships from the period of America's greatest maritime supremacy. Introduction. 109 halftones. 5 black-and-white line illustrations. Index. Total of 928pp. 5⅜ × 8½.
25115-2, 25116-0 Pa., Two-vol. set $17.90

TOWARDS A NEW ARCHITECTURE, Le Corbusier. Pioneering manifesto by great architect, near legendary founder of "International School." Technical and aesthetic theories, views on industry, economics, relation of form to function, "mass-production spirit," much more. Profusely illustrated. Unabridged translation of 13th French edition. Introduction by Frederick Etchells. 320pp. 6⅛ × 9¼. (Available in U.S. only)
25023-7 Pa. $8.95

THE BOOK OF KELLS, edited by Blanche Cirker. Inexpensive collection of 32 full-color, full-page plates from the greatest illuminated manuscript of the Middle Ages, painstakingly reproduced from rare facsimile edition. Publisher's Note. Captions. 32pp. 9⅜ × 12¼.
24345-1 Pa. $4.95

BEST SCIENCE FICTION STORIES OF H. G. WELLS, H. G. Wells. Full novel *The Invisible Man*, plus 17 short stories: "The Crystal Egg," "Aepyornis Island," "The Strange Orchid," etc. 303pp. 5⅜ × 8½. (Available in U.S. only)
21531-8 Pa. $4.95

AMERICAN SAILING SHIPS: Their Plans and History, Charles G. Davis. Photos, construction details of schooners, frigates, clippers, other sailcraft of 18th to early 20th centuries—plus entertaining discourse on design, rigging, nautical lore, much more. 137 black-and-white illustrations. 240pp. 6⅛ × 9¼.
24658-2 Pa. $5.95

ENTERTAINING MATHEMATICAL PUZZLES, Martin Gardner. Selection of author's favorite conundrums involving arithmetic, money, speed, etc., with lively commentary. Complete solutions. 112pp. 5⅜ × 8½.
25211-6 Pa. $2.95

THE WILL TO BELIEVE, HUMAN IMMORTALITY, William James. Two books bound together. Effect of irrational on logical, and arguments for human immortality. 402pp. 5⅜ × 8½.
20291-7 Pa. $7.50

THE HAUNTED MONASTERY and THE CHINESE MAZE MURDERS, Robert Van Gulik. 2 full novels by Van Gulik continue adventures of Judge Dee and his companions. An evil Taoist monastery, seemingly supernatural events; overgrown topiary maze that hides strange crimes. Set in 7th-century China. 27 illustrations. 328pp. 5⅜ × 8½.
23502-5 Pa. $5.95

CELEBRATED CASES OF JUDGE DEE (DEE GOONG AN), translated by Robert Van Gulik. Authentic 18th-century Chinese detective novel; Dee and associates solve three interlocked cases. Led to Van Gulik's own stories with same characters. Extensive introduction. 9 illustrations. 237pp. 5⅜ × 8½.
23337-5 Pa. $4.95

Prices subject to change without notice.

Available at your book dealer or write for free catalog to Dept. GI, Dover Publications, Inc., 31 East 2nd St., Mineola, N.Y. 11501. Dover publishes more than 175 books each year on science, elementary and advanced mathematics, biology, music, art, literary history, social sciences and other areas.